T0237287

Undergraduate Texts in Physics

Undergraduate Texts in Physics (UTP) publishes authoritative texts covering topics encountered in a physics undergraduate syllabus. Each title in the series is suitable as an adopted text for undergraduate courses, typically containing practice problems, worked examples, chapter summaries, and suggestions for further reading. UTP titles should provide an exceptionally clear and concise treatment of a subject at undergraduate level, usually based on a successful lecture course. Core and elective subjects are considered for inclusion in UTP.

UTP books will be ideal candidates for course adoption, providing lecturers with a firm basis for development of lecture series, and students with an essential reference for their studies and beyond.

More information about this series at https://link.springer.com/bookseries/15593

Hiqmet Kamberaj

Electromagnetism

With Solved Problems

 Springer

Hiqmet Kamberaj
Faculty of Engineering
International Balkan University
Skopje, North Macedonia

ISSN 2510-411X   ISSN 2510-4128 (electronic)
Undergraduate Texts in Physics
ISBN 978-3-030-96779-6   ISBN 978-3-030-96780-2 (eBook)
https://doi.org/10.1007/978-3-030-96780-2

This Springer imprint is published by the registered company Springer Nature Switzerland AG
The registered company address is: Gewerbestrasse 11, 6330 Cham, Switzerland

To the memory of my Mother and Father

Preface

Physics is part of any curriculum in science and engineering. The main objective of this course is to help students of engineering and other sciences in more advanced courses in these fields. The textbook will introduce the students to the fundamental concepts of physics and how different theories developed from physical observations and phenomena.

It starts with electrostatics in free space, introducing basic concepts, such as electrical charge Coulomb's law, and ideas of electric field and electric field lines (Chap. 1). Chapter 2 introduces the electric flux and Gauss's law. Electrostatic potential and electrostatic potential energy are introduced in Chap. 3. Chapter 4 presents the concepts of capacitance and dielectrics. Also, the electrostatics of a macroscopic medium and Maxwell's equations of the electrostatic field are discussed.

Chapter 5 introduces the concepts of electric current and Ohm's law. Chapter 6 continues with the magnetic field and its interactions with charges and currents. Then, Chap. 7 introduces the concept of magnetic field sources, where Biot-Savart law and Ampère's law are introduced. Chapter 8 describes the magnetism in medium, Faraday's law, and Maxwell's magnetic field equations. Then, Chap. 9 describes Maxwell's equations of electromagnetic fields. In particular, this chapter focuses on the potential vector and scalar of the electromagnetic field, electromagnetic field energy, and conservation laws. Then, the dynamics of charged particles in the electromagnetic field and averaging of microscopic properties to obtain the macroscopic Maxwell's equations are introduced. Chapter 10 describes some advanced topics on the induction law and alternating current circuit systems, aiming at understanding electromagnetism applications to wireless charging. Chapter 11 introduces some applications of the theory of electromagnetism in macromolecular solutions and wireless charging technology.

Chapter 12 introduces electromagnetic waves in vacuum and medium, coherence of electromagnetic waves, the polarization of electromagnetic waves, reflection and refraction of electromagnetic waves, and Fresnel's equations. Chapter 13 introduces electromagnetic wave equations in dispersive media. This chapter also describes the absorption, Lorentz's oscillator model of a dielectric, the wave equation of a conductor, the wave equation of a dilute plasma, and the magnetized plasma or dielectric.

Besides, Appendix introduces some mathematical background in vector analysis and vector differential operators.

The textbook is geared more towards examples and problem-solving techniques. The students will get a firsthand experience of how the theories in physics are applied to problems in engineering and science. The textbook is mainly aimed at undergraduate students in engineering and science. However, some chapters and sections are aimed at senior undergraduate students working in the final year thesis in theoretical and computational biophysics, physics, electrical and electronic engineering, and chemistry.

Skopje, North Macedonia   Hiqmet Kamberaj
January 2022

Acknowledgements I thank my family for their continuous support: Nera (my wife), Jon (my son) and Lina (my daughter).

Contents

Chapter 1
Electrostatics in Free Space

The chapter aims to introduce some basic concepts of the electrostatics, such as the concept of charge, interactions between the charged particles, and the electric field.

In this chapter, we will introduce the electrostatics in free space. First, we will introduce the concept of the charges, and then present Coulomb's law of the interactions between the charges. Next, we discuss the concept of electric field and electric field lines. Also, we will describe the motion of a charged particle in the presence of an electric field. The reader can also consider other literature (Holliday et al. 2011) for further reading.

1.1 Electrical Charges

There exist several simple experiments to demonstrate the existence of electrical charges and forces. For example,

1. When we comb our hair on a dry day, we find that the comb attracts pieces of paper.
2. The same effect of attracting pieces of paper occurs when materials such as glass or rubber are rubbed with silk or fur.

As a general rule, for every material behaving in that way, we can say that it is *electrified*, or it becomes *electrically charged*.

Benjamin Franklin (1706–1790) found that there exist two types of electric charges, namely *positive* and *negative*. The following experiment can be used to demonstrate his finding. Suppose that we rubber with fur a hard rubber rod. In addition, we rub a glass rod with silk material. Then, if the glass rod is brought near

© The Author(s), under exclusive license to Springer Nature Switzerland AG 2022
H. Kamberaj, *Electromagnetism*, Undergraduate Texts in Physics,
https://doi.org/10.1007/978-3-030-96780-2_1

the rubber rod, we will observe that the two attract each other. However, if we bring near each other two charged rubber rods or two charged glass rods, then the two repel each other. This experiment indicates the existence of two different states of electrification for the rubber and glass. Furthermore, it finds that like charges repel each other and unlike charges attract each other.

By convention, the electric charge on the glass rod is *positive*, and that on the rubber rod is *negative*. Based on that convention, any charged object repelled by another charged object must have the same sign of charge with it, and any charged object attracted by another charged object must have an opposite sign of charge. It is important to note that the electricity model of Franklin implies that electric charge is always conserved. That is, an electrified state (positive or negative) is due to the charge transfer from one object to the other. In other words, when an object gains some amount of positive/negative charge, then the other gains an equal amount of the electric charge of the opposite sign.

Robert Millikan (1868–1953), in 1909, discovered that electric charge always appears as a multiple integer of a fundamental amount of charge, called e such that the electric charge q, which is a standard symbol for the charge, is quantized as

$$q = Ne \qquad (1.1)$$

Here, N is an integer number, $N = 0, \pm 1, \pm 2, \ldots$.

1.2 Coulomb's Law

Based on an experiment performed by Coulomb, the electric force between two charged particles at rest is proportional to the inverse of the square of distance r between them and directed along the line joining the two particles. In addition, the electric force is proportional to the charges q_1 and q_2 on each particle. Also, the electric force is attractive if the charges are of opposite sign and repulsive if the charges have the same sign. That is known as Coulomb's Law.

Definition 1.1 Force is proportional to the product of the magnitudes of the charges and inversely proportional to the square of the distance between them. Mathematically, the law may be written as

$$F = k_e \frac{|q_1||q_2|}{r^2} \qquad (1.2)$$

In Eq. (1.2), k_e is the Coulomb constant. Note that, in SI, the unit of charge is the coulomb (C). Therefore, the Coulomb constant k_e in SI units has the value

$$k_e = 8.9875 \times 10^9 \text{ N} \cdot \text{m}^2/\text{C}^2 \qquad (1.3)$$

Often, the constant is written as

Fig. 1.1 Electric force
vectors of the interactions
between two charges
(positive or negative)

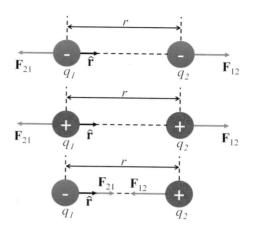

$$k_e = \frac{1}{4\pi\epsilon_0} \tag{1.4}$$

where ϵ_0 is the *permittivity of free space* given by

$$\epsilon_0 = 8.8542 \times 10^{-12} \ \mathrm{C^2/N \cdot m^2} \tag{1.5}$$

Coulomb's force is a vector; hence it has a magnitude expressed by Eq. (1.2) and a direction. Therefore, the Coulomb's law can be expressed in vector form concerning the electric force, $\mathbf{F}_{12}$, exerted by the charge q_1 (*positive or negative*) on another charge q_2 (*positive or negative*) as

Coulomb's force vector

$$\mathbf{F}_{12} = k_e \frac{q_1 q_2}{r^2} \hat{\mathbf{r}} \tag{1.6}$$

In Eq. (1.6), $\hat{\mathbf{r}}$ denotes a unit vector pointing from q_1 to q_2. Note that based on the Newton's third law, the electric force, $\mathbf{F}_{21}$, exerted by a charge q_2 (*positive or negative*) on a second charge q_2 (*positive or negative*) is

$$\mathbf{F}_{21} = -\mathbf{F}_{12} \tag{1.7}$$

Figure 1.1 illustrates graphically the direction of Coulomb's force vectors for different combinations of the pairs of positive and negative charges, namely negative-negative, positive-positive, and negative-positive charge-charge interactions.

Fig. 1.2 A system of N
interacting charges

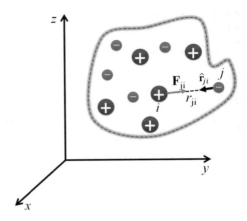

1.3   Coulomb's Law for a System of Charges

Consider a system of N charges (positive and negative), $q_1, q_2, \ldots, q_N$, as shown in
Fig. 1.2. The force exerted on any charge i from all other charges is

$$\mathbf{F}_i = \sum_{j=1 \neq i}^{N} \mathbf{F}_{ji} = \sum_{j=1 \neq i}^{N} k_e \frac{q_i q_j}{r_{ji}^2} \hat{\mathbf{r}}_{ji} \tag{1.8}$$

where $\hat{\mathbf{r}}_{ji}$ is a unit vector directed from q_j to q_i. In Eq. (1.8), $\mathbf{F}_{ji}$ is the force exerted on
i charge particle from j charge particle, and the some runs over all charges excluding
the charge i.

1.4   Electric Field

1.4.1   Force Fields

The field forces act through space, producing an effect even when no physical contact
between the objects occurs. As an example, we can mention the gravitational field.
Michael Faraday developed a similar approach to electric forces. That is, an electric
field exists in the region of space around any charged body, and when another charged
The
electric   body is inside this region of the electric field, an electric force acts on it.
field
Definition 1.2 *The electric field* $\mathbf{E}$ *at a point in space is defined as the electric force*
$\mathbf{F}_e$ *acting on a positive test charge* q_0 *placed at that point divided by the magnitude*
of the test charge:

$$\mathbf{E} = \frac{\mathbf{F}_e}{q_0} \tag{1.9}$$

Fig. 1.3 A positive test charge q_0 placed in the electric field, **E**, of another charge Q distributed uniformly in a sphere

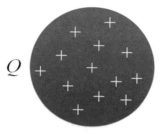

The vector **E** has the SI units of newtons per coulomb (N/C). Figure 1.3 illustrates the electric field **E** created by a positively charged sphere with total charge Q at the positive test charge q_0. Here, we have assumed that the test charge q_0 is small enough that it does not disturb the charge distribution of the sphere responsible for the electric field.

Note that **E** is the field produced by some charge *external* to the test charge, and it is not the field produced by the test charge itself. Also, note that the existence of an electric field is a property of its source. For example, every electron comes with its electric field. An electric field exists at a point if a test charge at rest at that point experiences an electric force. The electric field direction is the direction of the force on a positive test charge placed in the field. Once we know the magnitude and direction of the electric field at some point, the electric force exerted on any charged particle (either positive or negative) placed at that point can be calculated. The electric field exists at some point space, including the free space, independent of the existence of another test charge at that point.

To determine the direction of electric field, consider a point charge q located some distance r from a test positive charge q_0 located at a point P, as shown in Fig. 1.4. Coulomb's law defines the force exerted by q on q_0 as

Direction of electric field

$$\mathbf{F}_e = k_e \frac{q q_0}{r^2} \hat{\mathbf{r}} \tag{1.10}$$

where $\hat{\mathbf{r}}$ represents the usual unit vector directed from q toward q_0 (see Fig. 1.4). Electric field created by q (positive or negative) is

Fig. 1.4 A positive test
charge q_0 placed at distance
r from another charge q
(positive or negative)

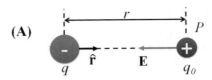

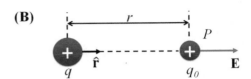

$$E = \frac{F_e}{q_0} = k_e \frac{q}{r^2}\hat{r} \tag{1.11}$$

From Eq. (1.11), when $q < 0$, then $\mathbf{E}$ is pointing opposite to vector $\hat{r}$, and hence
the electric field of a negative charge is pointing toward that charge, see Fig. 1.4a.
On the other hand, when $q > 0$, $\mathbf{E}$ and $\hat{r}$ are parallel, and hence the electric field of
a positive charge is pointing away from that charge, as shown in Fig. 1.4b.

1.4.2 Superposition Principle

According to superposition principle, at any point P, the total electric field due to a
set of discrete point charges, $q_1, q_2, \ldots, q_N$, positive and negative charges, is equal to
the sum of the individual charge electric field vectors (see Fig. 1.5). Mathematically,
we can write

$$\mathbf{E}(\mathbf{r}) = \sum_{i=1}^{N} \mathbf{E}_i = \sum_{i=1}^{N} k_e \frac{q_i}{|\mathbf{r} - \mathbf{r}_i|^2}\hat{r}_i \tag{1.12}$$

In Eq. (1.12), $|\mathbf{r} - \mathbf{r}_i|$ is the distance from q_i to the point P (the location of a test
charge), where $\mathbf{r}$ is the position vector of the point P with respect to some reference
frame, as indicated in Fig. 1.5, and $\mathbf{r}_i$ is the position vector of the charge i in that
reference frame. Furthermore, $\hat{r}_i$ is a unit vector directed from q_i toward P.

Note that in Eq. (1.12) the dependence of $\mathbf{E}$ on only position vector of point
P, $\mathbf{r}$, assumes a *static* configuration of the charges in space. That is, for some other
configuration distribution of charges in space, $\mathbf{E}$ at the same point P may be different.

Note that often for convenience, Eq. (1.12) is also written as

Fig. 1.5 Superposition of
the electric field created by
set of charges at the point P

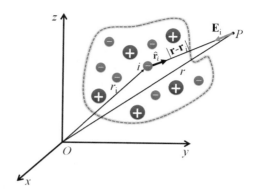

$$\mathbf{E}(\mathbf{r}) = \sum_{i=1}^{N} k_e \frac{q_i\,(\mathbf{r} - \mathbf{r}_i)}{|\,\mathbf{r} - \mathbf{r}_i\,|^3} \tag{1.13}$$

where

$$\hat{\mathbf{r}}_i = \frac{\mathbf{r} - \mathbf{r}_i}{|\,\mathbf{r} - \mathbf{r}_i\,|} \tag{1.14}$$

If the distances between charges in a set of charges are much smaller, compare with the distance of the set from a point where the electric field is to be calculated, then charge distribution is continuous.

To calculate the net electric field created by a continuous charge distribution in some volume V, we follow these steps. First, we divide the charge distribution into macroscopically small elements with small charge Δq_i, as shown in Fig. 1.6a. $\Delta q_i = \rho_i \Delta V$, where ρ_i is seen from a microscopic viewpoint as a uniform charge density within the volume element i, which represents one of the possible configurations of microscopic description. It is important to note that with "macroscopically small" we should understand a small volume in space with a characteristic microscopic configuration of the charges inside it that can, on average, macroscopically be represented as a point-like charge, Δq_i. Then, we calculate the electric field due to one of these macroscopically point charges, Δq_i, at some point P at distance $|\,\mathbf{r} - \mathbf{r}_i\,|$ from the charge element, Δq_i, as

$$\Delta\mathbf{E}(\mathbf{r}, \mathbf{r}_i) = k_e \frac{\Delta q_i}{|\,\mathbf{r} - \mathbf{r}_i\,|^2} \hat{\mathbf{r}}_i \tag{1.15}$$

where $\hat{\mathbf{r}}_i$ is a unit vector directed from the charge element Δq_i toward P. Here, $\mathbf{r}$ is position vector of point P in some reference frame, and $\mathbf{r}_i$ is the position vector of the macroscopically point charge Δq_i.

To evaluate the total electric field at P due to all charge elements distributed in the volume V, we apply the superposition principle:

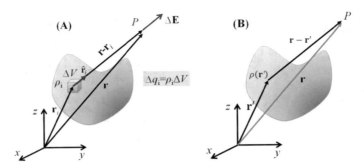

Fig. 1.6 Superposition of the electric field created by a continuous charge distribution at the point P. **a** Macroscopically small volume elements ΔV with charge $\Delta q_i = \rho_i \Delta V$, where ρ_i is the microscopic uniform charge distribution of a small volume ΔV, and **b** a continuous charge distribution in the limit when $\Delta V \rightarrow 0$ with a position dependent charge density ρ at the position $\mathbf{r}'$ with respect to some frame

$$\mathbf{E}(\mathbf{r}) \approx \sum_i k_e \frac{\Delta q_i}{|\mathbf{r} - \mathbf{r}_i|^2} \hat{\mathbf{r}}_i \qquad (1.16)$$

Using the limit of small element volume $\Delta V \rightarrow 0$,

$$\mathbf{E}(\mathbf{r}) = \lim_{\Delta V \rightarrow 0} \sum_i k_e \frac{\rho_i \Delta V}{|\mathbf{r} - \mathbf{r}_i|^2} \hat{\mathbf{r}}_i \qquad (1.17)$$

$$= k_e \int_V \frac{\rho(\mathbf{r}')(\mathbf{r} - \mathbf{r}')}{|\mathbf{r} - \mathbf{r}'|^3} d\mathbf{r}'$$

In Eq. (1.17), $\mathbf{r}'$ denotes the position vector of a point in the volume V with respect to reference frame, as shown in Fig. 1.6b, and the integration is performed over all volume extended by the charge distribution. Note that $\hat{\mathbf{r}}_i$ (or $\hat{\mathbf{r}}'$ when $\Delta V \rightarrow 0$) depends on the position of the volume element (or position vector $\mathbf{r}'$ when $\Delta V \rightarrow 0$, $\hat{\mathbf{r}}' = (\mathbf{r} - \mathbf{r}')/|\mathbf{r} - \mathbf{r}'|$), and hence, it cannot be put outside the sign of integral. Furthermore, $dV = d\mathbf{r}'$.

If we assume that ρ is constant, i.e., uniform charge distribution, then

$$\mathbf{E}(\mathbf{r}) = k_e \rho \int_V \frac{(\mathbf{r} - \mathbf{r}')}{|\mathbf{r} - \mathbf{r}'|^3} d\mathbf{r}' \qquad (1.18)$$

Note that if the charge Q is uniformly distributed in a volume V, then the charge density can be expressed as

$$\rho = \frac{Q}{V} \left(\frac{C}{m^3} \right) \tag{1.19}$$

For a charge Q uniformly distributed on a surface area A, a surface charge density σ is defined as

$$\sigma = \frac{Q}{A} \left(\frac{C}{m^2} \right) \tag{1.20}$$

If a charge Q is uniformly distributed along any line of length L, a linear charge density μ is as follows:

$$\mu = \frac{Q}{L} \left(\frac{C}{m} \right) \tag{1.21}$$

In general, for a charge Q nonuniformly distributed over a volume, surface, or line, we can define the charge densities as

$$\rho = \frac{dQ}{dV} \tag{1.22}$$
$$\sigma = \frac{dQ}{dA}$$
$$\mu = \frac{dQ}{dL}$$

1.5 Electric Field Lines

By definition, electric field lines are drawn to follow the same direction as the electric field vector at any point. Furthermore, the electric field vector is tangent to the line at every point along the field line.

Elec-
tric
field
lines

The electric field lines are such that $\mathbf{E}$ is tangent to the electric field line at each point. The number of lines per unit surface area passing a surface perpendicular to the lines is proportional to the magnitude $|\mathbf{E}|$ in that region. Furthermore, the lines are directed radially away from the positive point charge. Moreover, the lines are directed radially toward the negative point charge.

In Fig. 1.7, we show the electric field lines of a negative and positive point charge. It can be seen that for a negative point charge, $-q$, the electric field lines are drawn toward the charge (see Fig. 1.7a). On the other hand, for a positive point charge, $+q$, electric field lines are leaving the charge, as shown in Fig. 1.7b.

Fig. 1.7 **a** Electric field lines
of a negative point charge
$-q$; **b** Electric field lines of a
positive point charge $+q$

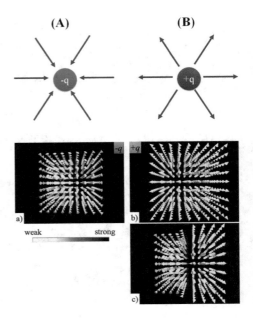

Fig. 1.8 **a** Electric field lines
of a negative point charge
$-q$; **b** Electric field lines of a
positive point charge $+q$; **c**
Electric field lines of the
resultant electric field. The
positive charge is located at
the point $(0, 3, 0)$ and the
negative charge at $(0, -3, 0)$

The following general rules for drawing electric field lines apply:

> The lines start from a positive charge and end on a negative charge. Also, the
> number of lines drawn, leaving a positive charge, or approaching a negative
> charge is proportional to the magnitude of the charge. Moreover, no two field
> lines can cross.

In Fig. 1.8, we show the electric field vector for a positive point charge $+q$ located
at the point $(0, 3, 0)$ (Fig. 1.8b) and a negative point charge $-q$ located at $(0, -3, 0)$
(Fig. 1.8a), colored according to the magnitude of the electric field **E** using a color
scaling, as depicted in Fig. 1.8. Besides, the electric field lines of the resultant electric
field are shown in Fig. 1.8c.

1.6 Motion in Uniform Electric Field

Suppose a charge particle of mass m and charge q is moving in a uniform electric
field **E**. Electric field **E** exerts on a particle placed in it the force

$$\mathbf{F} = q\mathbf{E} \tag{1.23}$$

If that force is equal to the resultant force exerted on the particle, it causes the particle to accelerate, based on Newton's second law:

$$m\mathbf{a} = q\mathbf{E} \tag{1.24}$$

The acceleration gained by the charge is given as

$$\mathbf{a} = \frac{q}{m}\mathbf{E} \tag{1.25}$$

Therefore, if $\mathbf{E}$ is uniform (that is, constant in magnitude and direction), then $\mathbf{a}$ is constant. Furthermore, if the particle has a positive charge, then its acceleration is in the direction of the electric field. On the other hand, if the particle has a negative charge, then its acceleration is in the direction opposite the electric field.

1.7 Exercises

Exercise 1.1 Suppose the electron and proton are in a hydrogen atom. Assume their average separation is approximately 5.3×10^{-11} m. What are the magnitudes of the electric and gravitational force between the two particles?

Solution 1.1 Using the Coulomb's law,

$$F_e = k_e \frac{|-e|\,|+e|}{r^2} \tag{1.26}$$

$$= \left(8.9875 \times 10^9 \text{ N} \cdot \text{m}^2/\text{C}^2\right) \frac{(1.60 \times 10^{-19} \text{ C})^2}{(5.3 \times 10^{-11} \text{ m})^2}$$

$$= 8.2 \times 10^{-8} \text{ N}$$

Using the Newton's law of gravitation,

$$F_g = G \frac{m_e m_p}{r^2} \tag{1.27}$$

$$= \left(6.70 \times 10^{-11} \frac{\text{N} \cdot \text{m}^2}{\text{kg}^2}\right) \frac{(9.11 \times 10^{-31} \text{ kg})(1.67 \times 10^{-27} \text{ kg})}{(5.3 \times 10^{-11} \text{ m})^2}$$

$$= 3.6 \times 10^{-47} \text{ N}$$

The ratio is

$$\frac{F_e}{F_g} = \frac{8.2 \times 10^{-8} \text{ N}}{3.6 \times 10^{-47} \text{ N}} \approx 2 \times 10^{39} \tag{1.28}$$

Exercise 1.2 Consider a configuration in space of three point charges at the corners of a right triangle. Their charges are $q_1 = q_3 = 5.0\,\mu C, q_2 = -2.0\,\mu C$, and $a = 0.10$ m, as shown in Fig. 1.9. Find the resultant force exerted on q_3.

Solution 1.2 Using the Coulomb's law, $\mathbf{F}_3 = \mathbf{F}_{13} + \mathbf{F}_{23}$, and

$$\mathbf{F}_3 = F_{3x}\mathbf{i} + F_{3y}\mathbf{j} \tag{1.29}$$

where

$$F_{3x} = F_{13x} + F_{23x} \tag{1.30}$$
$$F_{3y} = F_{13y} + F_{23y}$$

$$F_{13x} = F_{13}\cos\theta = F_{13}\frac{a}{\sqrt{2}a} = F_{13}/\sqrt{2} \tag{1.31}$$
$$F_{13y} = F_{13}\sin\theta = F_{13}\frac{a}{\sqrt{2}a} = F_{13}/\sqrt{2}$$

and

$$F_{13} = k_e \frac{|q_1||q_3|}{(\sqrt{2}a)^2} \tag{1.32}$$
$$= \left(8.9875 \times 10^9\,\frac{N\cdot m^2}{C^2}\right)\frac{(5.0 \times 10^{-6}\,C)(5.0 \times 10^{-6}\,C)}{2(0.10\,m)^2}$$
$$= 11\,N$$

Thus,

$$F_{13x} = 11\,N/\sqrt{2} = 7.9\,N; \quad F_{13y} = 11\,N/\sqrt{2} = 7.9\,N \tag{1.33}$$

Furthermore,

$$F_{23x} = -F_{23} \tag{1.34}$$
$$F_{23y} = 0\,N$$

and

Fig. 1.9 A system of three point charges

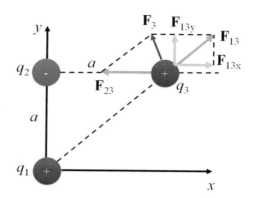

$$F_{23} = k_e \frac{|q_2||q_3|}{a^2} \tag{1.35}$$

$$= \left(8.9875 \times 10^9 \; \frac{\text{N} \cdot \text{m}^2}{\text{C}^2}\right) \frac{(-2.0 \times 10^{-6} \; \text{C})(5.0 \times 10^{-6} \; \text{C})}{(0.10 \; \text{m})^2}$$

$$= 9 \; \text{N}$$

Thus,

$$F_{23x} = -9 \; \text{N} \tag{1.36}$$

Then,

$$F_{3x} = F_{13x} + F_{23x} = -1.1 \; \text{N} \tag{1.37}$$
$$F_{3y} = F_{13y} = 7.9 \; \text{N}$$

and

$$\mathbf{F}_3 = (-1.1\mathbf{i} + 7.9\mathbf{j}) \; \text{N} \tag{1.38}$$

Exercise 1.3 It can be seen that the point charges lie along the x-axis, see also Fig. 1.10. The positive charge $q_1 = 15.0 \; \mu\text{C}$ is at $x = 2.00 \; \text{m}$, the positive charge $q_2 = 6.00 \; \mu\text{C}$ is at the origin, and the resultant force acting on q_3 is zero. What is the x coordinate of q_3?

Solution 1.3 We know $\mathbf{F}_3 = \mathbf{F}_{13} + \mathbf{F}_{23} = 0$, or projecting along the x-axis:

$$F_{3x} = F_{13x} + F_{23x} = 0 \tag{1.39}$$

Therefore,

Fig. 1.10 A system of three
point charges along the
x-axis

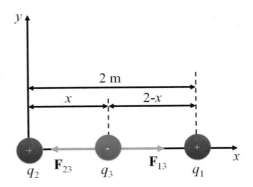

$$F_{13x} = -F_{23x} \tag{1.40}$$

Furthermore,

$$F_{13x} = +F_{13} = k_e \frac{|q_1||q_3|}{(2.0-x)^2} \tag{1.41}$$

$$F_{23x} = -F_{23} = -k_e \frac{|q_2||q_3|}{x^2}$$

Then,

$$k_e \frac{|q_1||q_3|}{(2.0-x)^2} = k_e \frac{|q_2||q_3|}{x^2} \tag{1.42}$$

or

$$(2.0-x)^2 |q_2| = x^2 |q_1| \tag{1.43}$$

After replacing the numerical values, we get

$$(4.0 - 4.0x + x^2)(6.0 \times 10^{-6} \text{ C}) = x^2(15.0 \times 10^{-6} \text{ C}) \tag{1.44}$$

Solving for x we get $x = 0.775$ m.

Exercise 1.4 A charge $q_1 = 7.0 \ \mu\text{C}$ is located at the origin, and a second
charge $q_2 = -5.0 \ \mu\text{C}$ is located on the x-axis, 0.30 m from the origin (see also
Fig. 1.11). What is the electric field at the point P with coordinates $(0, 0.40)$ m?

Fig. 1.11 A system of two point charges along the x-axis

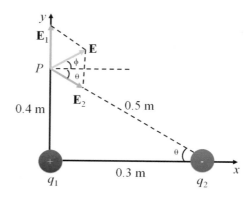

Solution 1.4 Using superposition principle, $\mathbf{E} = \mathbf{E}_1 + \mathbf{E}_2$, where

$$\mathbf{E}_1 = k_e \frac{q_1}{r_1^2} \hat{\mathbf{r}}_1 \tag{1.45}$$

$$= \left(8.9875 \times 10^9 \ \frac{\text{N} \cdot \text{m}^2}{\text{C}^2}\right) \frac{7.0 \times 10^{-6} \ \text{C}}{(0.40 \ \text{m})^2} \mathbf{j}$$

$$= \left(3.9 \times 10^5 \ \frac{\text{N}}{\text{C}}\right) \mathbf{j}$$

and

$$\mathbf{E}_2 = k_e \frac{q_2}{r_2^2} \hat{\mathbf{r}}_2 \tag{1.46}$$

$$= \left(8.9875 \times 10^9 \ \frac{\text{N} \cdot \text{m}^2}{\text{C}^2}\right) \frac{-5.0 \times 10^{-6} \ \text{C}}{(0.50 \ \text{m})^2} (-\cos\theta \mathbf{i} + \sin\theta \mathbf{j})$$

$$= (-1.8 \times 10^5 \ \text{N/C}) (-0.6\mathbf{i} + 0.8\mathbf{j})$$

Then,

$$\mathbf{E} \approx \left(1.1 \times 10^5 \mathbf{i} + 2.5 \times 10^5 \mathbf{j}\right) \ \text{N/C} \tag{1.47}$$

Exercise 1.5 Consider an electric dipole formed by a positive charge q and a negative charge $-q$ separated by some distance (see Fig. 1.12). Find the electric field $\mathbf{E}$ at P due to the charges, where P is a distance $y \gg a$ from the origin.

Fig. 1.12 An electric dipole
formed by two point charges
along the x axis

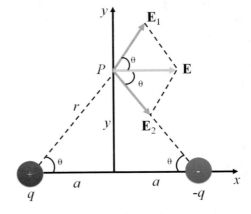

Solution 1.5 Using superposition principle,

$$\mathbf{E} = \mathbf{E}_1 + \mathbf{E}_2 \tag{1.48}$$

$$\mathbf{E}_1 = k_e \frac{q}{r^2} \hat{\mathbf{r}}_1 \tag{1.49}$$

$$= k_e \frac{q}{r^2} (\cos\theta \mathbf{i} + \sin\theta \mathbf{j})$$

$$= k_e \frac{q}{a^2 + y^2} \left(\frac{a}{\sqrt{a^2 + y^2}} \mathbf{i} + \frac{y}{\sqrt{a^2 + y^2}} \mathbf{j} \right)$$

$$= k_e \frac{q}{(a^2 + y^2)^{3/2}} (a\mathbf{i} + y\mathbf{j})$$

Similarly,

$$\mathbf{E}_2 = k_e \frac{-q}{r^2} \hat{\mathbf{r}}_2 \tag{1.50}$$

$$= k_e \frac{-q}{r^2} (-\cos\theta \mathbf{i} + \sin\theta \mathbf{j})$$

$$= k_e \frac{-q}{a^2 + y^2} \left(-\frac{a}{\sqrt{a^2 + y^2}} \mathbf{i} + \frac{y}{\sqrt{a^2 + y^2}} \mathbf{j} \right)$$

$$= -k_e \frac{q}{(a^2 + y^2)^{3/2}} (-a\mathbf{i} + y\mathbf{j})$$

Then,

$$\mathbf{E} = k_e \frac{2qa}{(a^2 + y^2)^{3/2}} \mathbf{i} \approx k_e \frac{2qa}{y^3} \mathbf{i} \tag{1.51}$$

Fig. 1.13 Electric field created by a rod of length ℓ having a uniform charge distribution Q

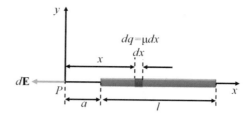

Exercise 1.6 A rod of length ℓ has a uniform positive charge per unit length μ and a total charge Q (see Fig. 1.13). What is the electric field at P located along the long axis of the rod and at distance a from one end?

Solution 1.6 We can chose a reference frame as shown in Fig. 1.13. Then the charge in a small element dx at distance x from P is $dq = \mu dx > 0$ and its electric field at P is

$$d\mathbf{E} = -k_e \frac{dq}{x^2}\mathbf{i} \qquad (1.52)$$

and the total electric field is

$$\mathbf{E} = -k_e \int_a^{a+\ell} \frac{dq}{x^2}\mathbf{i} \qquad (1.53)$$

$$= -k_e \mu \left(\int_a^{a+\ell} \frac{dx}{x^2} \right)\mathbf{i}$$

$$= -k_e \mu \left[-\frac{1}{x} \right]_a^{a+\ell} \mathbf{i}$$

$$= -k_e \mu \left(-\frac{1}{a+\ell} + \frac{1}{a} \right)\mathbf{i}$$

$$= -k_e \mu \frac{\ell}{a(a+\ell)}\mathbf{i}$$

$$= -k_e Q \frac{1}{a(a+\ell)}\mathbf{i}$$

If $a \gg \ell$, then

$$\mathbf{E} = -k_e Q \frac{1}{a^2}\mathbf{i} \qquad (1.54)$$

which is the electric field of a point charge Q at a distance a from its position.

Fig. 1.14 A configuration of
three charges with equal
magnitude:
$|q_1|=|q_2|=|q_3|$,
distributed along the line of a
circle with radius r

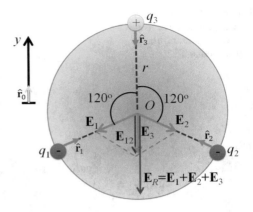

Exercise 1.7 Consider the configuration of three charges shown in Fig. 1.14.
They have equal magnitude, q, and are placed along the line of a circle separated
by an angle of 120°. Find the resultant electric field at the center of the circle,
if it has a radius of r.

Solution 1.7 The electric field $\mathbf{E}_1$ of the negative charge q_1 at the center of the circle
O is

$$\mathbf{E}_1 = -k_e \frac{q}{r^2} \hat{\mathbf{r}}_1 \qquad (1.55)$$

Similarly, the electric field $\mathbf{E}_2$ of the negative charge q_2 at the center of the circle O
is

$$\mathbf{E}_2 = -k_e \frac{q}{r^2} \hat{\mathbf{r}}_2 \qquad (1.56)$$

and of the positive charge q_3 is $\mathbf{E}_3$ given as

$$\mathbf{E}_3 = k_e \frac{q}{r^2} \hat{\mathbf{r}}_3 \qquad (1.57)$$

Choosing a vertical y-axis, as shown in Fig. 1.14, we can express the following unit
vectors in terms of the vertical unit axis vector $\hat{\mathbf{r}}_0$ as

$$\hat{\mathbf{r}}_1 = \cos 60° \hat{\mathbf{r}}_0 = \frac{1}{2} \hat{\mathbf{r}}_0 \qquad (1.58)$$

$$\hat{\mathbf{r}}_2 = \cos 60° \hat{\mathbf{r}}_0 = \frac{1}{2} \hat{\mathbf{r}}_0 \qquad (1.59)$$

$$\hat{\mathbf{r}}_3 = -\hat{\mathbf{r}}_0 \qquad (1.60)$$

Therefore, the resultant electric field at the center of the circle is

$$\mathbf{E}_R = \mathbf{E}_1 + \mathbf{E}_2 + \mathbf{E}_3 \tag{1.61}$$

$$= -k_e \frac{q}{r^2} \left(\frac{1}{2} + \frac{1}{2} + 1 \right) \hat{\mathbf{r}}_0$$

$$= -2k_e \frac{q}{r^2} \hat{\mathbf{r}}_0 \tag{1.62}$$

Exercise 1.8 Consider again the configuration of three charges shown in the previous exercise, having equal magnitude, q, and placed along the line of a circle separated by an angle of $120°$. Determine the sign of the charges for this configuration for which the resultant electric field at the center of the circle vanishes, if it has a radius of r.

Solution 1.8 In Fig. 1.15a and b, we show two possible configurations of charges that produce a zero resultant electric field at O.

For the first configuration, shown in Fig. 1.15a, we have

$$\mathbf{E}_1 = -k_e \frac{q}{r^2} \hat{\mathbf{r}}_1 = -k_e \frac{q}{2r^2} \hat{\mathbf{r}}_0 \tag{1.63}$$

$$\mathbf{E}_2 = -k_e \frac{q}{r^2} \hat{\mathbf{r}}_2 = -k_e \frac{q}{2r^2} \hat{\mathbf{r}}_0 \tag{1.64}$$

$$\mathbf{E}_3 = -k_e \frac{q}{r^2} \hat{\mathbf{r}}_3 = k_e \frac{q}{r^2} \hat{\mathbf{r}}_0 \tag{1.65}$$

Hence, the resultant field at the center of circle is

$$\mathbf{E}_R = \mathbf{E}_1 + \mathbf{E}_2 + \mathbf{E}_3 = 0 \tag{1.66}$$

On the other hand, for the configuration, shown in Fig. 1.15b, we have

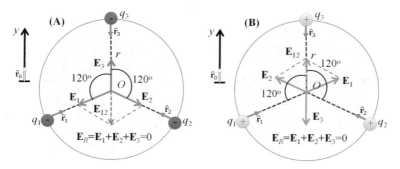

Fig. 1.15 A configuration of three charges with equal magnitude: $| q_1 |=| q_2 |=| q_3 |$, distributed along the line of a circle with radius r: **a** negative charges and **b** positive charges

$$\mathbf{E}_1 = k_e \frac{q}{r^2} \hat{\mathbf{r}}_1 = k_e \frac{q}{2r^2} \hat{\mathbf{r}}_0 \tag{1.67}$$

$$\mathbf{E}_2 = k_e \frac{q}{r^2} \hat{\mathbf{r}}_2 = k_e \frac{q}{2r^2} \hat{\mathbf{r}}_0 \tag{1.68}$$

$$\mathbf{E}_3 = k_e \frac{q}{r^2} \hat{\mathbf{r}}_3 = -k_e \frac{q}{r^2} \hat{\mathbf{r}}_0 \tag{1.69}$$

Then, the resultant field at the center of circle is

$$\mathbf{E}_R = \mathbf{E}_1 + \mathbf{E}_2 + \mathbf{E}_3 = 0 \tag{1.70}$$

In our calculations, $\hat{\mathbf{r}}_1, \hat{\mathbf{r}}_2, \hat{\mathbf{r}}_3, \hat{\mathbf{r}}_0$ are unit vectors as indicated in Fig. 1.15.

Exercise 1.9 A positive point charge q of mass m is released from rest in a uniform electric field $\mathbf{E}$ directed along the x-axis. Describe its motion.

Solution 1.9 The acceleration is

$$\mathbf{a} = \frac{q}{m} \mathbf{E} \tag{1.71}$$

since $\mathbf{E}$ is constant, then $\mathbf{a}$ is constant and different from zero only along x-axis:

$$a_x = \frac{q}{m} E_x \tag{1.72}$$

$$a_y = 0$$

$$a_z = 0$$

and the motion is along x-axis.
   From kinematics,

$$x_f = x_i + v_{xi}t + a_x t^2/2 \tag{1.73}$$

$$v_{xf} = v_{xi} + a_x t$$

$$v_{xf}^2 - v_{xi}^2 = 2a_x(x_f - x_i)$$

For $x_i = 0$ and $v_{xi} = 0$, we get

$$x_f = a_x t^2/2 = \frac{qEt^2}{2m} \tag{1.74}$$

$$v_{xf} = a_x t = \frac{qEt}{m} \tag{1.75}$$

and

$$v_{xf}^2 = 2a_x x_f \tag{1.76}$$

$$= 2\frac{qE}{m}x_f$$

$$= \left(\frac{qE}{m}\right)^2 t^2$$

The final kinetic energy

$$K_f = mv_{xf}^2/2 \tag{1.77}$$

$$= \frac{m}{2}\left(\frac{qE}{m}\right)^2 t^2$$

$$= \frac{(qE)^2}{2m}t^2$$

Work done by electric force is

$$W = \Delta K = K_f - K_i = K_f \tag{1.78}$$

$$= \frac{(qE)^2}{2m}t^2$$

$$= qEx_f = F_x x_f$$

Exercise 1.10 Consider two oppositely charged flat metallic plates creating an electric field in the region between them approximately uniform. Suppose an electron of charge $-e$ is projected horizontally into this field with an initial velocity $v_i\mathbf{i}$. Describe the motion of the electron.

Solution 1.10 Because the electric field $\mathbf{E}$ is in the positive y direction, the acceleration is

$$\mathbf{a} = -\frac{eE}{m}\mathbf{j} \tag{1.79}$$

and $v_{xi} = v_i$ and $v_{yi} = 0$, thus

$$x_f = v_i t \tag{1.80}$$

$$y_f = a_y t^2/2 = -\frac{eE}{2m}t^2 \tag{1.81}$$

$$v_{xf} = v_i \tag{1.82}$$

$$v_{yf} = a_y t = -\frac{eE}{m}t \tag{1.83}$$

From here, we get

$$y_f = -\frac{eE}{2m}\left(\frac{x_f}{v_i}\right)^2 \tag{1.84}$$

Hence, the trajectory is a parabola.

Reference

Holliday D, Resnick R, Walker J (2011) Fundamentals of physics. John Wiley and Sons, New York

Chapter 2
Gauss's Law

*This chapter aims to introduce Gauss's law and its applications
to electrostatics.*

In this chapter, we introduce the electric flux and Gauss's law. Also, the application of
Gauss's law to insulators and conductors will be discussed. As extra reading material,
the reader can also consider other literature (Holliday et al. 2011).

2.1 Electric Flux

2.1.1 Uniform Electric Field

The electric flux concept describes quantitatively the electric lines. The number of
field lines per unit area (also called *line density*) going through a rectangular surface of
area A, which is perpendicular to the field, is proportional to the magnitude of electric
field, $\mathbf{E}$, as shown in Fig. 2.1. Furthermore, the total number of lines penetrating
the surface is proportional to the product $|\mathbf{E}| A$. By definition, the product of the
magnitude of electric field $|\mathbf{E}|$ and surface area A perpendicular to the field is called
the *electric flux*:

$$\Phi_E = |\mathbf{E}| \ A \qquad (2.1)$$

Using Eq. (2.1), from the SI units of E and A, we derive the SI units of the electric
flux:

$$[E] = \left[\frac{N}{C}\right], \ [A] = [m^2] \qquad (2.2)$$

Uni-
form
electric
field
flux

Fig. 2.1 A uniform electric
field perpendicular to a
rectangular surface area A

Fig. 2.2 A uniform electric
field penetrating a
rectangular surface area A
with unit vector **n**
perpendicular to surface
forming an angle θ with **E**

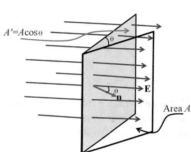

Thus, we obtain SI units of Φ_E:

$$[\Phi_E] = \left[\frac{\text{N} \cdot \text{m}^2}{\text{C}} \right] \tag{2.3}$$

Note that the electric flux is proportional to the number of electric field lines
penetrating some surface.

Moreover, consider the electric flux on any surface with an arbitrary orientation
with respect to electric field **E**, as shown in Fig. 2.2. Electric flux going through the
surface (with area A) not perpendicular to **E** is smaller than the product $| \text{ E } | A$. That
is, the number of lines that cross this area A is equal to the number of lines that cross
the area $A' = A \cos \theta$, which is a projection of A aligned perpendicular to the field.
Mathematically, the electric flux is given by (Fig. 2.2)

$$\Phi_E = | \text{ E } | A' = | \text{ E } | A \cos \theta \tag{2.4}$$

From the definition, Eq. (2.4), we can say that the maximum electric flux is
achieved when $\theta = 0°$; that is, the surface is perpendicular to **E**: $\Phi_E^{max} = | \text{ E } | A$
(see also Eq. (2.1)). Or, equivalently, when normal vector **n** to the surface is parallel

Fig. 2.3 A nonuniform
electric field penetrating an
arbitrary element surface
area ΔA_i with unit vector $\mathbf{n}_i$
perpendicular to surface
forming an angle θ_i with $\mathbf{E}_i$
at that point

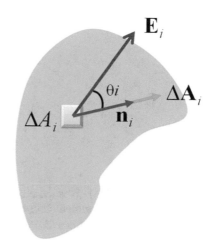

to $\mathbf{E}$. On the other hand, the minimum electric flux is achieved when $\theta = 90°$, that
is, the surface is parallel to $\mathbf{E}$: $\Phi_E^{min} = 0$. In this case, normal vector $\mathbf{n}$ to the surface
is perpendicular to $\mathbf{E}$. In general, denoting the vector $\mathbf{A} = A\mathbf{n}$, we can write

$$\Phi_E = \mathbf{E} \cdot \mathbf{A} \tag{2.5}$$

2.1.2 General Electric Field Flux

Let us consider the case of a general surface with a nonuniform electric field $\mathbf{E}_i$,
as shown in Fig. 2.3. We can partition the surface on small infinitesimal elements
ΔA_i, such that electric field is constant on every point of ΔA_i, then the electric flux
through ΔA_i is

$$\Delta \Phi_{E,i} = \mathbf{E}_i \cdot \Delta \mathbf{A}_i \tag{2.6}$$

The total flux can be approximated as

$$\Phi_E \approx \sum_i \mathbf{E}_i \cdot \Delta \mathbf{A}_i \tag{2.7}$$

Taking the limit when $\Delta \mathbf{A}_i \to 0$ on both sides of Eq. (2.7), we obtain the exact
electric flux through general surface (Fig. 2.3):

$$\Phi_E = \lim_{\Delta A_i \to 0} \sum_i \mathbf{E}_i \cdot \Delta \mathbf{A}_i = \int_{surface} \mathbf{E} \cdot d\mathbf{A} \tag{2.8}$$

Electric
flux
through
general
surface

Fig. 2.4 A nonuniform
electric field penetrating a
closed surface

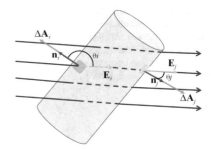

Fig. 2.4 A nonuniform electric field penetrating a closed surface

Note that Eq. (2.8) gives the electric flux through any surface. Often, the electric
flux is calculated through a closed surface.

Closed surface

A closed surface defines the surface that divides space into the inside and outside
regions, and to move from one region to another, one has to cross the surface. For
example, the surface of a sphere, ellipsoid, etc., are all closed surfaces.

Electric flux through a closed surface

Consider a nonuniform electric field penetrating a closed surface, for example, a
cylinder, as shown in Fig. 2.4. Every electric field line that passes through a surface
element ΔA_i is going to leave the closed surface at some other surface element ΔA_j.
The electric flux going through the surface element ΔA_i is

$$\Delta \Phi_{E,i} = \mathbf{E}_i \, \Delta \mathbf{A}_i = |\,\mathbf{E}_i\,| \, \Delta A_i \cos \theta_i \tag{2.9}$$

In Eq. (2.9), θ_i is the angle between electric field $\mathbf{E}_i$ and unit vector normal to surface
element $\mathbf{n}_i$:

$$\cos \theta_i = \begin{cases} > 0, \ 0 \leq \theta_i < 90° \\ = 0, \ \theta_i = 90° \\ < 0, \ 90° < \theta_i \leq 180° \end{cases} \tag{2.10}$$

Similarly, the electric flux going through the surface element ΔA_j is

$$\Delta \Phi_{E,j} = \mathbf{E}_j \, \Delta \mathbf{A}_j = |\,\mathbf{E}_j\,| \, \Delta A_j \cos \theta_j \tag{2.11}$$

Here, θ_j is the angle between $\mathbf{E}_j$ and unit vector normal to surface element $\mathbf{n}_j$.

If by convention, we define unit vector $\mathbf{n}_i$ as outward to the surface element,
then $\Delta \Phi_{E,j} > 0$ indicates that electric field lines leave the closed surface through
ΔA_j, and $\Delta \Phi_{E,i} < 0$ indicates that electric field lines are entering the closed surface
through ΔA_i.

Therefore, if the electric field lines are entering the closed surface, then their
electric flux contribution is negative. If the electric field lines are leaving the closed
surface, their contribution to electric flux is positive.

Since the total flux is approximated as

$$\Phi_E \approx \sum_i \mathbf{E}_i \cdot \Delta \mathbf{A}_i \tag{2.12}$$

therefore, the *net electric flux* through the surface is proportional to the net number of lines leaving the surface, or the number leaving the surface minus the number entering the surface.

As a consequence, if more lines are leaving than entering, the net flux is positive. On the other hand, if more lines are entering than leaving, the net flux is negative. Mathematically, we can write the net flux Φ_E through a closed surface as

$$\Phi_E = \oint_S \mathbf{E} \cdot d\mathbf{A} \tag{2.13}$$

$$= \oint_S \mathbf{E} \cdot \mathbf{n} \, dA$$

$$= \oint_S E_n \, dA$$

2.2 Gauss's Law

By definition, a Gaussian surface is called a closed surface. Gauss's law relates the electric flux through the closed surface and the charge enclosed by the surface.

For illustration consider a nonconducting sphere of radius r with an elementary positive charge $+q$ at the center of sphere. Note that the sphere's surface is equal to Gaussian surface, as indicated in Fig. 2.5. Electric field created by the charge $+q$ at any point at distance r from the charge is

$$\mathbf{E} = k_e \frac{q}{r^2} \hat{\mathbf{r}} \tag{2.14}$$

Gaussian surface

Fig. 2.5 A nonconducting sphere of radius r containing an elementary positive charge at the center, $+q$

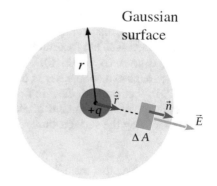

Gaussian surface

Fig. 2.6 Electric flux of an
elementary charge, q

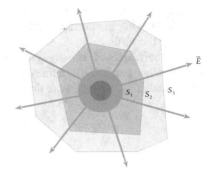

The electric flux through the sphere surface is

$$\Phi_E = \oint_S \mathbf{E} \cdot \mathbf{n}\, dA = \oint_S k_e \frac{q}{r^2} \hat{\mathbf{r}} \cdot \mathbf{n}\, dA \qquad (2.15)$$

$$= \oint_S k_e \frac{q}{r^2}\, dA = k_e \frac{q}{r^2} \oint_S dA = k_e \frac{q}{r^2}(4\pi r^2)$$

$$= 4\pi k_e q$$

or

$$\Phi_E = q4\pi \frac{1}{4\pi\epsilon_0} = \frac{q}{\epsilon_0} \qquad (2.16)$$

The result given by Eq. (2.16) indicates that Φ_E is independent of radius r. Furthermore, since both q and ϵ_0 are constants, then the electric flux is constant. Therefore,
the same electric flux is passing through the surface of any other sphere with radius
$R > r$, which has an elementary charge at the center.

Now, consider several closed surfaces surrounding a charge q, as shown in Fig. 2.6,
S_1 (spherical), S_2 and S_3 (nonspherical). The flux that passes through S_1 is given by
Eq. (2.16). Since flux is proportional to the number of electric field lines passing
Flux of through a surface, then the flux through S_2 and S_3 also is constant (see Eq. (2.16)).
a point   By definition, the net flux through any closed surface is independent of the shape
charge of that surface. The net flux through an arbitrary closed surface surrounding a point
charge q is given by Eq. (2.16).

Now, suppose a charge $+q$ is outside a closed surface (any shape), as shown in
Fig. 2.7. In that case, an electric field line that enters the surface leaves the surface
at another point. The number of electric field lines entering the surface equals the
number leaving the surface, thus

$$\Phi_E = 0 \qquad (2.17)$$

We can conclude that the net electric flux through a closed surface that surrounds no
charge is zero (Fig. 2.7).

Fig. 2.7 Electric flux of an elementary charge, q outside a closed surface area

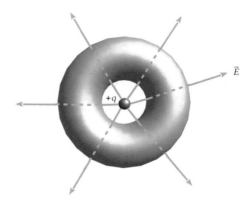

2.2.1 Gauss's Law for a System of Charges

Consider a system of N discrete point charges, $q_1, q_2, \ldots, q_N$, positive or negative, as shown in Fig. 2.8. Using the superposition principle of electric field, discussed in Chap. 1, we can write

$$\mathbf{E} = \sum_{i=1}^{N} \mathbf{E}_i \qquad (2.18)$$

Then, the net electric flux through a closed surface is

$$\Phi_E = \oint_S \mathbf{E} \cdot d\mathbf{A} \qquad (2.19)$$

$$= \oint_S \left(\sum_{i=1}^{N} \mathbf{E}_i \right) \cdot d\mathbf{A}$$

Fig. 2.8 Electric flux of a set of discrete charges through a Gaussian surface

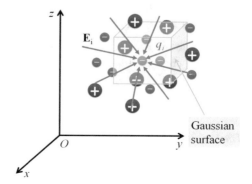

$$= \sum_{i=1}^{N} \oint_{S} \mathbf{E}_i \cdot d\mathbf{A}$$

where $\mathbf{E}_i$ is electric field of charge q_i.

The electric flux for the charge q_i through the closed surface is

$$\Phi_{E,i} = \oint_{S} \mathbf{E}_i \cdot d\mathbf{A} \tag{2.20}$$

$$= \begin{cases} \dfrac{q_i}{\epsilon_0}, & \text{if } q_i \text{ inside closed surface} \\ 0, & \text{if } q_i \text{ outside closed surface} \end{cases}$$

Electric flux of a discrete system of charges

Therefore, the net electric flux becomes

$$\Phi_E = \sum_{i=1}^{N} \Phi_{E,i} \tag{2.21}$$

$$= \sum_{i=1}^{N_{in}} \frac{q_i}{\epsilon_0} = \frac{Q_{in}}{\epsilon_0}$$

where Q_{in} is the net charge inside the closed surface and N_{in} denotes the number of charges inside the closed surface.

$$Q = \sum_{i=1}^{N_{in}} q_i \tag{2.22}$$

Gauss's law for a system of charges $q_1, q_1, \ldots, q_N$ says that the net electric flux through any close surface is given by Eq. (2.21). When using this equation, we should note that

1. The charge Q_{in} is the net charge inside the closed surface.
2. $\mathbf{E}$ represents the total electric field, which includes contributions from charges both inside and outside the surface.

For a continuous system, see also Fig. 2.9, we partition it into macroscopically elementary charges Δq_i, and then using the superposition principle of electric field,

$$\mathbf{E} \approx \sum_{i} \Delta \mathbf{E}_i \tag{2.23}$$

The net electric flux through a closed surface is approximately

Fig. 2.9 Electric flux of a continuous charge distribution with density ρ through a Gaussian surface. Δq_i and Δq_j are two macroscopically small elementary charges

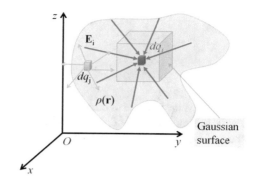

$$\Phi_E = \oint_S \mathbf{E} \cdot d\mathbf{A} \qquad (2.24)$$

$$\approx \oint_S \left(\sum_i \Delta\mathbf{E}_i \right) \cdot d\mathbf{A}$$

$$= \sum_i \oint_S \Delta\mathbf{E}_i \cdot d\mathbf{A}$$

where $\Delta\mathbf{E}_i$ is electric field of charge Δq_i. The electric flux for a charge Δq_i through the closed surface is

$$\Delta\Phi_{E,i} = \oint_S \Delta\mathbf{E}_i \cdot d\mathbf{A} \qquad (2.25)$$

$$= \begin{cases} \dfrac{\Delta q_i}{\epsilon_0}, & \text{if } \Delta q_i \text{ inside closed surface} \\ 0, & \text{if } \Delta q_i \text{ outside closed surface} \end{cases}$$

Therefore, the net electric flux is

$$\Phi_E \approx \sum_i \Delta\Phi_{E,i} \qquad (2.26)$$

$$= \sum_{i=1}^{N_{in}} \frac{\Delta q_i}{\epsilon_0}$$

Taking the limit when $\Delta q_i \to 0$ on both sides of Eq. (2.26), we get the net electric flux for a continuous system of charges:

Electric flux of a continuous system of charges

$$\Phi_E = \lim_{\Delta q_i \to 0} \sum_{i=1}^{N_{in}} \frac{\Delta q_i}{\epsilon_0} \qquad (2.27)$$

$$= \int_{V_{in}} \frac{dq}{\epsilon_0} = \frac{Q_{in}}{\epsilon_0}$$

where V_{in} is the volume of charges inside the Gaussian surface, Q_{in} is the total charge inside the closed surface, and N_{in} is the number of macroscopically elementary charges inside the closed surface.

2.3   Applications of Gauss's Law to Insulators

In the following, we are summarizing some tips for solving problems using Gauss's law. First, Gauss's law is useful in determining electric fields when the charge distribution is characterized by a high degree of symmetry. We should pay attention to ways of choosing the Gaussian surface over which the surface integral given by either Eq. (2.21) or Eq. (2.27) can be simplified and the electric field is determined.

In choosing the surface, we should always take advantage of the symmetry of the charge distribution so that we can remove **E** from the integral and solve it. Using this calculation, we can determine a surface that satisfies one or more of the following conditions:

1. The value of the electric field can be argued by symmetry to be constant over the surface.
2. The dot product can be expressed as a simple algebraic product $Ed A$ because **E** and $d\mathbf{A}$ are parallel.
3. The dot product is zero because **E** and $d\mathbf{A}$ are perpendicular.
4. The field is zero over the surface.

2.4   Conductors in Electrostatic Equilibrium

A good electrical conductor contains free electrons (with charge $-e$) moving inside the material. When there is no net motion of charge within a conductor, the conductor is in an electrostatic equilibrium, having the following properties:

1. There is no electric field inside the conductor; that is, the electric field is zero.
2. Any excess charge in an isolated conductor resides on its surface.
3. The electric field just outside a charged conductor is perpendicular to the surface of the conductor and has a magnitude σ/ϵ_0, where σ is the surface charge density at that point.
4. For an irregularly shaped conductor, the surface charge density is greatest at locations where the radius of curvature of the surface is the smallest.

Fig. 2.10 A conductor in a
uniform electric field **E**

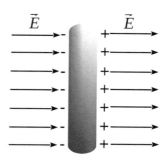

2.4.1 Property 1

Consider a conducting slab placed in an external field **E**, as shown in Fig. 2.10. If the
field were not zero inside the conductor, then free charges in the conductor would
accelerate under the action of the field. This motion of electrons would mean that the
conductor is not in electrostatic equilibrium, which is not true. Thus, a zero electric
field inside the conductor is a necessary condition for the existence of electrostatic
equilibrium.

In the absence of external field, free electrons distribute uniformly throughout
the conductor and are free to move about the conductor. On the other hand, if we
apply an external **E**, as shown in Fig. 2.10, the free electrons drift to the left. Thus,
a plane of negative charge will be seen on the left surface. As the electrons move to
the left, a plane of positive charge creates on the right surface. These charges create
an additional electric field inside the conductor that opposes the external field. As
the electrons move, the surface charge density increases until the magnitude of the
internal field is equal to that of the external field, and hence there is a net field of zero
inside the conductor. The time it takes for an excellent conductor to reach equilibrium
is of the order of 10^{-16} s, which for most purposes, can be considered instantaneous.

2.4.2 Property 2

Let's apply the Gauss's law to a conductor of arbitrary shape. First, the Gaussian
surface is a closed surface inside the conductor, as indicated in Fig. 2.11. Then, using
the Gauss's law, we write

$$\Phi_E = \oint_S \mathbf{E} \cdot d\mathbf{A} \tag{2.28}$$

$$= \oint_S 0 \cdot d\mathbf{A} = 0$$

$$= \frac{Q_{in}}{\epsilon_0}$$

Fig. 2.11 A Gaussian
surface inside the conductor

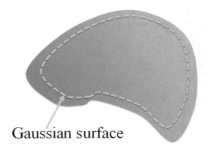

Gaussian surface

Combining Eqs. (2.27) and (2.28), we get $Q_{in} = 0$. Because the net charge inside
the arbitrary Gaussian surface is zero, any net charge on the conductor must reside
on its surface.

2.4.3 Property 3

First, a Gaussian surface in the shape of a small cylinder is drawn with end faces
parallel to the surface of the conductor, as indicated in Fig. 2.12. It can be seen
that a part of the cylinder is outside the conductor, and only a portion is inside the
conductor. Furthermore, the field vector is perpendicular to the conductor's surface
from the condition of electrostatic equilibrium. If **E** had a component parallel to the
conductor's surface, the free charges would move along the surface, and therefore,
the conductor would not be in equilibrium.

Therefore, we have

1. For the curved part of the cylindrical Gaussian surface there is no flux through
   this part of the Gaussian surface because **E** is parallel to the surface.
2. There is no flux through the flat face of the cylinder inside the conductor because
   here $E = 0$.
3. The net flux through the Gaussian surface is only through the flat face outside the
   conductor, where the field is perpendicular to the Gaussian surface.

Fig. 2.12 A cylinder
Gaussian surface
perpendicular to the
conductor

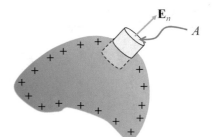

Applying Gauss's law to this surface, we obtain

$$\Phi_E = \oint_S \mathbf{E} \cdot \mathbf{n} \, dA \qquad (2.29)$$

$$= \int_{Base} E \, dA = EA$$

Using Eqs. (2.27) and (2.29), we obtain

$$EA = \frac{Q_{in}}{\epsilon_0} = \frac{\sigma A}{\epsilon_0} \qquad (2.30)$$

where σ is the surface charge density such that $Q_{in} = \sigma A$. Therefore, we get

$$E = \frac{\sigma}{\epsilon_0} \qquad (2.31)$$

2.4.4 Property 4

Consider an irregularly shaped conductor. Then, the surface charge density is highest at locations where the radius of curvature of the surface is smallest. To demonstrate that, suppose we partition irregular surface into small elements dA_i of equal angle at the center of the conductor.

Then, properties (1) and (3) require that $\sigma_i dA_i = constant$, and since dA_i depends on the radius of curvature, the smaller the radius of curvature, the smaller the dA_i, thus property (4) follows.

2.5 Exercises

Exercise 2.1 Determine the electric flux through a sphere that has a radius of 1.00 m and carries a charge of +1.00 μC at its center.

Solution 2.1 Magnitude of electric field at distance $r = 1.00$ m from the charge is

$$E = k_e \frac{q}{r^2} \qquad (2.32)$$

$$= (8.99 \times 10^9 \text{ N} \cdot \text{m}^2/\text{C}^2) \frac{1.00 \times 10^{-6} \text{ C}}{(1.00 \text{ m})^2}$$

$$= 8.99 \times 10^3 \text{ N/C}$$

The direction of **E** is radially and outward from the charge and perpendicular to the surface of sphere; thus, the electric flux penetrating the surface of sphere is

$$\Phi_E = E A \tag{2.33}$$
$$= (8.99 \times 10^3 \text{ N/C})(4\pi r^2)$$
$$= 1.13 \times 10^5 \text{ N} \cdot \text{m}^2/\text{C}$$

Exercise 2.2 Find the electric flux through a sphere that has a radius of 0.50 m and carries a charge of $+1.00 \ \mu\text{C}$ at its center.

Solution 2.2 Magnitude of electric field at distance $r = 0.50$ m from the charge is

$$E = k_e \frac{q}{r^2} \tag{2.34}$$
$$= (8.99 \times 10^9 \text{ N} \cdot \text{m}^2/\text{C}^2) \frac{1.00 \times 10^{-6} \text{ C}}{(0.50 \text{ m})^2}$$
$$= 3.60 \times 10^4 \text{ N/C}$$

The direction of **E** is radially and outward from the charge and perpendicular to the surface of sphere; thus, the electric flux penetrating the surface of sphere is

$$\Phi_E = E A \tag{2.35}$$
$$= (3.60 \times 10^4 \text{ N/C})(4\pi r^2)$$
$$= 1.13 \times 10^5 \text{ N} \cdot \text{m}^2/\text{C}$$

Exercise 2.3 Consider a uniform electric field **E** along the x-axis. What is the net electric flux through the surface of a cube of edges a, oriented as shown in Fig. 2.13?

Solution 2.3 We can distinguish the flux through six faces of the cube: two along x-axis, two along y-axis, and two along z-axis. The total electric flux is

$$\Phi_E = \int_1 \mathbf{E} \cdot d\mathbf{A} + \int_2 \mathbf{E} \cdot d\mathbf{A} + \int_3 \mathbf{E} \cdot d\mathbf{A} \tag{2.36}$$
$$+ \int_4 \mathbf{E} \cdot d\mathbf{A} + \int_5 \mathbf{E} \cdot d\mathbf{A} + \int_6 \mathbf{E} \cdot d\mathbf{A}$$

Fig. 2.13 A uniform electric field $\mathbf{E}$ oriented in the x direction passing through the surface of a cube of edges a

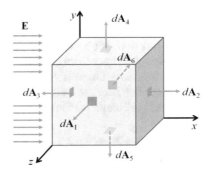

Since $\mathbf{E}$ is along x-axis

$$\int_1 \mathbf{E} \cdot d\mathbf{A} = \int_4 \mathbf{E} \cdot d\mathbf{A} \qquad (2.37)$$

$$= \int_5 \mathbf{E} \cdot d\mathbf{A}$$

$$= \int_6 \mathbf{E} \cdot d\mathbf{A} = 0$$

The total electric flux is

$$\Phi_E = \int_3 \mathbf{E} \cdot d\mathbf{A} + \int_2 \mathbf{E} \cdot d\mathbf{A} \qquad (2.38)$$

$$= \int_3 \mathbf{E} \cdot \mathbf{n}_1 dA + \int_2 \mathbf{E} \cdot \mathbf{n}_2 dA$$

Since E is constant

$$\Phi_E = -\int_3 E dA + \int_2 E dA = -Ea^2 + Ea^2 = 0 \qquad (2.39)$$

Exercise 2.4 Consider a spherical Gaussian surface enclosing a point charge q. Determine what happened to the net flux (a) if the charge is tripled; (b) if the radius of the sphere is doubled; (c) if the surface is changed to a cube; and (d) if the charge is moved to another location inside the surface.

Solution 2.4 1. The flux through the surface is tripled because the flux is proportional to the amount of charge inside the surface.

2. The flux does not change because electric field lines from the charge pass through
   the sphere, regardless of its radius.
3. The flux does not change when the shape of the Gaussian surface changes because
   electric field lines from the charge pass through the surface, regardless of its shape.
4. The flux does not change when the charge is moved to another location inside
   that surface because Gauss's law refers to the total charge enclosed, regardless of
   where the charge is located inside the surface.

Exercise 2.5 Calculate the electric field of an isolated point charge q using
Gauss's law.

Solution 2.5 Using the spherical symmetry of the problem, we choose a spherical
Gaussian surface of radius r centered on the point charge, as shown in Fig. 2.14.
Since the charge is positive, then the electric field is directed radially outward, and
hence it is normal to the surface at every point.

$$\Phi_E = \oint_S \mathbf{E} \cdot d\mathbf{A} \tag{2.40}$$
$$= \oint_S (\mathbf{E} \cdot \mathbf{n}) dA = \oint_S E_n dA$$
$$= E_n \oint_S dA = E_n (4\pi r^2)$$
$$= \frac{Q_{in}}{\epsilon_0} = \frac{q}{\epsilon_0}$$

or

$$E_n \equiv E = \frac{q}{4\pi\epsilon_0 r^2} \tag{2.41}$$
$$= k_e \frac{q}{r^2}$$

Fig. 2.14 The Gaussian
surface and an isolated point
charge q

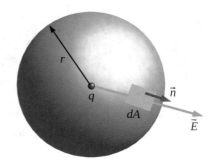

Fig. 2.15 Gaussian surfaces for calculation of the electric field created by an insulating solid sphere of radius a has a uniform volume charge density ρ and carries a total positive charge Q

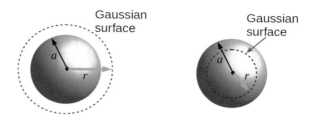

Exercise 2.6 Consider an insulating solid sphere of radius a, which has a uniform volume charge density ρ and carries a total positive charge Q. Determine the magnitude of the electric field at a point (a) outside the sphere, and (b) inside the sphere.

Solution 2.6 (a) Because the charge distribution is spherically symmetric, we again select a spherical Gaussian surface of radius r, concentric with the sphere (see Fig. 2.15), and

$$\Phi_E = \oint_S \mathbf{E} \cdot d\mathbf{A} \tag{2.42}$$

$$= \oint_S (\mathbf{E} \cdot \mathbf{n}) dA$$

$$= \oint_S E_n dA = E_n \oint_S dA$$

$$= E_n(4\pi r^2) = \frac{Q_{in}}{\epsilon_0}$$

$$= \frac{Q}{\epsilon_0}$$

or

$$E_n \equiv E = \frac{Q}{4\pi\epsilon_0 r^2} \tag{2.43}$$

$$= k_e \frac{Q}{r^2}$$

(b) We select a spherical Gaussian surface with $r < a$, concentric with the insulated sphere, as shown in Fig. 2.15. Let us denote the volume of this smaller sphere by V_0. To calculate Q_{in}, we use

$$Q_{in} = \rho V_0 = \rho \frac{4}{3} \pi r^3 \tag{2.44}$$

$$= \frac{Q}{\frac{4}{3} \pi a^3} \frac{4}{3} \pi r^3$$

$$= Q \frac{r^3}{a^3} \quad (< Q)$$

Using the Gauss's law,

$$\Phi_E = \oint_S \mathbf{E} \cdot d\mathbf{A} \tag{2.45}$$

$$= \oint_S (\mathbf{E} \cdot \mathbf{n}) dA$$

$$= \oint_S E_n dA = E_n \oint_S dA$$

$$= E_n (4\pi r^2) = \frac{Q_{in}}{\epsilon_0}$$

$$= \frac{1}{\epsilon_0} Q \frac{r^3}{a^3}$$

Or

$$E_n \equiv E = \frac{Q}{4\pi r^2 \epsilon_0} \frac{r^3}{a^3} \tag{2.46}$$

$$= k_e \frac{Qr}{a^3}$$

Exercise 2.7 Suppose a thin spherical shell of radius a has a total charge Q distributed uniformly over its surface (see Fig. 2.16). What is the electric field at points (a) outside and (b) inside the shell?

Solution 2.7 (a) For finding E outside the spherical shell, we select a spherical Gaussian surface of radius $r > a$, concentric with the spherical shell, as shown in Fig. 2.16, and

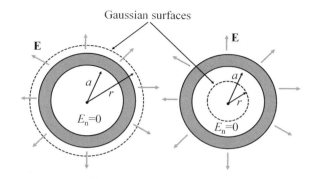

Fig. 2.16 A thin spherical shell of radius a with a total charge Q distributed uniformly over its surface, and the Gaussian surfaces

$$\Phi_E = \oint_S \mathbf{E} \cdot d\mathbf{A} \tag{2.47}$$

$$= \oint_S (\mathbf{E} \cdot \mathbf{n}) dA = \oint_S E_n dA$$

$$= E_n \oint_S dA = E_n(4\pi r^2) = \frac{Q_{in}}{\epsilon_0}$$

$$= \frac{Q}{\epsilon_0}$$

or

$$E_n \equiv E = \frac{Q}{4\pi\epsilon_0 r^2} \tag{2.48}$$

$$= k_e \frac{Q}{r^2}$$

(b) For finding E inside the spherical shell, we select a spherical Gaussian surface of radius r ($r < a$), concentric with the spherical shell (see also Fig. 2.16), and

$$\Phi_E = \oint_S \mathbf{E} \cdot d\mathbf{A} \tag{2.49}$$

$$= \oint_S (\mathbf{E} \cdot \mathbf{n}) dA = \oint_S E_n dA$$

$$= E_n \oint_S dA$$

$$= E_n(4\pi r^2)$$

$$= \frac{Q_{in}}{\epsilon_0} = 0$$

or

$$E_n \equiv E = 0 \tag{2.50}$$

Fig. 2.17 A line of positive
charge of infinite length and
constant charge per unit
length λ. Gaussian surface is
a cylinder of length ℓ and a
radius of the base r, where
$d\mathbf{A}$ is a small surface
element vector

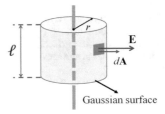

Gaussian surface

Exercise 2.8 What is the electric field at a distance r from a line of positive
charge of infinite length and constant charge per unit length λ (see Fig. 2.17)?

Solution 2.8 Because of the symmetry of the charge distribution, the vector $\mathbf{E}$ is
perpendicular to the line charge and directed outward, as shown in Fig. 2.17. Based
on symmetry of the charge distribution, a cylindrical Gaussian surface of radius r
and length ℓ can be selected that is coaxial with the line charge. From Gauss's law:

$$\Phi_E = \oint_S \mathbf{E} \cdot d\mathbf{A} \tag{2.51}$$

$$= \oint_S (\mathbf{E} \cdot \mathbf{n}) dA$$

$$= \oint_S E_n dA$$

$$= E_n \oint_S dA = E_n (2\pi r \ell)$$

$$= \frac{Q_{in}}{\epsilon_0} = \frac{\lambda \ell}{\epsilon_0}$$

or

$$E_n \equiv E = \frac{\lambda \ell}{2\pi \epsilon_0 r \ell} \tag{2.52}$$

$$= 2k_e \frac{\lambda}{r}$$

Exercise 2.9 Determine the electric field of a nonconducting, infinite plane of
positive charge with uniform surface charge density σ, as shown in Fig. 2.18.

Fig. 2.18 The infinite plane of positive charge with uniform surface charge density σ and a Gaussian cylinder surface

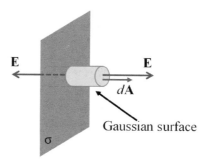

Solution 2.9 By symmetry, **E** must be perpendicular to the plane and must have the same magnitude at all points equidistant from the plane. The fact that the direction of **E** is away from positive charges indicates that the direction of **E** on one side of the plane must be opposite to its direction on the other side, as shown in the figure. We chose a Gaussian surface, which because of the symmetry is a small cylinder with an axis perpendicular to the plane and with both ends have an area A and are equidistant from the plane. From Gauss's law,

$$\Phi_E = \oint_A \mathbf{E} \cdot d\mathbf{A} \tag{2.53}$$

$$= \int_{Base1} (\mathbf{E} \cdot \mathbf{n}) dA + \int_{Base2} (\mathbf{E} \cdot \mathbf{n}) dA$$

$$= 2EA = \frac{Q_{in}}{\epsilon_0} = \frac{\sigma A}{\epsilon_0}$$

or

$$E = \frac{\sigma}{2\epsilon_0} \tag{2.54}$$

Exercise 2.10 Let us consider a solid conducting sphere of radius a carrying a net positive charge $2Q$ (see Fig. 2.19). Suppose a conducting spherical shell of inner radius b and outer radius c is concentric with the solid sphere and it carries a net negative charge $-Q$. Find the electric field in the regions labeled 1, 2, 3, and 4. What is the charge distribution on the shell when the entire system is in electrostatic equilibrium?

Solution 2.10 The charge distributions on the sphere and shell have spherical symmetry around their common center. To determine the electric field at different dis-

Fig. 2.19 The charge distributions on both the sphere and the shell are characterized by spherical symmetry around their common center

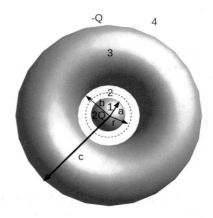

tances r from the center, we construct a spherical Gaussian surface for each of the four regions of interest.

1. Region 1: From Gauss's law,

$$\Phi_E = \oint_S \mathbf{E} \cdot d\mathbf{A} \tag{2.55}$$
$$= \frac{Q_{in}}{\epsilon_0}$$
$$= 0 \quad (Q_{in} = 0)$$

This implies that $E = 0$.

2. Region 2: From Gauss's law,

$$\Phi_E = \oint_S \mathbf{E} \cdot d\mathbf{A} \tag{2.56}$$
$$= \oint_S E\, dA = E(4\pi r^2)$$
$$= \frac{Q_{in}}{\epsilon_0}$$
$$= \frac{2Q}{\epsilon_0}$$

Thus, we obtain the electric field as

$$E = \frac{1}{4\pi\epsilon_0} \frac{2Q}{r^2} \tag{2.57}$$
$$= k_e \frac{2Q}{r^2}$$

3. Region 3: The electric field must be zero because the spherical shell is also a conductor in equilibrium. From Gauss's law,

$$\Phi_E = \oint_S \mathbf{E} \cdot d\mathbf{A} \tag{2.58}$$
$$= 0$$
$$= \frac{Q_{in}}{\epsilon_0}$$

Thus, we get $Q_{in} = 0$.

Therefore, the charge on the inner surface of the spherical shell is $-2Q$ to cancel the charge $+2Q$ on the solid sphere.

4. Region 4: From Gauss's law

$$\Phi_E = \oint_S \mathbf{E} \cdot d\mathbf{A} \tag{2.59}$$
$$= \oint_S Ed A = E(4\pi r^2) = \frac{Q_{in}}{\epsilon_0}$$
$$= \frac{2Q - Q}{\epsilon_0}$$

We obtain the electric field as

$$E = \frac{1}{4\pi \epsilon_0} \frac{Q}{r^2} \tag{2.60}$$
$$= k_e \frac{Q}{r^2}$$

Reference

Holliday D, Resnick R, Walker J (2011) Fundamentals of physics. John Wiley and Sons, New York

Chapter 3
Electrostatic Potential

This chapter aims to introduce electrostatic potential and the methods used to calculate it.

In this chapter, we will introduce electrostatic potential and discuss the methods used to calculate the electrostatic potential. Furthermore, we present the potential difference and equipotential surfaces. Moreover, we will discuss the calculation of the electrostatic potential for a single point charge and system of charges. As extra reading material, the reader can also consider other literature (Holliday et al. 2011; Jackson 1999; Griffiths 1999).

3.1 Electrostatic Potential Energy

Suppose a test charge q_0 is placed in the electric field $\mathbf{E}$ of some other charged object. Then, electric force on q_0 is $\mathbf{F} = q_0\mathbf{E}$. If $\mathbf{E}$ is created by a system of charges $q_1, \ldots, q_N$, then from superposition principle

$$\mathbf{E}_R = \sum_{i=1}^{N} \mathbf{E}_i \tag{3.1}$$

The electric force acting on q_0 is

$$\mathbf{F} = \sum_{i=1}^{N} q_0\mathbf{E}_i = q_0\mathbf{E}_R \tag{3.2}$$

© The Author(s), under exclusive license to Springer Nature Switzerland AG 2022
H. Kamberaj, *Electromagnetism*, Undergraduate Texts in Physics,
https://doi.org/10.1007/978-3-030-96780-2_3

Therefore, if the test charge q_0 moves in the electric field $\mathbf{E}$, the electrostatic forces (see also Eq. (3.2)) do work on q_0.

Suppose that q_0 moves in the field $\mathbf{E}$ by some external agent, then the work done by electric field is negative of the work done by external agent. Let $d\mathbf{s}$ be an infinitesimal displacement in the electric field, then work done by the field on test charge q_0 is calculated as

$$dW_e = \mathbf{F} \cdot d\mathbf{s} = q_0 \mathbf{E} \cdot d\mathbf{s} = -dU \tag{3.3}$$

**Elec-
tro-
static
poten-
tial
energy**

In Eq. (3.3), $-dU$ is the decrease of the potential energy of *charge-field* system. Therefore,

$$dU = -dW_e = -q_0 \mathbf{E} \cdot d\mathbf{s} \tag{3.4}$$

That is, the electric field's work $\mathbf{E}$ decreases electrostatic potential energy of the charge moving in the field.

3.2 Electric Potential

For a finite displacement of charge from A to some B, the change in potential energy of the system charge-field ΔU is

$$\Delta U = U_B - U_A = \int_A^B dU = -q_0 \int_A^B \mathbf{E} \cdot d\mathbf{s} \tag{3.5}$$

The integral in Eq. (3.5) is called *path integral* or *line integral*, and it is performed along the path that q_0 follows as it moves from A to B. Since electric force is *conservative force*, then the integral does not depend on the path taken from A to B.

**Elec-
tric
poten-
tial**

The quantity $\dfrac{U}{q_0}$ is independent of q_0, but it depends only on $\mathbf{E}$.

By definition, the ratio $\dfrac{U}{q_0}$ is called *electric potential* ϕ:

$$\phi = \frac{U}{q_0} \tag{3.6}$$

**Elec-
tric
poten-
tial
differ-
ence**

Equation (3.6) implies that electric potential, ϕ, is a scalar quantity.

Potential difference is the difference of electric potential between two points A and B:

$$\Delta\phi = \phi_B - \phi_A \tag{3.7}$$

$$= \frac{\Delta U}{q_0}$$

$$= -\int_A^B \mathbf{E} \cdot d\mathbf{s}$$

Note that potential difference, $\Delta\phi$, is different physical quantity than change in potential energy, ΔU:

$$\Delta U = q_0 \Delta\phi \tag{3.8}$$

Electric potential is a scalar quantity characterizing the electric field. It is independent of the test charge placed in the field. On the other hand, the potential energy refers to a charge-field system, and hence it is not independent of the test charge placed in the field.

Moreover, the change in potential energy, ΔU, of a test charge q_0 is minus the work done by the electric field on the charge (or it is equal to the work done on the charge by the external agent). The potential difference, $\Delta\phi = \phi_B - \phi_A$, is equal to the work per unit charge that an external agent does to move a test charge from A to B without accelerating it (or without changing the kinetic energy of the test charge).

Similar to potential energy, only differences in electric potential are meaningful. Often, it is avoided to work with potential differences; therefore, by convention, the value of the electric potential is considered zero at some convenient point in an electric field. Practically, it is established that the electric potential is zero at a location that is at an infinite distance from the system of charges producing the field.

Then, the electric potential at an arbitrary point P in an electric field is equal to the work needed to move a positive unit test charge from a point at the infinity to that point. Thus, the electric potential at any point P is

$$\phi_P = \phi_P - \phi_\infty = -\int_\infty^P \mathbf{E} \cdot d\mathbf{s} \tag{3.9}$$

From Eq. (3.9), SI units of electric potential is $N \cdot m/C = J/C$, which is defined as Volt (V):

$$1\,V \equiv 1\,\frac{J}{C} \tag{3.10}$$

SI units for electric field could also be

$$1\,\frac{N}{C} = 1\,\frac{V}{m} \tag{3.11}$$

Often, in atomic and nuclear physics, the units of energy are given in *electron volt* (eV). By definition, 1 eV is the energy an electron (or proton) gains or loses by moving it through a potential difference of 1 V:

$$1 \, \text{eV} = 1.60 \times 10^{-19} \, \text{C} \cdot (1 \, \text{V}) = 1.60 \times 10^{-19} \, \text{J} \tag{3.12}$$

where the charge of 1.60×10^{-19} C is a fundamental charge equal to the magnitude of the charge of an electron.

3.3   Potential Difference in a Uniform Electric Field

Consider the motion of a point charge q in a uniform electric field **E**, as shown in Fig. 3.1. As a result of interaction with the electric field, the charge moves from A to B, assuming that it is a positive test charge.

First, we express the electric potential difference between points A and B:

$$\Delta\phi = \phi_B - \phi_A = -\int_A^B \mathbf{E} \cdot d\mathbf{s} \tag{3.13}$$

We also define $\mathbf{E} = -E\mathbf{j}$, where $\mathbf{j}$ is a unit vector along y-axis. Moreover, $d\mathbf{s} = -ds\mathbf{j}$. Then, Eq. (3.14) can also be written as

$$\Delta\phi = -\int_A^B E(\mathbf{j} \cdot \mathbf{j}) ds = -E\int_A^B ds = -Ed \tag{3.14}$$

where $(-)$ sign indicates that $\phi_B < \phi_A$. Therefore, the electric field lines always point in the direction of decreasing electric potential.

Since the electric potential ϕ is a characteristic of the electric field **E**, for the same field **E** as in Fig. 3.1, and an arbitrary test charge q_0 (positive or negative) moving

Fig. 3.1 Movement of a charge q in a uniform electric field **E**

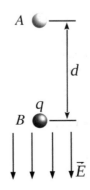

Fig. 3.2 Movement of a
charge q in a uniform electric
field $\mathbf{E}$ along a diagonal
direction from A to B

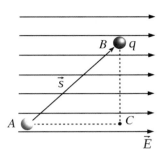

from A to B, the change in potential energy of the charge-field system is

$$\Delta U = q_0 \Delta \phi = -q_0 E d \tag{3.15}$$

If $q_0 > 0$, then $\Delta U < 0$; that is, a positive charge loses its potential energy when
it moves in the direction of an electric field, and the field does work on the charge.
Therefore, when released from the rest in an external electric field, a positive charge
particle gains acceleration in the direction of the external electric field.

On the other hand, if $q_0 < 0$, then $\Delta U > 0$; that is, a negative charge increases
its potential energy when it displaces along the electric field direction. Therefore,
if a negative charge releases from rest in the field $\mathbf{E}$, it accelerates in the opposite
direction to the external electric field.

In the following, we consider a more general case. Assume a charged particle
moves freely between any two points in a uniform electric field directed along the
x-axis, as indicated in Fig. 3.2. Let $\mathbf{s}$ be the displacement vector between A and B,
then

$$\Delta \phi = \phi_B - \phi_A \tag{3.16}$$

$$= -\int_A^B \mathbf{E} \cdot \mathbf{s}$$

$$= -\mathbf{E} \cdot \int_A^B d\mathbf{s}$$

$$= -\mathbf{E} \cdot \mathbf{s}$$

The change in the potential energy of a charge q_0 is

$$\Delta U = q_0 \Delta \phi = -q_0 \mathbf{E} \cdot \mathbf{s} \tag{3.17}$$

Fig. 3.3 Movement of a
charge q in a uniform electric
field $\mathbf{E}$ along a diagonal
direction from A to B

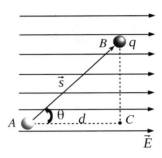

3.4 Equipotential Surface

Consider again the potential difference between the points A and B, see Fig. 3.3, as

$$\Delta\phi = \phi_B - \phi_A = -\mathbf{E}\cdot\mathbf{s} = -Es\cos\theta = -Ed \qquad (3.18)$$

On the other hand, the potential difference between the points C and A is

$$\Delta\phi = \phi_C - \phi_A = -Ed \qquad (3.19)$$

Combining Eqs. (3.18) and (3.19), we obtain the following potential difference relationship:

$$\phi_B - \phi_A = \phi_C - \phi_A \qquad (3.20)$$

Thus, $\phi_B = \phi_C$.

**Equipo-
tential
surface**   By definition, the *equipotential surface* is called any surface consisting of a continuous distribution of points having the same electric potential.

Using Eq. (3.8), since $\phi_B = \phi_C$ for any two points B and C in an equipotential surface

$$\Delta U = q_0(\phi_C - \phi_B) = 0 \qquad (3.21)$$

Hence, no work is done in moving a test charge between any two points on an equipotential surface.

The equipotential surfaces of a uniform electric field consist of a family of planes that are all perpendicular to the field.

3.5 Electric Potential of a Point Charge

Consider an isolated positive point charge q, as indicated in Fig. 3.4. Note that such a charge produces an electric field that is directed radially outward from the charge, as discussed in Chap. 1:

Fig. 3.4 Electric potential of a point charge q

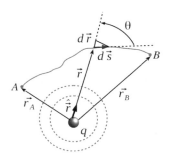

$$\mathbf{E} = k_e \frac{q}{r^2} \hat{\mathbf{r}} \tag{3.22}$$

Consider an arbitrary path from some point A to the point B, and a small displacement vector along that path $d\mathbf{s}$ at the position $\mathbf{r}$ relative to the charge q, as shown in Fig. 3.4. Using Eq. (3.7), the potential difference is

$$\phi_B - \phi_A = -\int_A^B \mathbf{E} \cdot d\mathbf{s} \tag{3.23}$$

$$= -\int_A^B k_e \frac{q}{r^2} \hat{\mathbf{r}} \cdot d\mathbf{s}$$

$$= -\int_A^B k_e \frac{q}{r^2} \cos\theta \, ds$$

$$= -\int_A^B k_e \frac{q}{r^2} dr$$

$$= k_e q \left(\frac{1}{r_B} - \frac{1}{r_A} \right)$$

where r_A and r_B are the distances of A and B relative to q, and $\hat{\mathbf{r}}$ is a unit vector along the radial direction. In Eq. (3.23), θ is angle between the small displacement vector $d\mathbf{s}$ along the path between A and B and the radial small displacement vector $d\mathbf{r}$, as in Fig. 3.4.

Equation (3.23) implies that the electric potential of any arbitrary charge q at a distance r from the charge is given as

$$\phi(r) = k_e \frac{q}{r} \tag{3.24}$$

which is a function of the distance r from the charge q.

Equation (3.24) indicates that if q is a positive charge, then the work done to bring a positively charged object from infinite toward the charge q (where $r = 0$) increases to infinite and it is positive. That is, the region around a positive charge q is analogous to a hill for a positively charged object approaching q. In contrast, for a negative charge q, the work is done to bring a positively charged object from infinite toward the charge q decreases, and it is negative. Hence, the region around a negative charge q is analogous to a hole for a positively charged object approaching q. Furthermore, when a charged object is infinitely distant from another charge, the electric potential surface is flat and has a value of zero.

3.6 Electric Potential of a System of Point Charges

Let $q_1, q_2, \ldots, q_N$ be a set of N static discrete charges, positive or negative, as shown in Fig. 3.5. Based on superposition principle, the electric potential resulting from those point charges at some point P, with position vector $\mathbf{r}$ with respect to the origin of the reference frame, is

$$\phi(\mathbf{r}) = k_e \sum_{i=1}^{N} \frac{q_i}{|\mathbf{r} - \mathbf{r}_i|} \tag{3.25}$$

where $\mathbf{r}_i$ is the position vector of the ith charge with respect to the origin O, as indicated in Fig. 3.5.

Equation (3.25) indicates that the total potential at any point P of a set of N point charges is the sum of the potentials due to the individual charges.

In particular, for a system of two charged particles, we denote by ϕ_1 the electric potential created by the charge q_1 at a point P at distance r from the charge q_1, which is taken to be at the origin O of a coordinate system:

$$\phi_1 = k_e \frac{q_1}{r} \tag{3.26}$$

The work done by an external agent to move the second charge q_2 from infinity to P without accelerating it (i.e., the kinetic energy remains constant) is

Fig. 3.5 A set of N static discrete charges $q_1, q_2, \ldots, q_N$. P is an observation point at position vector $\mathbf{r}$ from the origin O of a coordinate system. The position vectors of the charges with respect to the origin O are denoted by $\mathbf{r}_i$, $i = 1, 2, \ldots, N$, and $\mathbf{r} - \mathbf{r}_i$ gives the relative position vector of P with charge i

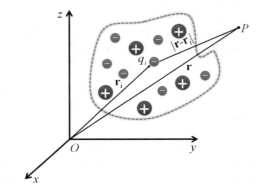

$$W = q_2 \phi_1 \qquad (3.27)$$

Therefore, from Eq. (3.27), the work is equal to the interaction potential energy U_{12} of the particles, when they are separated by a distance r_{12}:

$$U_{12} = k_e \frac{q_1 q_2}{r_{12}} \qquad (3.28)$$

For two particles with the same sign of charges, U_{12} is positive, which is in agreement with what we know from the previous discussion that positive work has to be done by an external agent on the system to move the two charges near one another. That is also in agreement with the view that charges with the same sign repel each other (see Chap. 1). On the other hand, if the charges have opposite sign, U_{12} is negative; that is, the external agent does a negative work to move the charges near each other against the attractive force between charges (in this case, it is the field which does the positive work).

When the system is composed of more than two charged particles, we can similarly obtain the total potential energy of the system. For that, to calculate U, we need to know the pair-wise potential between every pair of charges, then we sum up the terms algebraically:

$$U = \sum_{i=1}^{N} \sum_{j=1>i}^{N} k_e \frac{q_i q_j}{r_{ij}} \qquad (3.29)$$

$$= \frac{1}{2} \sum_{i=1}^{N} \sum_{j=1 \neq i}^{N} k_e \frac{q_i q_j}{r_{ij}}$$

where r_{ij} is the distance between the charges i and j:

$$r_{ij} = \sqrt{(x_i - x_j)^2 + (y_i - y_j)^2 + (z_i - z_j)^2} \qquad (3.30)$$

For a physical interpretation of the expression given by Eq. (3.29), imagine that q_i is fixed at the position, but all other $q_{j \neq i}$ are at the infinity. In that case, an external agent has to do work to move any of the charges q_j from infinity to its position near q_i given as $k_e \frac{q_i q_j}{r_{ij}}$. Furthermore, that is independent of the order in which the charges are displaced.

Note that Eq. (3.29) can also be written as

$$U = \frac{1}{2} \sum_{i=1}^{N} q_i \phi_i (\mathbf{r}_i) \qquad (3.31)$$

where $\phi_i(\mathbf{r}_i)$ is electric potential created by all other charges at the position where the ith charge is located:

$$\phi_i(\mathbf{r}_i) = \sum_{\substack{j=1 \neq i}}^{N} k_e \frac{q_j}{|\mathbf{r}_i - \mathbf{r}_j|} \tag{3.32}$$

It is interesting to note the behavior of the electric potential ϕ given by Eq. (3.24) for $r \to 0$. From Eq. (3.24), we can see that $\phi(r \to 0) \to \infty$. Now, let us consider an elementary charged particle in its own electric potential, then the so-called self-energy or the energy of the field created by that charge is given as

$$U_{self} = q\phi(r = 0) = \infty \tag{3.33}$$

Equation (3.33) is a contradiction because the self-energy of a static particle is finite and given as mc^2, based on the theory of relativity. Therefore, that result (Eq. (3.33)) will give a rest mass of a charged particle, for example, consider an electron, infinite. To solve that contradiction, we will assume that every charged particle is surrounded by a screening sphere such that within that sphere the classical electrostatic theory does not apply, and this theory is valid only outside that region. By simple consideration, we can calculate the radius of that sphere, which is also called classical radius, r_0, which identifies the boundary between the classical ($r \geq r_0$) and a higher accuracy region called quantum electrodynamics region ($r < r_0$). To calculate r_0, we equalize the self-energy at $r = r_0$ with the rest energy mc^2, where c is the speed of light in vacuum and m is the rest mass:

$$q\phi(r_0) = k_e \frac{q^2}{r_0} = mc^2 \tag{3.34}$$

Solving it for r_0, we obtain the classical radius as

$$r_0 = k_e \frac{q^2}{mc^2} \tag{3.35}$$

For example, if the charged particle is an electron, then r_0 is

$$r_0 = \left(8.9875 \times 10^9 \ \frac{\mathrm{N \cdot m^2}}{\mathrm{C^2}}\right) \frac{(-1.6 \times 10^{-19} \ \mathrm{C})^2}{(9.1 \times 10^{-31} \ \mathrm{kg})\left(3 \times 10^8 \ \frac{\mathrm{m}}{\mathrm{s}}\right)^2} \tag{3.36}$$

$$= 0.28 \times 10^{-14} \ \mathrm{m}$$

Considering that the size of atoms is of the order 10^{-10} m, then the classical electromagnetic theory could be a valid method to describe the electron-electron and electron-nucleus interactions within an atom; however, it may fail to describe the interactions within nuclei, which have a size of the order 10^{-15} m.

3.7   Electric Potential of a Continuous Charge Distribution

We can partition the volume into macroscopically small charge elements dq, as shown in Fig. 3.6. Then, we consider the potential due to macroscopically small charge element dq by treating this element as a point charge:

$$d\phi = k_e \frac{dq}{|\mathbf{r} - \mathbf{r}'|} \tag{3.37}$$

To obtain the total potential at point P, we integrate to include contributions from all elements of the charge distribution:

$$\phi = k_e \int \frac{dq}{|\mathbf{r} - \mathbf{r}'|} \tag{3.38}$$

If we assume that $dq = \rho \, dV$, where ρ is the charge density inside the macroscopically small volume $dV = d\mathbf{r}'$, then

$$\phi(\mathbf{r}) = k_e \int_V \rho(\mathbf{r}') \frac{dV}{|\mathbf{r} - \mathbf{r}'|} \tag{3.39}$$

which is a function of $\mathbf{r}$.

For a continuous charge distribution, the potential interaction energy can be written as

$$U = \frac{k_e}{2} \int_V \int_V \frac{\rho(\mathbf{r})\rho(\mathbf{r}')}{|\mathbf{r} - \mathbf{r}'|} \, d\mathbf{r} \, d\mathbf{r}' \tag{3.40}$$

Using Eq. (3.39), we can write Eq. (3.40) in the following convenient form:

Fig. 3.6 A continuous charge distribution in a volume V with density $\rho(\mathbf{r}')$ over a macroscopically small volume centered at point $\mathbf{r}'$. dq denotes an elementary charge

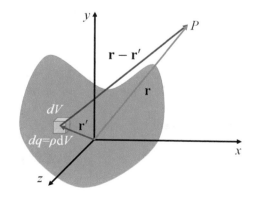

$$U = \frac{1}{2} \int_V \rho(\mathbf{r})\phi(\mathbf{r}) \, d\mathbf{r} \tag{3.41}$$

3.8   Differential form of Electric Potential

For an elementary displacement vector $d\mathbf{s}$, the potential difference is given as the following:

$$d\phi = -\mathbf{E} \cdot d\mathbf{s} \tag{3.42}$$

or

$$\mathbf{E} = -\frac{d\phi}{d\mathbf{s}} \tag{3.43}$$

It can also be written as

$$\mathbf{E} = -\nabla \phi \tag{3.44}$$

where ∇ is the gradient. Furthermore, we can write in Cartesian coordinates as

$$\mathbf{E} = -\left(\frac{\partial \phi(x, y, z)}{\partial x}\mathbf{i} + \frac{\partial \phi(x, y, z)}{\partial y}\mathbf{j} + \frac{\partial \phi(x, y, z)}{\partial z}\mathbf{k} \right) \tag{3.45}$$

where $\mathbf{i}, \mathbf{j}, \mathbf{k}$ are unit vectors along x, y, and z axes, respectively. Therefore,

$$E_x = -\frac{\partial \phi(x, y, z)}{\partial x} \tag{3.46}$$

$$E_y = -\frac{\partial \phi(x, y, z)}{\partial y}$$

$$E_z = -\frac{\partial \phi(x, y, z)}{\partial z}$$

In spherical coordinates, Eq. (3.44) can be written as

$$\mathbf{E} = -\left(\hat{\mathbf{r}}\frac{\partial \phi(r, \theta, \psi)}{\partial r} + \hat{\boldsymbol{\theta}}\frac{1}{r}\frac{\partial \phi(r, \theta, \psi)}{\partial \theta} + \hat{\boldsymbol{\psi}}\frac{1}{r \sin\theta}\frac{\partial \phi(r, \theta, \psi)}{\partial \psi} \right) \tag{3.47}$$

where $\hat{\mathbf{r}}, \hat{\boldsymbol{\theta}}, \hat{\boldsymbol{\psi}}$ are unit vectors along r, θ, and ψ directions, respectively. Therefore,

$$E_r = -\frac{\partial \phi(r, \theta, \psi)}{\partial r} \tag{3.48}$$

$$E_\theta = -\frac{1}{r}\frac{\partial \phi(r, \theta, \psi)}{\partial \theta}$$

$$E_\psi = -\frac{1}{r \sin\theta}\frac{\partial \phi(r, \theta, \psi)}{\partial \psi}$$

For a distribution of charges producing an electric field with a spherical symmetry (i.e., the electrostatic potential depends only on r, $\phi = \phi(r)$), then $E = E(r)$:

$$E_r = -\frac{\partial \phi(r)}{\partial r} \tag{3.49}$$

$$E_\theta = E_\psi = 0$$

3.9 Multipole Expansion

In Eq. (3.39), the charge density $\rho(\mathbf{r}')$ describes the distribution of charges localized in a finite volume V, and it is zero outside that volume. Often, we are interested for the potential ϕ outside the region of the volume V at the distances such that $|\mathbf{r}| \gg |\mathbf{r}'|$. Therefore, an expansion of the term $1/|\mathbf{r} - \mathbf{r}'|$ around $\mathbf{r}' = 0$ in Eq. (3.39) can be used to identify the value of the potential in different regions.

Using direct Taylor expansion of $1/|\mathbf{r} - \mathbf{r}'|$ around $\mathbf{r}' = 0$, we write

$$\frac{1}{|\mathbf{r} - \mathbf{r}'|} = \frac{1}{r} \tag{3.50}$$

$$+ \left(\nabla_{\mathbf{r}'} \frac{1}{|\mathbf{r} - \mathbf{r}'|} \right)_{\mathbf{r}'=0} \cdot \mathbf{r}$$

$$+ \frac{1}{2} \sum_{i,j=1}^{3} \frac{\partial^2}{\partial x_i' \partial x_j'} \left(\frac{1}{|\mathbf{r} - \mathbf{r}'|} \right)_{\mathbf{r}'=0} x_i x_j + \cdots$$

$$= \frac{1}{r}$$

$$+ \frac{\mathbf{r}' \cdot \mathbf{r}}{r^3}$$

$$+ \frac{1}{2} \sum_{i \neq j=1}^{3} \frac{\partial^2}{\partial x_i' \partial x_j'} \left(\frac{1}{|\mathbf{r} - \mathbf{r}'|} \right)_{\mathbf{r}'=0} x_i x_j$$

$$+ \frac{1}{2} \sum_{i=1}^{3} \frac{\partial^2}{\partial x_i' \partial x_i'} \left(\frac{1}{|\mathbf{r} - \mathbf{r}'|} \right)_{\mathbf{r}'=0} x_i^2 + \cdots$$

$$= \frac{1}{r}$$

$$+ \frac{\mathbf{r}' \cdot \mathbf{r}}{r^3}$$

$$+ \frac{1}{2} \sum_{i \neq j=1}^{3} \frac{3 x_i' x_j'}{r^5} x_i x_j - \frac{1}{2} \sum_{i}^{3} \frac{3(x_i')^2}{r^5} x_i^2 + \cdots$$

$$= \frac{1}{r}$$

$$+ \frac{\mathbf{r}' \cdot \mathbf{r}}{r^3}$$

$$+ \frac{1}{2} \sum_{i,j=1}^{3} \frac{3x_i' x_j' - \delta_{ij}(r')^2}{r^5} x_i x_j + \cdots$$

where δ_{ij} is the Kronecker's number: $\delta_{ij} = 1$ if $i = j$, otherwise $\delta_{ij} = 0$. Substituting Eq. (3.50) into Eq. (3.39), we obtain

$$\phi(\mathbf{r}) = k_e \frac{1}{r} \int_V \rho(\mathbf{r}')d\mathbf{r}' \tag{3.51}$$

$$+ k_e \frac{1}{r^3} \left(\int_V \rho(\mathbf{r}')\mathbf{r}'d\mathbf{r}' \right) \cdot \mathbf{r}$$

$$+ k_e \frac{1}{r^5} \left(\sum_{i,j=1}^{3} \frac{1}{2} \int_V \rho(\mathbf{r}') \left(3x_i'x_j' - \delta_{ij}(r')^2\right) d\mathbf{r}' \right) x_i x_j + \cdots$$

We can introduce the following physical quantities:

$$Q = \int_V \rho(\mathbf{r}')d\mathbf{r}' \quad \text{(monopole)} \tag{3.52}$$

$$\mathbf{p} = \int_V \rho(\mathbf{r}')\mathbf{r}'d\mathbf{r}' \quad \text{(dipole)}$$

$$Q_{ij} = \int_V \rho(\mathbf{r}') \left(3x_i'x_j' - \delta_{ij}(r')^2\right) d\mathbf{r}' \quad \text{(quadrupole)}$$

where the first expression gives the so-called electric "monopole", or the total charge in the volume V, the second term defines the electric dipole moment, and the third term is the trace-less quadrupole moment tensor element. Using the quantities in Eq. (3.52), we obtain the following convenient expression for the potential:

$$\phi(\mathbf{r}) = k_e \frac{Q}{r} + k_e \frac{\mathbf{p} \cdot \mathbf{r}}{r^3} + \frac{k_e}{2} \sum_{i,j=1}^{3} Q_{ij} \frac{x_i x_j}{r^5} + \Theta \left(\frac{1}{r^4}\right) \tag{3.53}$$

where the first term is the zeroth-order approximation, which gives the electrostatic potential that would have created a point charge equal to the total charge in the confined free space volume V as it was placed at the origin of the reference frame (i.e., $\mathbf{r}' = 0$); the second term gives the dipole electrostatic potential, which takes into account polarity of the charge distribution; the third term gives the quadrupole electrostatic potential. In typical calculations, higher-order terms in Eq. (3.53) (i.e., terms of order $\Theta(1/r^4)$) can be ignored, in particular, when the electrostatic potential is required at some point far from the origin (or, $r \gg r'$).

This formalism can straightforward apply to a system of static discrete charges q_i for $i = 1, 2, \ldots, N$. In that case, the charge density at any point $\mathbf{r}$ in space is written as

$$\rho(\mathbf{r}) = \sum_{i=1}^{N} q_i \delta (\mathbf{r} - \mathbf{r}_i) \tag{3.54}$$

where δ is the delta function such that

$$\delta (\mathbf{r} - \mathbf{r}_i) = \begin{cases} 1, \ \mathbf{r} = \mathbf{r}_i \\ 0, \ \mathbf{r} \neq \mathbf{r}_i \end{cases} \tag{3.55}$$

Using Eq. (3.52), we can calculate the dipole moment as

$$\mathbf{p} = \int_V \sum_{i=1}^{N} q_i \delta (\mathbf{r}' - \mathbf{r}_i) \, \mathbf{r}' \, d\mathbf{r}' = \sum_{i=1}^{N} q_i \mathbf{r}_i \tag{3.56}$$

Combining Eqs. (3.54) and (3.52), we can also calculate the total charge of system (or monopole) as

$$Q = \int_V \sum_{i=1}^{N} q_i \delta \left(\mathbf{r}' - \mathbf{r}_i\right) \, d\mathbf{r}' = \sum_{i=1}^{N} q_i \tag{3.57}$$

which is an expected result. Furthermore, the quadrupole of the discrete system of static charges is

$$Q_{ij} = \int_V \left(\sum_{k=1}^{N} q_k \delta (\mathbf{r}' - \mathbf{r}_k) \right) \left(3x_i' x_j' - \delta_{ij} (r')^2 \right) d\mathbf{r}' \tag{3.58}$$

$$= \sum_{k=1}^{N} q_k \left(3x_{ki} x_{kj} - \delta_{ij} r_k^2 \right)$$

Then, Eq. (3.53) can be used to calculate the electric potential at any point P as in the case of the continuous charge distribution.

Note that dipole moment depends on the origin of the coordinate system, and it is independent of that origin only if the system is neutral (see the Exercises). Furthermore, it is straightforward to show that the quadrupole has a trace equal to zero: $\sum_{i=1}^{3} Q_{ii} = 0$ (see also exercises).

We can calculate the electric field using Eq. (3.44). In the following we are evaluating the gradient of each term in Eq. (3.53). The first term is rather simple to be evaluated:

$$\nabla \phi(0) = k_e Q \nabla \left(\frac{1}{r} \right) = -k_e Q \frac{\mathbf{r}}{r^3} \tag{3.59}$$

The gradient of the second term gives

$$\nabla \left(k_e \frac{\mathbf{p} \cdot \mathbf{r}}{r^3} \right) = k_e \left(\nabla_x \left(\frac{\mathbf{p} \cdot \mathbf{r}}{r^3} \right) \mathbf{i} + \nabla_y \left(\frac{\mathbf{p} \cdot \mathbf{r}}{r^3} \right) \mathbf{j} + \nabla_z \left(\frac{\mathbf{p} \cdot \mathbf{r}}{r^3} \right) \mathbf{k} \right) \qquad (3.60)$$

$$= k_e \left(\frac{r^2 p_x - 3(\mathbf{p} \cdot \mathbf{r})x}{r^5} \mathbf{i} \right.$$

$$+ \frac{r^2 p_y - 3(\mathbf{p} \cdot \mathbf{r})y}{r^5} \mathbf{j}$$

$$\left. + \frac{r^2 p_z - 3(\mathbf{p} \cdot \mathbf{r})z}{r^5} \mathbf{k} \right)$$

$$= k_e \frac{(\mathbf{r} \cdot \mathbf{r})\mathbf{p} - 3(\mathbf{p} \cdot \mathbf{r})\mathbf{r}}{r^5}$$

where $\mathbf{i}$, $\mathbf{j}$, and $\mathbf{k}$ are unit vectors along the x-, y-, and z-axis, respectively. Furthermore, the gradient of the third term is

$$\nabla \left(\frac{k_e}{2} \sum_{i,j=1}^{3} Q_{ij} \frac{x_i x_j}{r^5} \right) = \frac{k_e}{2} \left(\sum_{k=1}^{3} \frac{\partial}{\partial x_k} \left(\sum_{i,j=1}^{3} Q_{ij} \frac{x_i x_j}{r^5} \right) \hat{\mathbf{x}}_k \right) \qquad (3.61)$$

$$= \frac{k_e}{2} \left(\sum_{i,j=1}^{3} Q_{ij} \sum_{k=1}^{3} \frac{\partial}{\partial x_k} \left(\frac{x_i x_j}{r^5} \right) \hat{\mathbf{x}}_k \right)$$

$$= \frac{k_e}{2} \left(\sum_{i,j=1}^{3} Q_{ij} \sum_{k=1}^{3} \frac{r^2(\delta_{ik}x_j + x_i \delta_{jk}) - 5x_i x_j x_k}{r^7} \hat{\mathbf{x}}_k \right)$$

which are all terms of order $\Theta(1/r^4)$ that are ignored from our discussion. In Eq. (3.60), $\hat{\mathbf{x}}_k$ denotes a unit vector along one of the axes ($k = x, y, z$).

Combining Eqs. (3.44), (3.59), and (3.60), we obtain multipole expansion of the electric field as follows:

$$\mathbf{E}(\mathbf{r}) = k_e Q \frac{\mathbf{r}}{r^3} \qquad (3.62)$$

$$+ k_e \frac{3(\mathbf{p} \cdot \mathbf{r})\mathbf{r} - (\mathbf{r} \cdot \mathbf{r})\mathbf{p}}{r^5}$$

$$+ \Theta \left(\frac{1}{r^4} \right)$$

For convenience, we use the expression of the unit vector $\hat{\mathbf{r}}$ along the direction from the origin of reference frame toward the point P:

$$\hat{\mathbf{r}} = \frac{\mathbf{r}}{r} \qquad (3.63)$$

and Eq. (3.62) is written in the following convenient form:

$$\mathbf{E}(\mathbf{r}) = k_e \frac{Q}{r^2}\hat{\mathbf{r}} + k_e \frac{3(\mathbf{p} \cdot \hat{\mathbf{r}})\hat{\mathbf{r}} - \mathbf{p}}{r^3} + \Theta\left(\frac{1}{r^4}\right) \tag{3.64}$$

In Eq. (3.64), the first term gives electric field created by a point charge placed at the origin of the reference frame with a charge equal to the charge distributed in all free space, and the second term gives the electric dipole moment field of a dipole at the origin of coordinate system. Note that if the dipole has its center some position $\mathbf{r}_0$ related to the origin of coordinate system, then Eq. (3.64) can be modified as

$$\mathbf{E}(\mathbf{r}) = k_e \frac{Q}{r^2}\hat{\mathbf{r}} + k_e \frac{3(\mathbf{p} \cdot \hat{\mathbf{r}})\hat{\mathbf{r}} - \mathbf{p}}{|\mathbf{r} - \mathbf{r}_0|^3} + \Theta\left(\frac{1}{r^4}\right) \tag{3.65}$$

3.10 Electric Potential of a Charged Conductor

We can show that the surface of a charged conductor in equilibrium is an equipotential surface. For that, consider two points A and B on the surface of a charged conductor, as shown in Fig. 3.7. Along a surface path connecting these points, $\mathbf{E}$ is always perpendicular to the displacement $d\mathbf{s}$; that is, $\mathbf{E} \cdot d\mathbf{s} = 0$. Therefore,

$$\phi_B - \phi_A = -\int_A^B \mathbf{E} \cdot d\mathbf{s} = 0 \tag{3.66}$$

which implies that $\phi_B = \phi_A$.

Therefore, ϕ is constant everywhere on the surface of a charged conductor in equilibrium. That is, the surface of any charged conductor in electrostatic equilibrium is an equipotential surface. Besides, the electric field is zero inside the conductor, that is

$$E_r = -d\phi/dr = 0 \tag{3.67}$$

Fig. 3.7 Electric potential on the surface of a charged conductor in electrostatic equilibrium

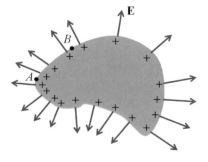

Fig. 3.8 A conducting
sphere of radius R and total
positive charge Q. Two
Gaussian surfaces are drawn
for $r > R$ and $r < R$

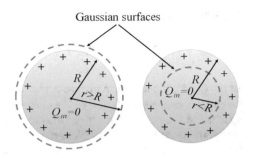

which indicates that the electric potential ϕ is constant everywhere inside the conductor, including the surface. Therefore, the potential ϕ inside the conductor in equilibrium is equal to its value at the surface.

As an example, let us consider a solid metal conducting sphere of radius R and total positive charge Q (see also Fig. 3.8). Using Gauss's law with Gaussian surfaces as indicated in Fig. 3.8, we can calculate the electric flux as

$$\Phi_E = \oint_S \mathbf{E} \cdot d\mathbf{A} = E4\pi r^2 = \frac{Q}{\epsilon_0} \tag{3.68}$$

Equation (3.68) implies that the magnitude of electric field is

$$E = \begin{cases} k_e \dfrac{Q}{r^2}, & r \geq R \\ 0, & r < R \end{cases} \tag{3.69}$$

In addition, electric field vector is given as

$$\mathbf{E} = \begin{cases} k_e \dfrac{Q}{r^2}\hat{\mathbf{r}}, & r \geq R \\ 0, & r < R \end{cases} \tag{3.70}$$

Using Eq. (3.14), we can calculate the potential difference between a point B at the infinity and a point A at the surface of the sphere with radius R as

$$\phi_\infty - \phi_R = -\int_R^\infty k_e \frac{Q}{r^2} dr = -k_e \frac{Q}{R} \tag{3.71}$$

Therefore, we get

$$\phi_R = k_e \frac{Q}{R} \tag{3.72}$$

If B is an interior point, $r < R$, then

$$\phi_r - \phi_R = -\int_R^r \mathbf{E} \cdot d\mathbf{r} \tag{3.73}$$

$$= -\int_R^r 0 \cdot dr = 0$$

Then, we obtain

$$\phi_r = k_e \frac{Q}{R} \tag{3.74}$$

On the other hand, if B is an outside point, $r > R$, then

$$\phi_\infty - \phi_r = -\int_r^\infty \mathbf{E} \cdot d\mathbf{r} \tag{3.75}$$

$$= -\int_r^\infty k_e \frac{Q}{r^2} dr = -k_e \frac{Q}{r}$$

Therefore,

$$\phi_r = k_e \frac{Q}{r} \tag{3.76}$$

For a nonspherical conductor, the surface charge density σ is not uniformly dis-tributed along the surface. That is, higher surface charge density is obtained at the regions with small radius of curvature or where the surface is convex. In contrast, σ is low at the small radius of curvature regions or where the surface is concave. Furthermore, we know that the electric field just outside the conductor is given as $E_n = \sigma/\epsilon_0$. Therefore, the electric field is large near convex points having small radii of curvature and reaches very high values at sharp points.

3.10.1 Cavity Within a Conductor

Let us consider a cavity of no charge within a conductor of any shape, as shown in Fig. 3.9. Based on Gauss's law ($Q_{in} = 0!$), the electric field inside the cavity is zero independent of the charge distribution on the outside surface of the conductor. Furthermore, the field within the cavity is zero, even if the conductor is in an external electric field because every point inside the conductor is at the same electric potential. For instance, consider the points A and B inside the conductor and on the surface of the cavity. Then, for those points:

$$\phi_A = \phi_B \tag{3.77}$$

On the other hand, using Eq. (3.14), we have

$$\phi_B - \phi_A = -\int \mathbf{E} \cdot d\mathbf{s} \tag{3.78}$$

Fig. 3.9 A conductor of arbitrary shape containing a cavity with no charges inside the cavity

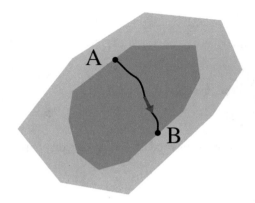

where $\mathbf{E}$ is electric field inside cavity. Since $\phi_A = \phi_B$, $\int \mathbf{E} \cdot d\mathbf{s} = 0$, or $\mathbf{E} = 0$.

3.11   Exercises

Exercise 3.1 Assume a battery produces a specified potential difference of 12 V between conductors attached to the battery terminals. It is connected between two parallel plates (see Fig. 3.10) at a separation of $d = 0.30$ cm. Assuming that the electric field between the plates is uniform, what is the electric field between the plates?

Solution 3.1 The field points from the positive plate (A) to the negative plate (B). Furthermore, the positive plate is at a higher electric potential than the negative plate. The potential difference between the plates must be equal to the potential difference between the battery terminals:

$$\phi_B - \phi_A = -Ed \tag{3.79}$$

The magnitude of E is

$$E = \frac{\mid \phi_B - \phi_A \mid}{d} = \frac{12 \text{ V}}{0.30 \times 10^{-2} \text{ m}} = 4.0 \times 10^3 \text{ V/m} \tag{3.80}$$

Fig. 3.10 A battery of 12 V is connected between two parallel plates at separation $d = 0.30$ cm

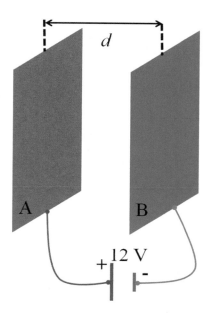

Exercise 3.2 Consider a proton released at rest in a uniform electric field of magnitude of 8.0×10^4 V/m and is directed along the positive x-axis, as indicated in Fig. 3.11. The proton undergoes a displacement of 0.50 m in the direction of **E**. What is (a) the electric potential difference of points A and B? (b) the proton's potential energy change for this displacement?

Solution 3.2 For proton $q_0 = | e | = 1.60 \times 10^{-19}$ C.

(a) Potential difference is

$$\Delta\phi = -Ed = -(8.0 \times 10^4 \text{ V/m})(0.50 \text{ m}) = -4.0 \times 10^4 \text{ V} \qquad (3.81)$$

(b) Change in potential energy is

$$\Delta U = q_0 \Delta\phi \qquad (3.82)$$
$$= (1.60 \times 10^{-19} \text{ C})(-4.0 \times 10^4 \text{ V})$$
$$= -6.4 \times 10^{-15} \text{ J}$$

$(-)$ sign indicates that the proton's potential energy decreases during the displacement along the electric field direction. That is, proton gains kinetic energy and at the same time loses electric potential energy.

Fig. 3.11 A proton
undergoing a displacement
of d in the direction of $\mathbf{E}$

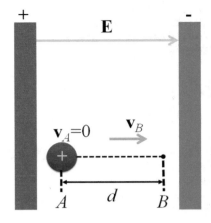

Exercise 3.3 A charge $q_1 = 2.00\ \mu$C is located at the origin, and a charge $q_2 = -6.00\ \mu$C is located at $(0, 3.00)$ m (see also Fig. 3.12). (a) What is the total electric potential created by these charges at the point P with coordinates $(4.00, 0)$ m? (b) What is the potential energy change of a charge of $3.00\ \mu$C when it moves from infinity to point P?

Solution 3.3 (a) Total electric potential at P is

$$\phi_P = k_e \left(\frac{q_1}{r_1} + \frac{q_2}{r_2} \right) \tag{3.83}$$
$$= \left(8.99 \times 10^9\ \frac{\text{N} \cdot \text{m}^2}{\text{C}^2} \right) \left(\frac{2.00 \times 10^{-6}\ \text{C}}{4.00\ \text{m}} + \frac{-6.00 \times 10^{-6}\ \text{C}}{5.00\ \text{m}} \right)$$
$$= -6.29 \times 10^3\ \text{V}$$

(b) For the charge at infinity: $U_i = 0$, then charge is at point P,

$$U_P = q_3 \phi_P = -18.9 \times 10^{-3}\ \text{J} \tag{3.84}$$

Fig. 3.12 A system of
interacting charges

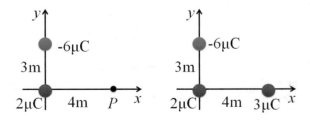

Then the potential energy $\Delta U = U_P - U_i = -18.9 \times 10^{-3}$ J.

The work done by an external agent to move the charge from point P back to infinity is

$$W = -\Delta U = 18.9 \times 10^{-3} \text{ J} \tag{3.85}$$

which is positive work.

Exercise 3.4 A charge $q_1 = 2.00 \ \mu\text{C}$ is located at the origin, and a charge $q_2 = -6.00 \ \mu\text{C}$ is located at $(0, 3.00)$ m. Find the total potential energy.

Solution 3.4 Total potential energy is

$$U = k_e \sum_{i=1}^{3} \sum_{j>i}^{3} \frac{q_i q_j}{r_{ij}} \tag{3.86}$$

$$= k_e \left(\frac{q_1 q_2}{r_{12}} + \frac{q_1 q_3}{r_{13}} + \frac{q_2 q_3}{r_{23}} \right)$$

$$= -5.48 \times 10^{-2} \text{ J}$$

Exercise 3.5 Consider an electric dipole consisting of two charges of equal magnitude and opposite sign at a distance $2a$ from each other (see Fig. 3.13). The dipole vector is taken parallel to x-axis and it is centered at the origin. (a) Calculate the electric potential at point P. (b) Calculate ϕ and E_x at a point far from the dipole.

Fig. 3.13 An electric dipole

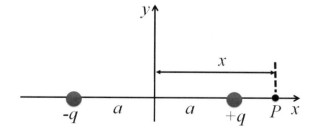

Solution 3.5   (a)   Total electric potential

$$\phi_P = k_e \frac{q_1}{r_1} + k_e \frac{q_2}{r_2} \tag{3.87}$$

$$= k_e \left(\frac{q}{x - a} - \frac{q}{x + a} \right)$$

$$= 2k_e \frac{qa}{x^2 - a^2}$$

The electric field

$$E_x = -\frac{d\phi}{dx} = 4k_e \frac{qax}{(x^2 - a^2)^2} \tag{3.88}$$

(b)   For $x \gg a$, we have $x^2 - a^2 \approx x^2$, thus

$$\phi_P \approx 2k_e \frac{qa}{x^2} \tag{3.89}$$

and the electric field

$$E_x = -\frac{d\phi}{dx} \approx 4k_e \frac{qa}{x^3} \tag{3.90}$$

Exercise 3.6   (a) Define the expression of the electric potential at point P
located at the central axis perpendicular to a uniformly charged ring of radius a
and total charge Q (see Fig. 3.14). (b) What is the expression for the magnitude
of the electric field at point P?

Solution 3.6   (a)   Orient the ring with its plane perpendicular to an x-axis and its
center is at the origin. Then take point P to be at a distance x from the center of
ring. The charge element dq is at a distance $\sqrt{x^2 + a^2}$ from point P. Hence, we
can express ϕ as

Fig. 3.14 A uniformly
charged ring of radius a and
total charge Q

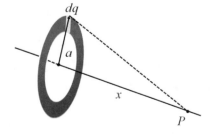

$$\phi = \int k_e \frac{dq}{r} \tag{3.91}$$

$$= \int k_e \frac{dq}{\sqrt{x^2 + a^2}}$$

$$= k_e \frac{1}{\sqrt{x^2 + a^2}} \int dq$$

$$= k_e \frac{Q}{\sqrt{x^2 + a^2}}$$

(b) From symmetry, we see that along the x-axis $\mathbf{E}$ can have only an x component. Then, from

$$\mathbf{E} = -\nabla \phi \tag{3.92}$$

we get

$$E_x = -\frac{d\phi}{dx} \tag{3.93}$$

$$= -k_e Q \frac{d}{dx} (x^2 + a^2)^{-1/2}$$

$$= k_e \frac{Qx}{(x^2 + a^2)^{3/2}}$$

Exercise 3.7 Find (a) the electric potential and (b) the magnitude of electric field along the perpendicular central axis of a uniformly charged disk of radius a and surface charge density σ (see Fig. 3.15).

Solution 3.7 (a) Orient the disk with its plane perpendicular to an x-axis and its center is at the origin. Then take point P to be at a distance x from the center of disk. Take a ring with width dr and radius r, then charge element $dq = \sigma(2\pi r dr)$

Fig. 3.15 A uniformly charged disk of radius a and surface charge density σ

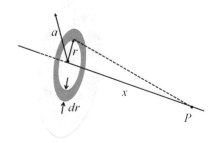

is at a distance $\sqrt{x^2 + r^2}$ from point P. Hence, we can express $d\phi$ as in the previous exercise.

$$d\phi = 2\pi k_e \frac{\sigma r \, dr}{\sqrt{x^2 + r^2}} \tag{3.94}$$

Then, the total ϕ is

$$\phi = \int_0^a 2\pi k_e \frac{\sigma r \, dr}{\sqrt{x^2 + r^2}} \tag{3.95}$$

$$= \pi k_e \sigma \int_0^a \frac{d(x^2 + r^2)}{\sqrt{x^2 + r^2}}$$

$$= \pi k_e \sigma \left[2\sqrt{x^2 + r^2} \right]_0^a$$

$$= 2\pi k_e \sigma \left[\sqrt{x^2 + a^2} - x \right]$$

(b) From symmetry, we see that along the x-axis $\mathbf{E}$ can have only an x component. Then, from

$$\mathbf{E} = -\nabla \phi \tag{3.96}$$

we get

$$E_x = -\frac{d\phi}{dx} \tag{3.97}$$

$$= 2\pi k_e \sigma \left[1 - \frac{x}{\sqrt{x^2 + a^2}} \right]$$

Exercise 3.8 Show that the electric dipole moment of a system of discrete charges depends on the origin of the coordinate system and it is independent of that only if the system of charges is neutral.

Solution 3.8 Consider that the origin of the coordinate system is shifted by $\mathbf{r}_0$, then the position vector of a charge i, $\mathbf{r}_i$ related to the origin O of the old system can be related to the position vector $\mathbf{r}'_i$ of that charge in the new coordinate system as: $\mathbf{r}'_i = \mathbf{r}_i - \mathbf{r}_0$, see also Fig. 3.16. Using Eq. (3.56), we can write the dipole moment in the new reference frame as

$$\mathbf{p}' = \sum_{i=1}^N q_i \mathbf{r}'_i \tag{3.98}$$

Fig. 3.16 A discrete set of static charges relative to two reference frames, which have a shift on their origins given by $\mathbf{r}_0$

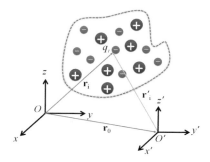

$$= \sum_{i=1}^{N} q_i \, (\mathbf{r}_i - \mathbf{r}_0)$$

$$= \sum_{i=1}^{N} q_i \mathbf{r}_i - \mathbf{r}_0 \sum_{i=1}^{N} q_i$$

$$= \mathbf{p} - \mathbf{r}_0 Q$$

where $Q = \sum_{i=1}^{N} q_i$. That result implies that dipole moment depends on the origin. However, if the system of charges is neutral, that is, $Q = 0$, then $\mathbf{p} = \mathbf{p}'$.

Exercise 3.9 Calculate the electric dipole moment and the quadrupole moment of the system of four charges with a configuration as shown in Fig. 3.17. $q = 3 \, \mu C$ gives the magnitude of each charge.

Solution 3.9 Using Eq. (3.56), we can write the dipole moment in the given reference frame as

Fig. 3.17 A configuration of four charges with $q = 3 \, \mu C$

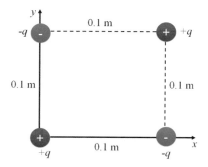

$$\mathbf{p} = \sum_{i=1}^{4} q_i \mathbf{r}_i \tag{3.99}$$

Then, each component of the dipole moment vector is

$$p_x = \sum_{i=1}^{4} q_i x_i \tag{3.100}$$

$$= (+3.0 \ \mu\text{C})(0.0 \ \text{m}) + (-3.0 \ \mu\text{C})(0.1 \ \text{m})$$
$$+ (3.0 \ \mu\text{C})(0.1 \ \text{m}) + (-3.0 \ \mu\text{C})(0.0 \ \text{m}) = 0.0 \ (\text{C} \cdot \text{m})$$

$$p_y = \sum_{i=1}^{4} q_i y_i \tag{3.101}$$

$$= (+3.0 \ \mu\text{C})(0.0 \ \text{m}) + (-3.0 \ \mu\text{C})(0.0 \ \text{m})$$
$$+ (3.0 \ \mu\text{C})(0.1 \ \text{m}) + (-3.0 \ \mu\text{C})(0.1 \ \text{m}) = 0.0 \ (\text{C} \cdot \text{m})$$

$$p_z = 0.0 \ (\text{C} \cdot \text{m}) \tag{3.102}$$

Now, we can calculate the components of the quadrupole moment as

$$Q_{xx} = \sum_{i=1}^{N} q_i \left(3x_i x_i - r_i^2\right) \tag{3.103}$$

$$= q \left(-3(0.1)^2 + (0.1)^2 + 3(0.1)^2 - 2(0.1)^2 - 3(0.0)^2 + (0.1)^2\right)$$
$$= (3.0 \ \mu\text{C}) \left(-0.03 + 0.01 + 0.03 - 0.02 + 0.01\right) = 0.0 \ (\text{C} \cdot \text{m}^2)$$

$$Q_{xy} = \sum_{i=1}^{N} q_i 3x_i y_i \tag{3.104}$$

$$= (3.0 \ \mu\text{C})3(0.1)(0.1) = 0.09 \times 10^{-6} \ (\text{C} \cdot \text{m}^2)$$

$$Q_{xz} = 0.0 \ (\text{C} \cdot \text{m}^2) \tag{3.105}$$

$$Q_{yx} = \sum_{i=1}^{N} q_i 3 y_i x_i \tag{3.106}$$

$$= (3.0 \ \mu\text{C})3(0.1)(0.1) = 0.09 \times 10^{-6} \ (\text{C} \cdot \text{m}^2)$$

$$Q_{yy} = \sum_{i=1}^{N} q_i \left(3 y_i y_i - r_i^2\right) \tag{3.107}$$

$$= q \left(-3(0.0)^2 + (0.1)^2 + 3(0.1)^2 - 2(0.1)^2 - 3(0.1)^2 + (0.1)^2\right)$$
$$= (3.0 \ \mu\text{C}) \left(0.01 + 0.03 - 0.02 - 0.03 + 0.01\right) = 0.0 \ (\text{C} \cdot \text{m}^2)$$

$$Q_{yz} = 0.0 \ (\text{C} \cdot \text{m}^2) \tag{3.108}$$

$$Q_{zz} = Q_{zy} = Q_{zx} = 0.0 \ (\text{C} \cdot \text{m}^2) \tag{3.109}$$

Exercise 3.10 Show that the quadrupole moment tensor has a trace equal to zero.

Solution 3.10 Using Eq. (3.56), we can write the diagonal of the quadrupole as

$$Q_{ii} = \sum_{k=1}^{N} q_k \left(3x_{ki}x_{ki} - r_k^2 \right) \tag{3.110}$$

Then, we calculate the trace of the quadrupole moment as follows:

$$\text{Trace}(\mathbf{Q}) = \sum_{i=1}^{3} Q_{ii} \tag{3.111}$$

$$= \sum_{i=1}^{3} \sum_{k=1}^{N} q_k \left(3x_{ki}x_{ki} - r_k^2 \right)$$

$$= \sum_{k=1}^{N} q_k \sum_{i=1}^{3} \left(3x_{ki}x_{ki} - r_k^2 \right)$$

$$= \sum_{k=1}^{N} q_k \left(3r_k^2 - 3r_k^2 \right) = 0$$

References

Holliday D, Resnick R, Walker J (2011) Fundamentals of physics. John Wiley and Sons, New York
Jackson JD (1999) Classical electrodynamics, 3rd edn. John Wiley and Sons, New York
Griffiths DJ (1999) Introduction to electrodynamics, 3rd edn. Prentice Hall, Hoboken

Chapter 4
Capacitance and Dielectrics

This chapter aims to introduce capacitance and molecular theory of dielectrics.

In this chapter, we will introduce capacitance and dielectrics. Then, we discuss the electrostatics of macroscopic media and introduce a molecular theory of dielectrics. Also, we will introduce electric polarization, and then derive Maxwell's equations for an electrostatic field in both free space and dielectric media. As extra reading material, the reader can also consider other literature (Holliday et al. 2011; Griffiths 1999).

4.1 Capacitance

We first start with the definition of a capacitor. For that, consider two charged conductors with charges of equal magnitude but of opposite sign, as shown in Fig. 4.1. Suppose they are combined, forming a so-called *capacitor*. The conductors are also called plates. Note that a potential difference $\Delta\phi$ exists between the conductors due to the presence of two different types of charges in each conductor.

Moreover, because the unit of potential difference is the volt, a potential difference is often called a *voltage*. Here, we denoted it ΔV, which is equal to absolute value of $\Delta\phi$, $\Delta V \equiv |\Delta\phi|$.

The capacity of an electric circuit to accumulate electric charge at a particular value of ΔV is called the *capacity*. Based on the experimental results, the amount of charge Q on a capacitor depends linearly on the potential difference ΔV between the two conductors. Furthermore, the constant of proportionality depends on the shape and distance between the conductors, as in the following demonstrated for a planar capacitor.

Side notes: Capacitor / Voltage / Capacity

© The Author(s), under exclusive license to Springer Nature Switzerland AG 2022
H. Kamberaj, *Electromagnetism*, Undergraduate Texts in Physics,
https://doi.org/10.1007/978-3-030-96780-2_4

Fig. 4.1 A capacitor formed
by two conductors carrying
charges Q and $-Q$

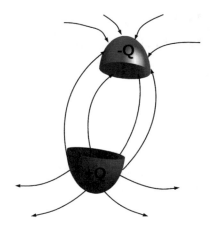

We can write this relationship, mathematically, as

$$Q = C\Delta V \tag{4.1}$$

Here, Q is the amount of charge in each capacitor, that is, $Q = |+Q| = |-Q|$, and

Capac-itance

hence, $C > 0$.

Therefore, the capacitance C of a capacitor (see also Eq. (4.1)) represents the ratio
of the magnitude of charge on each of conductors to the magnitude of the potential
difference between them:

$$C \equiv \frac{Q}{\Delta V} \tag{4.2}$$

From the definition (Eq. (4.2)), capacitance is always a positive quantity.
Furthermore, the potential difference ΔV in Eq. (4.2) is a positive quantity
($\Delta V > 0$). Moreover, the potential difference ΔV increases linearly with the
stored charge Q, and hence the ratio $Q/\Delta V$ is constant for a given capacitor
C.

Therefore, capacitance measures the ability of a capacitor to store charge
and electric potential energy, and hence it depends on the shape and distance
of the capacitor's plates. The capacitance has SI units of coulombs per volt.
The SI unit of capacitance is named *farad* (F), in honor of Michael Faraday:

$$1\,\mathrm{F} = 1\frac{\mathrm{C}}{\mathrm{V}} \tag{4.3}$$

The capacitance of electronic device ranges from picofarads to microfarads, and hence, the capacitance is labeled in "mF" for microfarads, "mmF" for micromicrofarads, or "pF" for picofarads.

4.2 Calculating Capacitance

To demonstrate the calculation of the capacitance of a capacitor, we consider a capacitor formed from a pair of parallel plates; each plate is connected to one terminal of a battery acting as a source of potential difference, as shown in Fig. 4.2. Thus, the voltage is ΔV.

We assume that the capacitor is initially uncharged, and the battery establishes an electric field in the connecting wires when the connections are made. The directions of the electric field lines are explained in Fig. 4.2 (on the left); that is,

$$\phi_+ - \phi_A = -\int_A^{(+)} \mathbf{E} \cdot d\mathbf{s} \tag{4.4}$$

where $d\mathbf{s}$ is a small displacement vector along the left wire, and

$$\phi_- - \phi_B = -\int_B^{(-)} \mathbf{E} \cdot d\mathbf{s} \tag{4.5}$$

where $d\mathbf{s}$ is a small displacement vector along the right wire. Initially, when $Q = 0$ on both plates $\phi_A = \phi_B = 0$, and hence $\mathbf{E} \neq 0$. Therefore, on the plate connected to the negative terminal of the battery, the electric field exerts a force on electrons, which are in the wire just outside this plate; the electrons accelerate to move onto the plate and hence starts charging the plate negatively. On the other hand, the electric field exerts forces on electrons of the side (which is closer to the wire) of the plate

Fig. 4.2 A capacitor formed by two parallel plates connected to the terminals of a battery at the potential difference ΔV

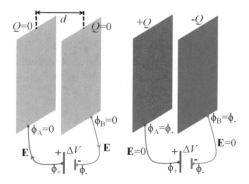

connected to the positive terminal of the battery. It accelerates the electrons to move onto the wire. Hence, leaving on the plate more positive charges in comparison to negative one (electrons); therefore, this plate is charged positively.

Those accelerations continue until the plates, the wires, and the terminals are all at the same electric potential. That is also illustrated in Fig. 4.2 (on the right). Once the equilibrium point is attained, no potential differences exist between the terminals and the plates on both sides, and hence; as a result, no electric field is present in the wire, $\mathbf{E} = 0$, and the movement of electrons stops. The right plate carries a negative charge, $-Q$, and left plate a positive charge, $+Q$. When the equilibrium establishes, the potential difference between the capacitor plates becomes that between the terminals of the battery.

To calculate the capacitance of a pair of oppositely charged conductors by an amount of charge Q, we calculate the potential difference ΔV as

$$\Delta V =| \phi_+ - \phi_- | \tag{4.6}$$

Then, the capacitance is evaluated using the expression given by Eq. (4.2). Note that for simple geometry of the capacitor, these calculations are relatively easy. In the following, we will discuss spherical shape and planar shapes of conductors.

4.2.1 Spherical Conductors

Let us consider first the capacitance of an isolated spherical conductor of radius R and charge Q concentric with a hollow sphere of infinite radius, which forms the second conductor making up the capacitor. We derived the electric potential of the sphere of radius R as

$$\phi_R = k_e \frac{Q}{R} \tag{4.7}$$

Since at infinity $\phi_\infty = 0$, we obtain

$$C = \frac{Q}{\Delta V} = \frac{Q}{k_e Q/R} = \frac{R}{k_e} = 4\pi\epsilon_0 R \tag{4.8}$$

where $\Delta V =| \phi_R - \phi_\infty |$. Equation (4.8) indicates that the capacitance of an isolated charged sphere depends only on its radius R, and it is independent of both the charge Q on the sphere and potential difference ΔV.

Fig. 4.3 The electric field
between the plates of a
parallel-plate capacitor is
uniform near the center but
nonuniform near the edges

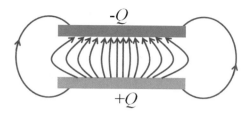

4.2.2 Parallel-Plate Capacitors

Now, let us consider a capacitor composed of two parallel conductor plates of equal
area A, which are at a distance d, see also Fig. 4.3. One of the plates carries a
charge $+Q$, and the other $-Q$. Note that charges of like sign repel one another and
that charges of opposite signs attract one another (see also Chap. 1). As a battery is
charging a capacitor, electrons flow into the negative plate and out of the positive
plate (see Fig. 4.2).

Note that the electric field between the plates of a parallel-plate capacitor is uniform near the center but nonuniform near the edges. When the capacitor plates are
large, the accumulated charges can distribute themselves over a substantial area, and
hence the amount of charge stored on each plate Q, for a given potential difference
ΔV, increases as the plate area increases to ensure a constant surface charge density
σ. A simple argument can be used for that: because the electric field just outside
the conductor is perpendicular to the surface of the conductor and with magnitude
$E = \sigma/\epsilon_0$, where E is proportional to constant ΔV, then σ is constant. Thus, we
expect the capacitance C to be proportional to the plate area A.

Above we derived a relationship between the electric field between the plates and
magnitude of potential difference, given as

$$E = \frac{\Delta V}{d} \qquad (4.9)$$

From Eq. (4.9), we see that when d decreases, E increases, for fixed ΔV. If we
move the plates closer together (that is, d decreases), We also consider the situation
before the charges have moved in response to that change, such that no charges have
moved. Hence, the electric field between the plates is the same but extends over a
shorter distance between plates. That situation corresponds to a new capacitor with
a potential difference between the plates that is different from the terminal voltage
of the battery. Now, across the wires connecting the battery to the capacitor exists a
potential difference (see also Fig. 4.2 for an illustration).

Based on the arguments that we discussed for a situation in Fig. 4.2, that potential
difference creates an electric field in the wires that drives more charges onto the
plates, which in turn increases the potential difference between the plates of the
capacitor. When it becomes equal to the potential difference between the terminals

of the battery (Fig. 4.2), the potential difference across the wires falls back to zero. Then, the flow of charge stops.

This simple experiment shows that decreasing the distance between the plates, d, yields an increase in the amount of charge stored on the capacitor. The opposite is also true; if d increases, the charge Q decreases. Therefore, this experiment indicates that the capacitance is inversely proportional to d.

To verify these physical arguments, we start writing the surface charge density on either plate as

$$\sigma = \frac{Q}{A} \tag{4.10}$$

If the plates are very close together (in comparison with their length and width), we can assume that the electric field is uniform between the plates and is zero elsewhere. The value of the electric field between the plates is

$$E = \frac{\sigma}{\epsilon_0} = \frac{Q}{\epsilon_0 A} \tag{4.11}$$

Because the field between the plates is uniform, the magnitude of the potential difference between the plates equals Ed; therefore,

$$\Delta V = Ed = \frac{Qd}{\epsilon_0 A} \tag{4.12}$$

The capacitance is

$$C = \frac{Q}{\Delta V} = \frac{Q}{\dfrac{Qd}{\epsilon_0 A}} \tag{4.13}$$

Or

$$C = \frac{\epsilon_0 A}{d} \tag{4.14}$$

Equation (4.14) implies that the capacitance of a capacitor made up of two parallel plates is proportional to the area A of each plate and inversely proportional to the plate separation d. That is in agreement with our expectations from the conceptual argument given above.

4.3 Combination of Capacitors

Often, in an electric circuit, two or more capacitors are combined. If that is the case, for convenience, we can calculate the equivalent capacitance of certain combinations using the following methods. Note that there exist circuit symbols for capacitors and batteries, as well as the color codes used for them, as shown in Fig. 4.4. As we can see,

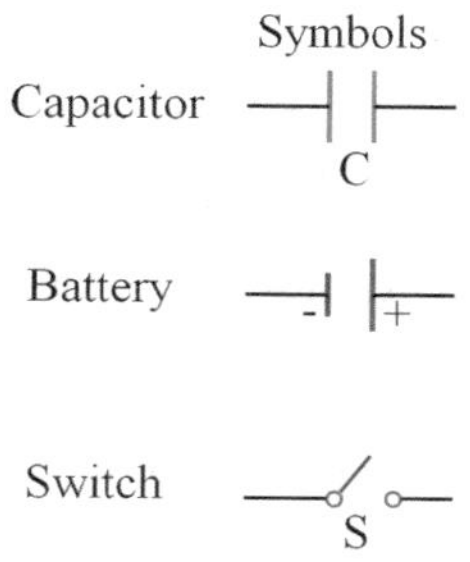

Fig. 4.4 The circuit symbols
and color code for capacitors,
batteries, and switches

Symbols

Capacitor

C

Battery

Switch

S

the symbol for the capacitor manifests the geometry of the most common capacitor
model of a pair of parallel plates. Also, the longer vertical line in a circuit symbol
with appropriate color indicates that the positive polarity of the battery is at the higher
potential.

4.3.1 Parallel Combination

Figure 4.5 presents a combination of two capacitors connected in parallel. Also,
we show a circuit diagram for this combination of capacitors, as often seen in an
electric circuit. Note from Fig. 4.5 that the left plates of the capacitors connect to the
positive terminal of the battery using conducting wires; therefore, those plates, after
equilibrium of the electric potential establishes, are at the same electric potential as
the positive terminal of the battery. For the same reason, the right plate connecting
to the negative terminal of the battery has equal electric potential with the negative
terminal after the equilibrium of the electric potential establishes. As a result, the
potential differences across each capacitor connected in parallel are the same and
equal to the voltage applied to the battery; that is, $\Delta V_1 = \Delta V_2 = \Delta V$.

Applying the model described above in Fig. 4.2, when two capacitors are initially
connected in a circuit, as shown in Fig. 4.5, electrons migrate between the wires
and the plates. As a result, the left plates charge positively, and the right plates

Fig. 4.5 Two capacitors
connected in parallel in an
electric circuit

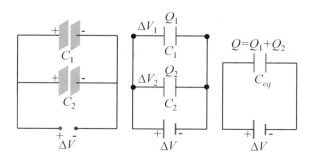

negatively. In other words, the internal chemical energy stored in the battery is the source of that migration; that is, the internal chemical energy of the battery converts into electric potential energy associated with the surface charges in the plates of the capacitors at a separation d. During the process of the electrons migration, the voltage across the capacitors becomes equal to that across the battery terminals and then charge transfer stops. When that establishes in the circuit, the capacitors load to their maximum charge capacity.

In the following, we show a few steps to calculate the equivalent capacitance, C_{eq}, of the combinations of C_1 and C_2. For that, we denote by Q_1 and Q_2 the maximum charges on each capacitor, respectively, and by Q the total charge stored by the two capacitors:

$$Q = Q_1 + Q_2 \qquad (4.15)$$

Q is also the charge stored in the capacitor C_{eq}. The voltages applied across each capacitor are the same, see also Fig. 4.5, and hence the charges in each capacitor are

$$Q_1 = C_1 \Delta V \qquad (4.16)$$
$$Q_2 = C_2 \Delta V$$

Replacing these two capacitors by one equivalent capacitor with a capacitance C_{eq} is equivalent to a new electric circuit having one capacitor, as shown in Fig. 4.5; that is, the effect of capacitor C_{eq} on the circuit is exactly the same as the effect of the combination of two capacitors. In other words, the equivalent capacitor stores the same charge Q when connected to the same battery with a voltage across the battery terminals ΔV, and hence the voltage across the equivalent capacitor is ΔV. Therefore, we have

$$Q = C_{eq} \Delta V \qquad (4.17)$$

Then, we obtain

$$C_{eq} \Delta V = C_1 \Delta V + C_2 \Delta V = (C_1 + C_2) \Delta V \qquad (4.18)$$

or

$$C_{eq} = C_1 + C_2 \qquad (4.19)$$

Note that the above description can be extended to three or more capacitors connected in parallel. For example, if N capacitors, $C_1, C_2, \ldots, C_N$, connect in parallel in an electric circuit, we find the equivalent capacitance to be

$$C_{eq} = C_1 + C_2 + \cdots + C_N \qquad (4.20)$$

4.3.2 Series Combination

Next, we consider an electric circuit in which two capacitors are combined in series, as shown in Fig. 4.6. That is known as a series combination of capacitors. In that combination, the left plate of capacitor 1 connects to one of the terminals of a battery (for example, the positive terminal in Fig. 4.6) and the right plate of capacitor 2 connects to the other terminal (for example, the negative terminal in Fig. 4.6). Furthermore, the other two plates, from each capacitor, connect each other via a conducting wire and to nothing else, as shown in Fig. 4.6. Two capacitors connected that way form an isolated conductor that is initially uncharged and must continue to have a net charge zero.

In the following, we will analyze the combination of two capacitors in series. When the two capacitors are initially uncharged and just connect to a battery in the circuit, then the electrons transfer from the left plate of C_1 and into the right plate of C_2. That is, during the process, a negative charge (electrons) stores on the right plate of C_2 and the same amount of negative charge leaves the left plate of C_2 as electrons migrating from that plate to the conducting wire leave behind the left plate having an excess positive charge. Therefore, we can say that the negative charge leaving the left plate of C_2 transfers via the conducting wire and stores on the right plate of C_1. As a result, the right plates, when the equilibrium establishes, accumulate a charge $-Q$, and the left plates a charge $+Q$. That indicates that the charges on capacitors connected as in Fig. 4.6 are the same.

It can be seen that the ΔV across the battery terminals is split between two capacitors:

$$\Delta V = \Delta V_1 + \Delta V_2 \qquad (4.21)$$

In Eq. (4.21), ΔV_1 and ΔV_2 are the potential across C_1 and C_2, respectively. In general, the total potential difference across any number of capacitors connected in series is the sum of the potential differences across the individual capacitors. Now, consider an equivalent capacitor, C_{eq}, with same effect on the circuit as the series combination of the capacitors. After it is fully charged, the equivalent capacitor must have a charge of $-Q$ on its right plate and a charge of $+Q$ on its left plate. Using the definition of capacitance to the equivalent circuit in Fig. 4.6, we have

Fig. 4.6 Two capacitors connected in series in an electric circuit

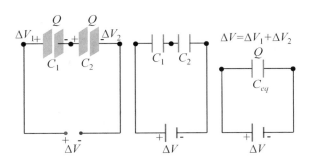

$$\Delta V = \frac{Q}{C_{eq}} \tag{4.22}$$

For each capacitor we can write

$$\Delta V_1 = \frac{Q}{C_1}; \quad \Delta V_2 = \frac{Q}{C_2} \tag{4.23}$$

Then,

$$\frac{Q}{C_{eq}} = \frac{Q}{C_1} + \frac{Q}{C_2} \tag{4.24}$$

or

$$\frac{1}{C_{eq}} = \frac{1}{C_1} + \frac{1}{C_2} \tag{4.25}$$

When this analysis is applied to three or more capacitors connected in series, for instance N capacitors $C_1, C_2, \ldots, C_N$, the equivalent capacitance is given as

$$\frac{1}{C_{eq}} = \frac{1}{C_1} + \frac{1}{C_2} + \frac{1}{C_3} + \cdots + \frac{1}{C_N} \tag{4.26}$$

4.4 Energy Storage in the Electric Field

To transfer an amount of charge from one plate of a capacitor to the other during the process of charging the capacitor, an external work is done against the electric field. That work stores in the capacitor in the form of the potential energy. For that, let q be the charge on the capacitor at some instant during the charging process when the potential difference across the capacitor is $\Delta V = q/C$. At that instant, one of the plates is carrying a charge $+q$ and the other $-q$. To transfer an increment of charge dq from the plate with charge $-q$ (which is at a lower electric potential) to the plate carrying charge $+q$ (which is at a higher electric potential) an elementary work is done against the electric field:

$$dW = \Delta V dq = \frac{q}{C} dq \tag{4.27}$$

To calculate the total work required to charge the capacitor from $q = 0$ to final charge Q, we integrate Eq. (4.27) as follows:

$$W = \int_0^Q \frac{q}{C} dq = \frac{1}{2} \frac{Q^2}{C} \tag{4.28}$$

This work done to charge the capacitor stores in the capacitor as an electric potential energy U. Therefore, $U = W$. Also, we can express the potential energy U in the following forms:

$$U = \frac{1}{2}\frac{Q^2}{C} \tag{4.29}$$

$$= \frac{1}{2}Q\Delta V \tag{4.30}$$

$$= \frac{1}{2}C(\Delta V)^2 \tag{4.31}$$

Note that all expressions given by Eqs. (4.29)–(4.31) are equivalent; that is, they can all be used to calculate the potential energy stored in a capacitor depending on what is known. We can consider the energy stored in a capacitor as being stored in the electric field created between the plates as the capacitor is charged. This description is reasonable from the viewpoint that the electric field is proportional to the charge Q stored on a capacitor. For a capacitor of two parallel plates, the potential difference is related to the electric field through a simple relationship $\Delta V = Ed$. Furthermore, its capacitance is $C = \epsilon_0\frac{A}{d}$. Then, we obtain

$$U = \frac{1}{2}\left(\epsilon_0\frac{A}{d}\right)(Ed)^2 = \frac{1}{2}\epsilon_0(Ad)E^2 \tag{4.32}$$

Since the volume is Ad, then the energy density is given

$$u_E = \frac{U}{Ad} = \frac{1}{2}\epsilon_0 E^2 \tag{4.33}$$

This expression is generally valid. That is, the energy density in any electric field is proportional to the square of the magnitude of the electric field at a given point.

4.5 Electrostatics of Macroscopic Media and Dielectrics

Until now, we have introduced the electric potential and electric field in the presence of other charges or conductors. Therefore, there was no need for distinguishing between microscopic and macroscopic fields. In fact, in the conductors, the surface charge densities imply a macroscopic description. However, other media may exist, where their effect on the electric charge movement is not negligible. In that case, the electrical response of the medium to the external charges and fields must be taken into account.

4.5.1   Dielectrics

There exist many materials that do not allow electric charges to move freely within them, or may allow such motion to occur only very slowly. Those materials are used to block the flow of electrical current, and to form the insulators. For example, they can create insulating layers between the plates of a capacitor. Those materials are known as *dielectric* materials. As an application, the use of the dielectric material for a capacitor reduces its size for a given capacitance or increases its working voltage.

Note that a dielectric material subject to a high enough electric field becomes a conductor; that is, the dielectric material experiences a *dielectric breakdown*. Thus, there exists a maximum voltage for dielectric capacitors to work. For example, there is a maximum power that a coaxial cable can adequately function in high-power applications such as radio transmitters; similarly, for microcircuits there are maximum voltages, which can be handled.

4.5.2   Comparison Between Dielectric Materials and Conductors

To know about the differences between dielectric and conducting materials, we can consider their behavior in electric fields. In particular, we have shown in Fig. 4.7 a conducting and dielectric sheet between the parallel plates in which a potential difference exists. That is, there are an equal amount of opposite charges on the two plates.

In the conducting sheet, the conducting electrons are free to move, and they establish a surface charge which exactly cancels the electric field within the conductor, as shown in Fig. 4.7. That is, the surface charge density of the plates and conducting sheet is the same but with opposite sign. On the other hand, the electrons in the dielectric material are bound to atoms, and the external electric field causes only a displacement of the electronic configuration of atoms (see Fig. 4.7). However, it is sufficient to produce some surface charge with density σ_{ind} (called an induced charge). We say that the dielectric is polarized. Note that the surface charge is not

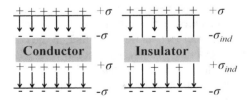

Fig. 4.7 Conductors and insulators in an external electric field. σ denotes the charge density of the plates of the capacitor creating the external field, and σ_{ind} denotes the induced charge density in the surface of insulator

Fig. 4.8 **a** Random orientation of dipoles in absence of external electric field, and **b** orientation of the dipoles along the external electric field

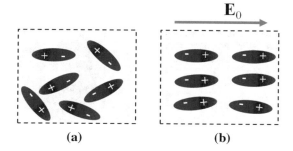

(a)   (b)

able to cancel the external electric field within the sheet; however, it does reduce. In the following, we will introduce a simplified molecular theory of dielectrics to understand the behavior of dielectric materials in the presence of an external electric field.[1] A more complicated, but more precise theory, will be introduced in the following sections, accounting for electric polarization of the ponderable media.[2]

4.5.3 Molecular Theory of Dielectrics

Consider a dielectric material placed in the electric field between the plates of a capacitor. The dielectric material is made up of polar molecules. Note that dielectric material can also be made up of the nonpolar molecules, which can be polarized in the presence of the external electric fields. In the dielectric made up of polar molecules, the permanent dipoles of molecules are also called permanent dipoles. For dielectrics made up of nonpolar molecules, the dipoles created due to polarization are called induced dipoles.

The permanent dipoles in the dielectric arising from the polar molecules of the dielectric material are randomly distributed and oriented in the absence of an electric field, as shown in Fig. 4.8. When an external field $\mathbf{E}_0$ applies, for example, created by charges on the capacitor plates, the forces exerted on the dipoles produce torques, causing them to align almost in the direction of the field. We can say that the dielectric is polarized; that is, macroscopic charge separation occurs. The degree of alignment of molecules with an electric field depends on temperature and the magnitude of the field. In general, the alignment increases with decreasing temperature and with increasing electric field. If the molecules of the dielectric are nonpolar, then the electric field due to the plates produces some charge separation and an induced dipole moment. These induced dipole moments tend to align with the external field, and the dielectric is polarized.

---

[1] See Holliday et al. (2011).
[2] See Jackson (1999).

Fig. 4.9 a Orientation of the
dipoles of a dielectric in the
external electric field; and **b**
Induced charge density σ_{ind}
in the surface of dielectric

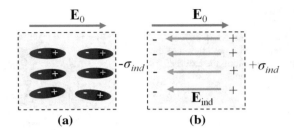

It is essential to say that we can polarize with an external field, both a dielectric
made up of polar molecules or nonpolar molecules. Consider a dielectric material
placed between the plates of a capacitor so that it is in a uniform electric field $\mathbf{E}_0$, as
shown in Fig. 4.9. Furthermore, the external electric field, $\mathbf{E}_0$, directed to the right,
polarizes the dielectric material. The net effect of $\mathbf{E}_0$ is the formation of a positive
surface-induced charge density σ_{ind} on the right face and an equal negative surface-
induced charge density $-\sigma_{ind}$ on the left side.

These surface-induced charges on the dielectric create an induced electric field
$\mathbf{E}_{ind}$ in the direction opposite the external field $\mathbf{E}_0$. Therefore, the net electric field
$\mathbf{E}$ in the dielectric is

$$\mathbf{E} = \mathbf{E}_0 + \mathbf{E}_{ind} \tag{4.34}$$

Projecting Eq. (4.34) along the direction of the external field, we obtain

$$E = E_0 - E_{ind} \tag{4.35}$$

Moreover, we suppose that the external field is created by two conducting plates
(electrodes), where each carries a surface charge σ per unit area. Also, the surface
charge on the dielectric is σ_{ind} per unit area. Now, assuming that the electric field is
everywhere uniform and normal to the plates, the field outside the dielectric is equal
to the field created by plates and is given as

$$E_0 = \frac{\sigma}{\epsilon_0} \tag{4.36}$$

Similarly the induced field within the dielectric is

$$E_{ind} = \frac{\sigma_{ind}}{\epsilon_0} \tag{4.37}$$

Equation (4.37) can further be written as

$$E = \frac{\sigma - \sigma_{ind}}{\epsilon_0} \qquad (4.38)$$

$$= \frac{\sigma - \sigma_{ind}}{\sigma} \frac{\sigma}{\epsilon_0}$$

$$= \epsilon_r E_0 \qquad (4.39)$$

where

$$\epsilon_r = \frac{\sigma - \sigma_{ind}}{\sigma} \qquad (4.40)$$

is called *relative permittivity of the material*, which is smaller than one: $\epsilon_r < 1$. Often,

$$\varepsilon = \frac{1}{\epsilon_r} \qquad (4.41)$$

is called *dielectric constant* and it is greater than one: $\varepsilon > 1$, then

$$E = \frac{E_0}{\varepsilon} \qquad (4.42)$$

Equation (4.42) indicates that the magnitude of electric field decreases in presence of dielectric material. Consider a dielectric completely filling the space between the plates of a parallel-plate capacitor. From the definition the capacitance C is

$$C = \frac{Q}{\Delta V} \qquad (4.43)$$

Let E be the magnitude of electric field between the plates of the capacitor, then, the magnitude of potential difference is $\Delta V = Ed$, and hence, $C = \frac{Q}{Ed}$. Using the relation $E = \frac{E_0}{\varepsilon}$, then

$$C = \frac{Q}{\frac{E_0}{\varepsilon}d} \qquad (4.44)$$

$$= \varepsilon \frac{Q}{E_0 d}$$

$$= \varepsilon C_0$$

where C_0 is the capacitance in absence of dielectric media. Equation (4.44) implies that the capacitance increases by the factor ε in the presence of the dielectric material completely filling the region between the plates.

4.5.4 Energy Stored in Capacitor

The energy stored in the absence of the dielectric is

$$U_0 = \frac{Q_0^2}{2C_0} \tag{4.45}$$

After the battery is removed and the dielectric inserted, the charge on the capacitor remains the same. Hence, the energy stored in the presence of the dielectric is

$$U = \frac{Q_0^2}{2C} \tag{4.46}$$

Using the relation $C = \varepsilon C_0$, then

$$U = \frac{Q_0^2}{2\varepsilon C_0} \tag{4.47}$$

or

$$U = \frac{U_0}{\varepsilon} \tag{4.48}$$

Because $\varepsilon > 1$, the final energy is less than the initial energy (see also Eq. (4.48)) $\Delta U = U - U_0 < 0$. We can account for the "missing" energy by noting that the dielectric, when inserted, gets pulled into the device. An external agent must do negative work to keep the dielectric from accelerating.

This work is simply the difference

$$W_a = U - U_0 \tag{4.49}$$

Alternatively, the positive work done on the external agent by the system is

$$W = -W_a = U_0 - U \tag{4.50}$$

In summary, a dielectric provides the following advantages:
1. Increase in capacitance;
2. Increase in maximum operating voltage;
3. Possible mechanical support between the plates, which allows the plates to be close together without touching, thereby decreasing d and increasing C.

4.6 Electric Polarization

Consider an electric field applied to a medium made up of a large number of particles, such as atoms or molecules. The charges bound in molecules will then respond to the external electric field, and they will follow the perturbed motion to align with the external field. Thus, the charge density within the molecules will be distorted. The dipole moments[3] of each molecule will be different in comparison to the dipole moments in the absence of the applied electric field. That is, in the absence of the external field, the average dipole moments over all molecules of the substance are zero because the dipole vectors are oriented randomly. In contrast, in the presence of the applied electric field, the net dipole moment of the substance is different from zero. Therefore, in the medium, there is an average dipole moment per unit volume, which is called *electric polarization* **P**, given as

$$\mathbf{P}(\mathbf{r}) = \sum_i n_i \langle \mathbf{p}_i \rangle \tag{4.51}$$

In Eq. (4.51), $\mathbf{p}_i$ is the dipole moment of the molecule type i in the medium, $\langle \cdots \rangle$ denotes the average over a small volume around $\mathbf{r}$, and n_i is the average number per unit volume of the molecule type i at the position $\mathbf{r}$.

If the net charge of the molecule i is Q_i, and there is a macroscopic excess or free charge, the charge density at the macroscopic level is

$$\rho(\mathbf{r}) = \sum_i n_i \langle Q_i \rangle + \rho_{free} \tag{4.52}$$

Note that, in general, average charge of a molecule i is zero, $\langle Q_i \rangle = 0$, and hence, the charge density ρ is equal to the macroscopic excess or free charge, ρ_{free}.

In the following, we will consider the case of a continuous charge distribution, as in Fig. 3.6 (Chap. 3), and see the medium from a macroscopic viewpoint. The potential at some point P at the position $\mathbf{r}$ from a macroscopic small volume element dV at the position $\mathbf{r}'$ is the sum of the potential created by the charge of dV, $dq = \rho(\mathbf{r}')dV$ and the dipole moment of dV is $\mathbf{P}(\mathbf{r}')dV$, assuming that there are no higher macroscopic multipole moment densities:

$$d\phi(\mathbf{r}, \mathbf{r}') = k_e \left(\frac{\rho(\mathbf{r}')dV}{|\mathbf{r} - \mathbf{r}'|} + \frac{\mathbf{P}(\mathbf{r}') \cdot (\mathbf{r} - \mathbf{r}')dV}{|\mathbf{r} - \mathbf{r}'|^3} \right) \tag{4.53}$$

In Eq. (4.53), P is outside the volume dV. To obtain the electric potential, we integrate over all space by treating the element volume dV as macroscopically infinitesimal, and hence $dV = d\mathbf{r}'$:

---

[3] The electric field of a molecule is characterized by the multipole moments of the molecule. However, here, we assume that the dominant molecular multipole with the external field is a dipole.

$$\phi(\mathbf{r}) = k_e \int d\mathbf{r}' \left(\frac{\rho(\mathbf{r}')}{|\mathbf{r} - \mathbf{r}'|} + \mathbf{P}(\mathbf{r}') \cdot \nabla_{\mathbf{r}'} \left(\frac{1}{|\mathbf{r} - \mathbf{r}'|} \right) \right) \tag{4.54}$$

In Eq. (4.54), the second term is the dipole contribution to the potential. This term can be integrated by parts as the following:

$$k_e \int d\mathbf{r}' \mathbf{P}(\mathbf{r}') \cdot \nabla_{\mathbf{r}'} \left(\frac{1}{|\mathbf{r} - \mathbf{r}'|} \right) = k_e \int d\mathbf{r}' \nabla_{\mathbf{r}'} \cdot \left(\frac{\mathbf{P}(\mathbf{r}')}{|\mathbf{r} - \mathbf{r}'|} \right) \tag{4.55}$$

$$- k_e \int d\mathbf{r}' \left(\frac{1}{|\mathbf{r} - \mathbf{r}'|} \right) \nabla_{\mathbf{r}'} \cdot \mathbf{P}(\mathbf{r}')$$

$$= -k_e \int d\mathbf{r}' \left(\frac{1}{|\mathbf{r} - \mathbf{r}'|} \right) \nabla_{\mathbf{r}'} \cdot \mathbf{P}(\mathbf{r}')$$

because the first integral is equal to zero. Therefore, Eq. (4.54) can be simplified as

$$\phi(\mathbf{r}) = k_e \int d\mathbf{r}' \frac{\rho(\mathbf{r}') - \nabla_{\mathbf{r}'} \cdot \mathbf{P}(\mathbf{r}')}{|\mathbf{r} - \mathbf{r}'|} \tag{4.56}$$

Equation (4.56) indicates that potential $\phi(\mathbf{r})$ is created by the effective charge distribution

$$\rho_{eff} = \rho(\mathbf{r}') - \nabla_{\mathbf{r}'} \cdot \mathbf{P}(\mathbf{r}') \tag{4.57}$$

where the first term is the macroscopic excess charge density in the dielectric and the second term is the polarization-charge density in the dielectric medium, which for a nonuniform polarization can either increase or decrease the net charge within a small volume.

Electric displacement

We can define the electric displacement vector $\mathbf{D}$ as

$$\mathbf{D} = \epsilon_0 \mathbf{E} + \mathbf{P} \tag{4.58}$$

Note that a relationship between the vectors $\mathbf{D}$ and $\mathbf{E}$ is important for obtaining a solution for the electric potential or field. If we exclude from the discussion the ferroelectricity of the macroscopic medium and assume a linear response of the medium to an external electric field, then the induced polarization $\mathbf{P}$ is parallel to $\mathbf{E}$:

$$\mathbf{P} = \epsilon_0 \chi_e \mathbf{E} \tag{4.59}$$

Here, the proportionality constant is independent of the direction; that is, the medium is isotropic. Furthermore, we have assumed that the electric field does not become extremely large. In Eq. (4.59), χ_e is the electric susceptibility of the medium. Combining Eqs. (4.58) and (4.59), we obtain

$$\mathbf{D} = \epsilon_0 (1 + \chi_e) \mathbf{E} \tag{4.60}$$

Denoting the relative electric permittivity of the medium as

$$\epsilon_r = \frac{1}{1 + \chi_e} \tag{4.61}$$

then, the dielectric constant is defined as

$$\epsilon = \frac{1}{\epsilon_r} = 1 + \chi_e \tag{4.62}$$

Therefore,

$$\mathbf{D} = \epsilon \epsilon_0 \mathbf{E} \tag{4.63}$$

If the dielectric is anisotropic and nonuniform, then ϵ is dependent on the position, $\epsilon(\mathbf{r})$. Furthermore, if the isotropic and uniform dielectric medium does not fill all of the space where the electric field is applied, then we have to consider the boundary conditions on both $\mathbf{D}$ and $\mathbf{E}$ at the interface between media. The boundary conditions include the normal components of $\mathbf{D}$ and tangential components of $\mathbf{E}$ on each side of the interface:

$$(\mathbf{D}_2 - \mathbf{D}_1) \cdot \mathbf{n} = \sigma \tag{4.64}$$
$$(\mathbf{E}_2 - \mathbf{E}_1) \times \mathbf{n} = 0$$

In Eq. (4.64), $\mathbf{n}$ is an outward unit normal vector to the surface; that is, it is directed from region 1 to region 2. Furthermore, σ is the macroscopic surface charge density at the boundary surface, excluding the polarization charge.

4.7 Set of Maxwell Equations for Electrostatic Field

4.7.1 Maxwell Equations for Free Space Electrostatic Field

First, we introduce the set of Maxwell equations for the electrostatic field in free space. Using Gauss's Law (see Chap. 2), we can write the electric flux of electric field created by continuous charge distribution in a volume V enclosed by the surface A as

$$\oint_A \mathbf{E} \cdot d\mathbf{A} = \frac{Q_{in}}{\epsilon_0} \tag{4.65}$$

Note that in Eq. (4.65) $\mathbf{E}$ is the electrostatic field created by all charges in space, and Q_{in} is the electric charge inside the volume V enclosed by the surface A. The left-hand side of Eq. (4.65) can be written in the following form using Gauss formula:

$$\oint_A \mathbf{E} \cdot d\mathbf{A} = \int_V \nabla \cdot \mathbf{E} \, dV \tag{4.66}$$

where V is the volume enclosed by the surface A. In addition, the right-hand side of Eq. (4.65) can be written as

$$\frac{Q_{in}}{\epsilon_0} = \int_V \frac{\rho(\mathbf{r})}{\epsilon_0} \, dV \tag{4.67}$$

Combining Eqs. (4.65), (4.66) and (4.67), we get

$$\int_V \nabla \cdot \mathbf{E} \, dV = \int_V \frac{\rho(\mathbf{r})}{\epsilon_0} \, dV \tag{4.68}$$

where $\nabla \cdot \mathbf{E}$ is the divergence of the vector $\mathbf{E}$, which produces a scalar.

Comparing both sides of Eq. (4.68), we obtain the first Maxwell equation in free space:

$$\nabla \cdot \mathbf{E}(\mathbf{r}) = \frac{\rho(\mathbf{r})}{\epsilon_0} \tag{4.69}$$

where both $\mathbf{E}$ and ρ can be functions of the position $\mathbf{r}$.

Using the expression of the electrostatic potential difference in free space, Eq. (4.10) (Chap. 3), we have

$$\Delta\phi = -\int_A^B \mathbf{E} \cdot d\mathbf{s} \tag{4.70}$$

where A and B are two points in free space, and $d\mathbf{s}$ is an infinitesimal displacement along the curve joining points A and B. If we consider a closed path, that is, $A = B$, then $\Delta\phi = \phi_B - \phi_A = \phi_A - \phi_A = 0$, and hence

$$\oint_{\mathcal{L}} \mathbf{E} \cdot d\mathbf{s} = 0 \tag{4.71}$$

where $\mathcal{L}$ is a closed path. Using Stokes' formula, we can write

$$\oint_{\mathcal{L}} \mathbf{E} \cdot d\mathbf{s} = \oint_{\mathcal{A}} (\nabla \times \mathbf{E}) \cdot d\mathbf{A} \tag{4.72}$$

where $\mathcal{A}$ is the an arbitrary surface enclosed by the path $\mathcal{L}$. Combining Eqs. (4.71) and (4.72), we obtain

$$\oint_{\mathcal{A}} (\nabla \times \mathbf{E}) \cdot d\mathbf{A} = 0 \tag{4.73}$$

Equation (4.73) implies that

$$\nabla \times \mathbf{E} = 0 \tag{4.74}$$

Equation (4.74) is the second Maxwell equation of the electrostatic field in free space.

4.7.2 Maxwell Equations for Dielectric Media Electrostatic Field

We mentioned that in the dielectric medium, an average over macroscopically small volumes, which are microscopically large, is necessary to obtain the Maxwell equations of the macroscopic phenomena.

The first observation is that Eq. (4.74) holds microscopically, that is

$$\nabla \times \mathbf{E}_{micro} = 0 \tag{4.75}$$

When averaging is made of the homogeneous Eq. (4.75), we obtain

$$\nabla \times \mathbf{E} = 0 \tag{4.76}$$

Equation (4.76) indicates that Eq. (4.74) holds for the averaged macroscopic electric field $\mathbf{E}$.

Using Eq. (4.57) for the effective charge density in the medium, Eq. (4.69) becomes

$$\nabla \cdot \mathbf{E}(\mathbf{r}) = \frac{\rho(\mathbf{r}) - \nabla \cdot \mathbf{P}(\mathbf{r})}{\epsilon_0} \tag{4.77}$$

Rearranging Eq. (4.77), we get

$$\nabla \cdot (\epsilon_0 \mathbf{E}(\mathbf{r}) + \mathbf{P}(\mathbf{r})) = \rho(\mathbf{r}) \tag{4.78}$$

Using the definition of the electric displacement vector given by Eq. (4.58), we write Eq. (4.78) as

$$\nabla \cdot \mathbf{D}(\mathbf{r}) = \rho(\mathbf{r}) \tag{4.79}$$

Note that Eqs. (4.76) and (4.79) are the macroscopic Maxwell equations in the dielectric medium, which are the counterparts of Eqs. (4.69) and (4.74).

4.8 Potential Energy of Electrostatic Field

Often, it is practical to interpret the electrostatic potential energy that emphasizes the interactions between the charges of a system as the energy stored in the electric field surrounding the charges. In that way, we emphasize the electric field instead of electric potential.

For that, we can use Eq. (3.41) (Chap. 3), and the first Maxwell's equation in the free space as $\rho = \epsilon_0 (\nabla \cdot \mathbf{E})$, then we write:

$$U = \frac{1}{2} \int_V \epsilon_0 (\nabla \cdot \mathbf{E}) \phi(\mathbf{r}) \, d\mathbf{r} \tag{4.80}$$

Furthermore, using Eq. (3.44) (Chap. 3), we obtain

$$U = -\frac{\epsilon_0}{2} \int_V (\nabla^2 \phi(\mathbf{r})) \phi(\mathbf{r}) \, d\mathbf{r} \tag{4.81}$$

If we integrate by parts in Eq. (4.81), we get

$$U = \frac{\epsilon_0}{2} \int_V (\nabla \phi(\mathbf{r}))^2 \, d\mathbf{r} = \frac{\epsilon_0}{2} \int_V |\mathbf{E}|^2 \, d\mathbf{r} \tag{4.82}$$

The integrand in Eq. (4.82) can be identified as the energy density, u:

$$u = \frac{\epsilon_0}{2} |\mathbf{E}|^2 \tag{4.83}$$

It is worth noting that the form of the right-hand side of Eq. (4.83) implies that $u \geq 0$; therefore, the total electrostatic potential energy $U \geq 0$. However, the electrostatic potential of the system of two charges discussed in Chap. 3 (see Eq. (3.28)) implies that when the two charges have opposite sign, then electrostatic potential, U, is negative. The reason for that contradiction is that the expression of U given by Eqs. (4.82) and (4.83) includes the self-energy term to the energy density; while Eq. (3.28), or more general Eq. (3.29), given in Chap. 3, does not.

In the presence of the dielectric medium, $\rho = \nabla \cdot \mathbf{D}$, Eq. (4.80) can be written as

$$U = \frac{1}{2} \int_V (\nabla \cdot \mathbf{D}) \phi(\mathbf{r}) \, d\mathbf{r} \tag{4.84}$$

Integrating Eq. (4.84) by parts, we get

$$U = -\frac{1}{2} \int_V \mathbf{D} \cdot (\nabla \phi(\mathbf{r})) \, d\mathbf{r} \tag{4.85}$$

Using the differential form of electric potential (Eq. (3.44), Chap. 3), we finally get

$$U = \frac{1}{2} \int_V \mathbf{D} \cdot \mathbf{E} \, d\mathbf{r} \tag{4.86}$$

4.9 Exercises

Exercise 4.1 Consider a two parallel plates capacitor, each with an area $A = 2.00 \times 10^{-4}$ m^2. The distance between the plates is $d = 1.00$ mm. Determine the capacitance.

Solution 4.1 From the formula of capacitance:

$$C = \epsilon_0 \frac{A}{d} \tag{4.87}$$

$$= (8.85 \times 10^{-12} \text{ C}^2/\text{N} \cdot \text{m}^2) \left(\frac{2.00 \times 10^{-4} \text{ m}^2}{1.00 \times 10^{-3} \text{ m}} \right)$$

$$= 1.77 \times 10^{-12} \text{ F} = 1.77 \text{ pF}$$

Exercise 4.2 Consider a solid cylindrical shape conductor with radius a and charge Q, which is coaxial with a cylindrical shell of negligible thickness, radius $b > a$, and opposite charge $-Q$, as shown in Fig. 4.10. Find the capacitance of this cylindrical capacitor if its length is ℓ.

Solution 4.2 First, we calculate the difference of potential between the two cylinders:

$$\Delta\phi = \phi_b - \phi_a = - \int_a^b \mathbf{E} \cdot d\mathbf{s} \tag{4.88}$$

where $\mathbf{E}$ is electric field inside $a < r < b$, which for a cylindrical charge distribution is

Fig. 4.10 A solid cylindrical conductor of a radius and charge Q is coaxial with a cylindrical shell of negligible thickness, radius $b > a$, and charge $-Q$

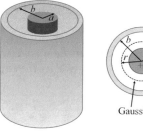
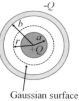

$$E_r = 2k_e \frac{\lambda}{r} \tag{4.89}$$

where $\lambda = Q/\ell$ is linear charge density.

Thus,

$$\phi_b - \phi_a = -\int_a^b E_r dr = -2k_e \lambda \int_a^b \frac{dr}{r} \tag{4.90}$$

$$= -2k_e \lambda \ln \left(\frac{b}{a} \right)$$

From $C = Q/\Delta V$ with $\Delta V = \mid \phi_b - \phi_a \mid$, we get

$$C = \frac{Q}{2k_e \lambda \ln \left(\dfrac{b}{a} \right)} \tag{4.91}$$

$$= \frac{Q}{2k_e \dfrac{Q}{\ell} \ln \left(\dfrac{b}{a} \right)}$$

$$= \frac{\ell}{2k_e \ln \left(\dfrac{b}{a} \right)}$$

Exercise 4.3 A spherical capacitor consists of a spherical conducting shell of radius b and charge $-Q$ concentric with a smaller conducting sphere of radius a and charge Q (see Fig. 4.11). Find the capacitance of this device.

Solution 4.3 The difference on the potential

$$\phi_b - \phi_a = -\int_a^b \mathbf{E} \cdot d\mathbf{s} \tag{4.92}$$

The electric field outside a spherically symmetric charge distribution has a radial symmetry, and it is given by the expression

$$\mathbf{E} = k_e \frac{Q}{r^2} \hat{\mathbf{r}} \tag{4.93}$$

Thus,

Fig. 4.11 A spherical capacitor consists of a spherical conducting shell of radius b and charge $-Q$ concentric with a smaller conducting sphere of radius a and charge Q

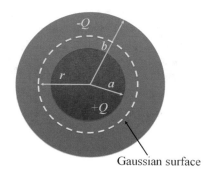

Gaussian surface

$$\phi_b - \phi_a = -\int_a^b E_r \, dr \tag{4.94}$$

$$= -k_e Q \int_a^b \frac{dr}{r^2} = k_e Q \left[\frac{1}{r}\right]_a^b$$

$$= k_e Q \left(\frac{1}{b} - \frac{1}{a}\right)$$

From $C = Q/\Delta V$ with

$$\Delta V = |\phi_b - \phi_a| \tag{4.95}$$

$$= k_e Q \frac{(b - a)}{ab}$$

we get

$$C = \frac{ab}{k_e(b - a)} \tag{4.96}$$

In the limit of $b \to \infty$:

$$C = \lim_{b \to \infty} \frac{ab}{k_e b \left(1 - \dfrac{a}{b}\right)} \tag{4.97}$$

$$= \frac{a}{k_e}$$

$$= 4\pi\epsilon_0 a$$

Exercise 4.4 A capacitor consists of two parallel plates in the form of disks with radius $r = 2$ mm at a separation $d = 2$ mm. Find the capacitance of this device.

Solution 4.4 From the formula of capacitance:

$$C = \epsilon_0 \frac{A}{d} \tag{4.98}$$

$$= (8.85 \times 10^{-12} \ \mathrm{C^2/N \cdot m^2}) \left(\frac{\pi (2 \times 10^{-3})^2 \ \mathrm{m^2}}{2.00 \times 10^{-3} \ \mathrm{m}} \right)$$

$$= 55.6 \times 10^{-15} \ \mathrm{F} = 0.0556 \ \mathrm{pF}$$

Exercise 4.5 A capacitor consists of two parallel plates connected to the terminals of a battery at the voltage $\Delta V = 10$ V. If the capacitor stores a net charge of $Q = 60 \ \mu\mathrm{C}$, calculate its capacitance.

Solution 4.5 From the formula of capacitance:

$$C = \frac{Q}{\Delta V} \tag{4.99}$$

we obtain

$$C = \frac{60 \ \mu\mathrm{C}}{10 \ \mathrm{V}} = 6 \times 10^{-6} \ \mathrm{F} = 6 \ \mu\mathrm{F} \tag{4.100}$$

Exercise 4.6 Consider a two parallel plates capacitor, each with an area $A = 4.00 \times 10^{-4} \ \mathrm{m^2}$. The distance between the plates is $d = 1.50$ mm. Determine the capacitance if the space between the plates is filled with a dielectric material of type glass with $\epsilon = 6$.

Solution 4.6 From the formula of capacitance, we find the capacitance in absence of dielectric material:

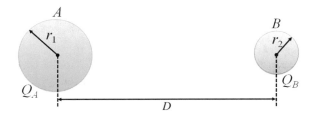

Fig. 4.12 Two long cylindrical conductors of radius r_1 and r_2, respectively, are parallel and separated by a distance d

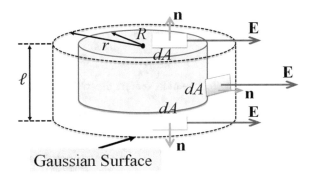

Fig. 4.13 A cylindrical Gaussian surface with radius r and length ℓ around a long cylindrical conductors of radius R and uniform charge Q

Gaussian Surface

$$C_0 = \epsilon_0 \frac{A}{d} \tag{4.101}$$

$$= (8.85 \times 10^{-12} \text{ C}^2/\text{N} \cdot \text{m}^2) \left(\frac{4.00 \times 10^{-4} \text{ m}^2}{1.50 \times 10^{-3} \text{ m}} \right)$$

$$= 23.6 \times 10^{-13} \text{ F} = 2.36 \text{ pF}$$

In presence of dielectric, we get

$$C = \epsilon C_0 = 14.16 \text{ pF} \tag{4.102}$$

Exercise 4.7 Two long cylindrical conductors of radius r_1 and r_2, respectively, are parallel and separated by a distance d, as shown in Fig. 4.12. Supposing that $r_1, r_2 \ll d$ and the charge on cylinder A is Q and the charge on cylinder B is $-Q$, calculate the capacitance per unit length of that parallel-wire line.

Solution 4.7 Assume the length of cylinders is ℓ. First, we have to calculate the electric field created by a cylindrical conductor of radius R and length ℓ with a uniform charge Q. For that, we can use Gauss's law with a Gaussian surface chosen as in Fig. 4.13. Then,

$$\oint_A \mathbf{E} \cdot \mathbf{n} \, dA = \frac{Q}{\epsilon_0} \tag{4.103}$$

Because of the symmetry, electric field $\mathbf{E}$ is radial, therefore, the basis of the cylinder give a zero contribution to the net electric flux. We obtain

$$E 2\pi r \ell = \frac{Q}{\epsilon_0} \tag{4.104}$$

or,

$$E = \frac{Q}{2\pi \epsilon_0 r \ell} \tag{4.105}$$

Therefore, the electric field vectors created by the two cylinders are

$$\mathbf{E}_A = \frac{Q_A}{2\pi \epsilon_0 r \ell} \hat{\mathbf{r}}_{AB} \tag{4.106}$$

$$\mathbf{E}_B = -\frac{Q_B}{2\pi \epsilon_0 r \ell} \hat{\mathbf{r}}_{AB} \tag{4.107}$$

where $Q_A = Q$ and $Q_B = -Q$, and $\hat{\mathbf{r}}_{AB}$ is a unit vector pointing from A to B. The capacitance of the two conductors of this two-wire line is

$$C = \frac{Q}{\Delta V_{AB}} \tag{4.108}$$

where ΔV_{AB} is the difference in potential between the two cylinders:

$$\Delta V_{AB} = \phi_A - \phi_B = \Delta V_{AB}^A + \Delta V_{AB}^B \tag{4.109}$$

where ΔV_{AB}^A is the voltage drop due to the charge Q_A in A and ΔV_{AB}^B is the voltage drop due to the charge Q_B on conductor B. Then, using the principle of superposition the voltage drop from conductor B to conductor A due to the charges is sum of the voltage drops by each charge individually. Therefore,

$$\Delta V_{AB}^A = -\int_B^A \mathbf{E}_A \cdot d\mathbf{s} = \int_D^{r_1} \frac{Q_A}{2\pi \epsilon_0 r \ell} \, dr = -\frac{Q}{2\pi \epsilon_0 \ell} \ln \frac{D}{r_1} \tag{4.110}$$

Similarly,

$$\Delta V_{AB}^B = -\int_B^A \mathbf{E}_B \cdot d\mathbf{s} = \int_D^{r_2} \frac{Q_B}{2\pi \epsilon_0 r \ell} \, dr = \frac{Q}{2\pi \epsilon_0 \ell} \ln \frac{r_2}{D} \tag{4.111}$$

$d\mathbf{s}$ is a small element length vector pointing from B to A.
Then, the net voltage drop is given as

$$\Delta V_{AB} = \Delta V_{AB}^A + \Delta V_{AB}^B = -\frac{Q}{2\pi\epsilon_0\ell}\left[\ln\frac{D}{r_1} - \ln\frac{r_2}{D}\right] \tag{4.112}$$

The capacitance is calculated as

$$C = \frac{Q}{|\Delta V_{AB}|} = \frac{2\pi\epsilon_0\ell}{\ln\dfrac{D}{r_1} - \ln\dfrac{r_2}{D}} \tag{4.113}$$

Then, the capacitance per unit length is calculated as

$$c = \frac{C}{\ell} = \frac{2\pi\epsilon_0}{\ln\dfrac{D}{r_1} - \ln\dfrac{r_2}{D}} \tag{4.114}$$

Exercise 4.8 Consider a two parallel plates capacitor with capacitance $C = 2.0$ pF connected to a battery of voltage $\Delta V = 1.2$ V. Determine the energy stored in the capacitor.

Solution 4.8 Using the formula of energy stored in a capacitor

$$U = \frac{1}{2}C(\Delta V)^2 = \frac{1}{2}(2.0 \times 10^{-12} \text{ F})(1.2 \text{ V})^2 = 1.44 \times 10^{-12} \text{ J} \tag{4.115}$$

Knowing that
$$1 \text{ eV} = 1.602 \times 10^{-19} \text{ J} \tag{4.116}$$

we get
$$U = 9 \text{ MeV} \tag{4.117}$$

Exercise 4.9 The voltage across a capacitor with capacitance $C = 1.0$ pF is $\Delta V = 0.6$ V. Determine the how much does energy stored in the capacitor increase if the voltage increases to $\Delta V = 1.2$ V.

Solution 4.9 Using the formula of energy stored in a capacitor

Fig. 4.14 An electric circuit
of four capacitors
$C_1 = 1.0$ pF, $C_2 = 0.5$ pF,
$C_3 = 1.5$ pF and
$C_4 = 2.0$ pF connected to a
battery of $\Delta V = 2.0$ V

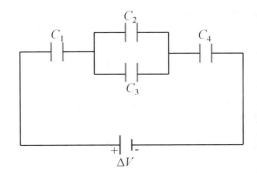

$$U = \frac{1}{2}C(\Delta V)^2 \tag{4.118}$$

For $\Delta V = 0.6$ V, we find

$$U_1 = \frac{1}{2}(1.0 \times 10^{-12} \text{ F})(0.6 \text{ V})^2 = 18 \times 10^{-10} \text{ J} \tag{4.119}$$

For $\Delta V = 1.2$ V, we find

$$U_2 = \frac{1}{2}(1.0 \times 10^{-12} \text{ F})(1.2 \text{ V})^2 = 72 \times 10^{-10} \text{ J} \tag{4.120}$$

Therefore, the increase is

$$\frac{U_2}{U_1} = 4 \text{ times} \tag{4.121}$$

Exercise 4.10 Consider the electric circuit of four capacitors $C_1 = 1.0$ pF,
$C_2 = 0.5$ pF, $C_3 = 1.5$ pF and $C_4 = 2.0$ pF connected to a battery of $\Delta V = 2.0$ V, as shown in Fig. 4.14. Calculate the charge stored in each capacitor and
the voltage across capacitors.

Solution 4.10 Since C_2 and C_3 are connected in parallel, they are equivalent to one
capacitor, C_{23}:

$$C_{23} = C_2 + C_3 = 2.0 \text{ pF} \tag{4.122}$$

Then, C_1, C_{23} and C_4 are connected in series, and hence, they are equivalent to C:

$$\frac{1}{C} = \frac{1}{C_1} + \frac{1}{C_{23}} + \frac{1}{C_4} \tag{4.123}$$

Replacing the numerical values, we get

$$\frac{1}{C} = \frac{1}{1.0 \text{ pF}} + \frac{1}{2.0 \text{ pF}} + \frac{1}{2.0 \text{ pF}} = (1.0 + 0.5 + 0.5) \text{ pF}^{-1} \qquad (4.124)$$

or $C = 0.5$ pF. Then, the charge stored in C is

$$Q = C\Delta V = (0.5 \text{ pF})(2.0 \text{ V}) = 1.0 \text{ pC} \qquad (4.125)$$

which is also the charge stored in C_1, C_{23} and C_4. Therefore, the voltage across each of these capacitors is

$$\Delta V_1 = \frac{Q}{C_1} = \frac{1.0 \text{ pC}}{1.0 \text{ pF}} = 1.0 \text{ V} \qquad (4.126)$$

$$\Delta V_{23} = \frac{Q}{C_{23}} = \frac{1.0 \text{ pC}}{2.0 \text{ pF}} = 0.5 \text{ V} \qquad (4.127)$$

$$\Delta V_4 = \frac{Q}{C_4} = \frac{1.0 \text{ pC}}{2.0 \text{ pF}} = 0.5 \text{ V} \qquad (4.128)$$

To find the charge stored in C_2 and C_3, first note that the voltage across C_2 and C_3 is the same; that is,

$$\Delta V_2 = \Delta V_3 = \Delta V_{23} = 0.5 \text{ V} \qquad (4.129)$$

Thus, the charges stored in C_2 and C_3 are calculated as

$$Q_2 = C_2\Delta V_2 = (0.5 \text{ pF})(0.5 \text{ V}) = 0.25 \text{ pC} \qquad (4.130)$$
$$Q_3 = C_3\Delta V_3 = (1.5 \text{ pF})(0.5 \text{ V}) = 0.75 \text{ pC} \qquad (4.131)$$

References

Holliday D, Resnick R, Walker J (2011) Fundamentals of physics. John Wiley and Sons, New York
Griffiths DJ (1999) Introduction to electrodynamics, 3rd edn. Prentice Hall, Hoboken
Jackson JD (1999) Classical electrodynamics, 3rd edn. John Wiley and Sons, New York

Chapter 5
Electric Current

The chapter aims to introduce the electric current and its theoretical microscopic view, including Ohm's law.

In this chapter, we will introduce the electric current. We introduced charges at rest, or electrostatics, so far. In this chapter, we will consider the phenomena associated with electric charges in motion. We will introduce the *electric current*, or simply *current*, which describes the rate of charge flow through some region of space. Also, in this chapter, we will introduce resistance and Ohm's law. As extra reading material, the reader can also consider other literature (Holliday et al. 2011).

5.1 Electric Current

Consider the motion in a system of electric charges, as presented in Fig. 5.1. A current will exist, if there is a net flow of charge through a region. To define current, we consider the charges moving as in Fig. 5.1 and a surface of area A perpendicular to the direction of motion of the charges.

By definition, the ratio of the amount of charge ΔQ that passes through the surface area A in a time interval Δt is the average current I_{av}:

Average electric current

$$I_{av} = \frac{\Delta Q}{\Delta t} \tag{5.1}$$

which represents the charge that passes through A per unit time. If the charge flow rate, $\Delta Q/\Delta t$ varies in time, then the current varies in time.

H. Kamberaj, *Electromagnetism*, Undergraduate Texts in Physics, https://doi.org/10.1007/978-3-030-96780-2_5

Fig. 5.1 A system of
electric charges in motion.
Charge passing through a
cross-sectional area A

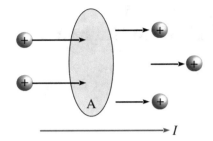

Then, the instantaneous current I is define as

$$I = \lim_{\Delta t \to 0} \frac{\Delta Q}{\Delta t} = \frac{dQ}{dt} \tag{5.2}$$

Note that the instantaneous current I is simply called electric current or current.
　In the SI units, the current has a unit of the ampere (A):

$$1\,\text{A} = 1\,\frac{\text{C}}{\text{s}} \tag{5.3}$$

Equation (5.2) implies that a current of 1 A is equivalent to a charge of 1 C passing
through the surface area in 1 s.

5.1.1　Direction of Electric Current

Since the charges passing through the surface can be positive or negative, or both,
by convention, the direction of the current flow is assigned to the direction of the
positive charge displacement. In electric conductors, the current is due to the motion
of negatively charged electrons; therefore, the direction of the current flow is opposite
the direction of the movement of electrons. On the other hand, for a beam of positively
charged ions (such as protons in an accelerator), the current flows in the same direction
as the motion of the positive ions. Note that there are situations when the current
is due to the flow of both positive and negative charges, such as in the gases and
electrolytes.

5.1.2　Charge Carrier

In Fig. 5.2a, we show a conducting wire in which the two ends connect to form a loop,
all points on the loop are at the same electric potential, for example, for the points A

Fig. 5.2 a A conducting loop, and **b** a conducting wire connected to a battery with voltage ΔV

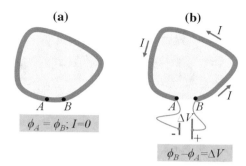

and B, $\phi_A = \phi_B$. Hence, the electric field within the conductor and at its surface is zero ($\phi_B - \phi_A = -\int_A^B \mathbf{E} \cdot d\mathbf{s}$); that is, there is no net drift of charge along the wire, and therefore, there is no current, $I = 0$. Moreover, the current in a conductor is zero even when the conductor has some excess charge.

Now, if the ends of the conduction wire connect to the terminals of a battery with a potential difference ΔV (see also Fig. 5.2b), not all points on the loop are at the same potential. For example, the potential difference between the points A and B of the conductor is $\phi_B - \phi_A = \Delta V$. The battery applies a potential difference between the ends A and B of the loop. Hence, an electric field is created within the wire. Furthermore, that electric field applies forces on the conduction electrons of the wire, and therefore, they accelerate around the loop, creating a current, I, with a flow direction as shown in Fig. 5.2b. Usually, we will refer to the moving charge (positive or negative) as a *mobile charge carrier*. Therefore, the mobile charge carriers in the conductor of Fig. 5.2 are electrons, and they move opposite to the direction of the current I flow. In general, for a metal, the mobile charge carrier is electron.

5.2 Microscopic Model of Current

In the following, we describe a microscopic model of conduction in a conductor to relate the current to the motion of the charge carriers. In particular, we will consider the current in a conductor with a cross-sectional area A, as shown in Fig. 5.3. Consider a section of the conductor with a length Δx. The volume of that section is

$$\Delta V = A \Delta x = A v_d \Delta t \tag{5.4}$$

Suppose that n is the volume number density of mobile charge carriers (or the charge carrier density), then, the total number of carriers in the volume ΔV is

$$N = n A \Delta x = n A v_d \Delta t \tag{5.5}$$

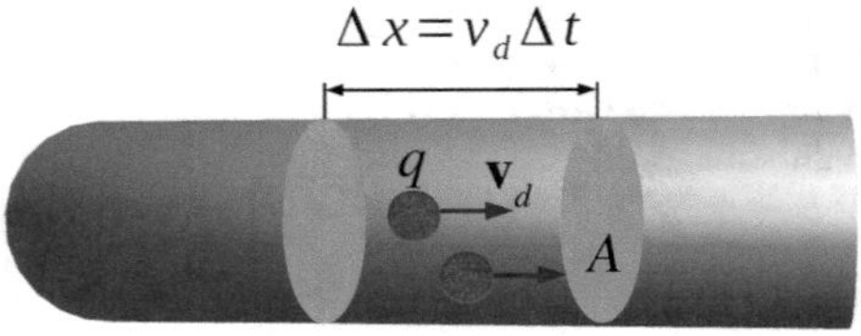

Fig. 5.3 Current in a
conductor of cross-sectional
area A

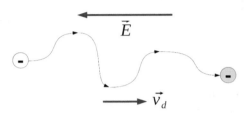

Fig. 5.4 A conductor in an
external electric field in
which the charge carriers are
free electrons

Therefore, the charge ΔQ in this volume is

$$\Delta Q = Nq = qnA\Delta x = qnAv_d\Delta t \qquad (5.6)$$

From Eq. (5.1), the average current in the conductor is

$$I_{av} = \frac{\Delta Q}{\Delta t} = qnAv_d \qquad (5.7)$$

**Drift
speed**

By definition, the *drift speed* represents the average speed of the charge carriers,
denoted as v_d. To understand the drift speed, we will consider a conductor, and
hence the charge carriers are free electrons. For an isolated conductor, the potential
difference across it is zero, as described above for Fig. 5.3, thus these electrons move
randomly as the motion of molecules of the gas in a container. If we apply a potential
difference across the conductor utilizing a battery, as also described above in Fig. 5.4,
an electric field sets up in the conductor. That field exerts an electric force on the
electrons, accelerating them in a given direction. That directed movement of electrons
produces a current, as shown in Fig. 5.4. It is important to note that the electrons do
not move in straight lines along the conductor. Indeed, they collide regularly with the
atoms of the conductor, and hence their resultant motion is a complicated movement,
considered here as a spiral motion. However, the collision just slows down the motion,
because the electrons move slowly along the conductor (in a direction opposite to **E**)
with a drift velocity $\mathbf{v}_d$, as shown in Fig. 5.4.

The atom-electron collisions in a conductor is considered an effective *internal friction* (or *drag force*) exerted by the atoms on the electrons. The amount of energy transferred to the atoms of the conductor from the electrons during collisions increases the vibration energy of the atoms, and hence an increase in the temperature of the conductor.

5.3 Resistance and Ohm's Law

Note that the electric field inside a conductor is zero, which is valid only if the conductor is in static equilibrium. However, when charges move in a conductor, they produce a current as a result of an electric field, which is maintained by connecting the conductor to a battery. The charges move because of the electric field, and hence a non-electrostatic situation exists in the conductor.

Current density

Let I be the current flowing in a conductor of cross-sectional area A. The ratio of the current I with cross-sectional surface area A defines the current density J in the conductor or the current per unit area:

$$J = \frac{I}{A} \tag{5.8}$$

Since the current $I = nqv_d A$, the current density is

$$J = nqv_d \tag{5.9}$$

Note that Eq. (5.8) implies that J has SI units of A/m^2. Furthermore, Eq. (5.9) is valid only if the current density is uniform, and only if the surface of the cross-sectional surface is perpendicular to the direction of the current flow.

The current density vector is defined as

$$\mathbf{J} = nq\mathbf{v}_d \tag{5.10}$$

Equation (5.10) indicates that current density is in the direction of charge motion for positive charge carriers ($q > 0$) and opposite the direction of motion for negative charge carriers ($q < 0$). Therefore, it is similar to the current I; however, current I is not a vector quantity but the current density $\mathbf{J}$ is a vector.

5.3.1 Ohm's Law

A current density $\mathbf{J}$ and an electric field $\mathbf{E}$ are established in a conductor if a potential difference ΔV is applied across the conductor. If a constant potential difference is maintained by connecting the two ends of the conductor to the terminals of a battery,

Ohm's Law

then the current is constant.

Ohm's law states that the current density is linearly proportional to the electric field:

$$\mathbf{J} = \sigma \mathbf{E} \tag{5.11}$$

In Eq. (5.11), the constant of proportionality σ is known as the conductivity of the conductor. Note that not all materials follow Eq. (5.11). The conducting materials that obey Ohm's law are often called ohmic, and the ratio between the current density and the electric field is a constant σ that is independent of the electric field producing the current.

Consider a straight conducting wire with uniform cross-sectional area A and length ℓ, as shown in Fig. 5.5. Suppose a potential difference $\Delta V = \phi_b - \phi_a > 0$ is maintained across the wire. Then, a uniform electric field and a current are created in the wire. We can relate the potential difference to the field as the following:

$$\Delta V = E\ell \tag{5.12}$$

The magnitude of the current density, based on Ohm's law given by Eq. (5.11), is

$$J = \sigma E = \sigma \frac{\Delta V}{\ell} \tag{5.13}$$

Resistance

Using Eq. (5.8), we obtain

$$\Delta V = \left(\frac{\ell}{\sigma A} \right) I \equiv RI \tag{5.14}$$

where

$$R = \frac{\ell}{\sigma A} \tag{5.15}$$

Fig. 5.5 A segment of straight wire of uniform cross-sectional area A and length ℓ

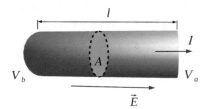

is called *resistance*.

Thus, the resistance R can be written as

$$R = \frac{\Delta V}{I} \tag{5.16}$$

or in a different known form as:

$$\Delta V = RI \tag{5.17}$$

Equation (5.16) implies that resistance has SI units of volts per ampere, V/A. One ohm (Ω) is defined as one volt per ampere:

$$1\,\Omega = \frac{1\,V}{1\,A} \tag{5.18}$$

By definition, the inverse of conductivity is called resistivity ρ: **Resis-
                                                                                tivity**

$$\rho = \frac{1}{\sigma} \tag{5.19}$$

From Eq. (5.19), ρ has the units ohm-meters ($\Omega \cdot m$). Using this definition and Eq. (5.15), we can express the resistance for a uniform block of a conducting material as

$$R = \rho \frac{\ell}{A} \tag{5.20}$$

Every ohmic material has a characteristic resistivity ρ that depends on the properties of the material and on temperature. Note that ρ is an important property in selection of the conducting materials for applications used in electronic devices.

5.3.2   *Classical Model for Electrical Conduction*

In the following, we will describe a classical model of electrical conduction in conductors. Note that Paul Drude first proposed that model in 1900. The model explains Ohm's law, and it shows that resistivity can be related to the motion of electrons in a conductor. Drude model has its limitations because it is a classical model; however, it introduces concepts that can be applied to more sophisticated models.

We will consider a conductor formed by atoms positioned in a regular array and a set of electrons that can freely move, which are called *conduction electrons*. If the atoms are not part of a solid, the conduction electrons are bound to atoms and are not free to move; however, when atoms condense into a solid, then conduction electrons become mobile. In there is no external electric field, the conduction electrons move in random directions in all the conductors with average speeds of the order of 10^6 m/s, similar to the motion of molecules in gas inside a container. Often, the conduction

Fig. 5.6 The random motion of electrons in a conductor in absence of electric field, and drift velocity zero along the horizontal

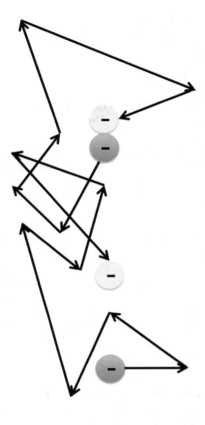

electrons in a conductor are assumed to form an *electron gas*. In the absence of the electric field, the drift velocity of the free electrons is zero, and hence there is no current, as presented in Fig. 5.6. In other words, on average, the same number electrons move in one direction as in the opposite direction, and thus the net charge flow is zero.

When an electric field is applied, as shown Fig. 5.7, the free electrons drift gently in the opposite direction of the electric field. Besides, the free electrons still undergo a random motion, as described in Fig. 5.6. Now, the average drift speed v_d is much smaller (typically 10^{-4} m/s) than average speed between collisions (typically 10^6 m/s). Therefore, the electric field **E** modified the random motion and made the electrons to drift in the opposite direction to the field. It is important to emphasize that there is slight curvature in the trajectories of the free electrons, as indicated in Fig. 5.7 because of the acceleration of the electrons between collisions. That is because the electric field applies a force on the free electrons.

In the classical model, we assume that the motion of an electron after a collision is independent of its motion before the collision. Besides, the excess energy gained by the electrons in the electric field is transferred to the atoms of the conductor during

Fig. 5.7 Motion of electrons in a conductor in presence of electric field, **E**

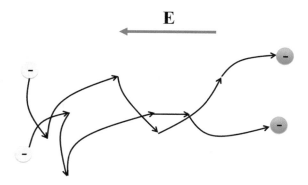

their collision. The energy transferred to the atoms increases the vibration energy of atoms, and therefore, the temperature of the conductor increases. Note that the temperature increase of a conductor because of the resistance can be used efficiently, such as in electric toasters and other familiar appliances. To derive a mathematical model, we will consider a free electron of mass m_e and charge q $(-e)$ in an electric field **E**. The electric field applies a force on the electron:

$$\mathbf{F} = q\mathbf{E} \tag{5.21}$$

Therefore, the electron gains an acceleration:

$$\mathbf{a} = \frac{\mathbf{F}}{m_e} = \frac{q\mathbf{E}}{m_e} \tag{5.22}$$

The acceleration, as given by Eq. (5.22), occurs for only a short time between collisions, which results only on a small drift velocity for the electron. We denote by t the time between two collisions, and $\mathbf{v}_i$ is the initial velocity of the electron, which is the velocity just after the first collision. Then, the velocity of the electron after time t (just before the second collision) is

$$\mathbf{v}_f = \mathbf{v}_i + \mathbf{a}t = \mathbf{v}_i + \frac{q\mathbf{E}}{m_e}t \tag{5.23}$$

Taking the time average of $\mathbf{v}_f$ over all possible times t and all possible values of $\mathbf{v}_i$:

$$\bar{\mathbf{v}}_f = \bar{\mathbf{v}}_i + \frac{q\mathbf{E}}{m_e}\bar{t} \tag{5.24}$$

Assuming that the initial velocities are uniformly distributed over all possible directions, we obtain

$$\bar{\mathbf{v}}_i = 0 \tag{5.25}$$

In Eq. (5.24), $\bar{t}$ is the average time interval (τ) between successive collisions:

$$\tau = \bar{t} \tag{5.26}$$

The average value of $\bar{\mathbf{v}}_f$ equals the drift velocity:

$$\mathbf{v}_d = \frac{q\mathbf{E}}{m_e}\tau \tag{5.27}$$

Then, the current density is

$$\mathbf{J} = nq\mathbf{v}_d = nq\frac{q\mathbf{E}}{m_e}\tau \tag{5.28}$$

and its magnitude:

$$J = nq^2\frac{E}{m_e}\tau \tag{5.29}$$

In Eq. (5.29), n is the number of charge carriers per unit volume.

Using the expression of Ohm's law, Eq. (5.12), we obtain the following relationships for conductivity:

$$\sigma = \frac{nq^2}{m_e}\tau \tag{5.30}$$

and resistivity from Eq. (5.21) becomes

$$\rho = \frac{1}{\sigma} = \frac{m_e}{nq^2\tau} \tag{5.31}$$

Equation (5.31) indicates that in this classical model, conductivity and resistivity do not depend on the strength of the electric field, which is a characteristic of the conductors obeying Ohm's law. The average time between collisions τ is related to the average distance between collisions ℓ and the average speed $\bar{v}$ as:

$$\tau = \frac{\ell}{\bar{v}} \tag{5.32}$$

5.3.3   Resistance and Temperature

It is found that the resistivity of a metal varies approximately linearly with temperature in a limited range of temperatures as follows:

$$\rho = \rho_0\left(1 + \alpha(T - T_0)\right) \tag{5.33}$$

where ρ is the resistivity at some temperature T (in degrees Celsius), ρ_0 is the resistivity at some reference temperature T_0 (usually taken to be $20\,°C$), and α is the temperature coefficient of resistivity. It is easy to obtain the temperature coefficient of resistivity as

$$\alpha = \frac{1}{\rho_0} \frac{\rho - \rho_0}{T - T_0} = \frac{1}{\rho_0} \frac{\Delta\rho}{\Delta T} \tag{5.34}$$

The unit for α is degrees Celsius^{-1} $[(°C)^{-1}]$. Because resistance is proportional to resistivity, we can write the variation of resistance as

$$R = R_0 \left(1 + \alpha(T - T_0)\right) \tag{5.35}$$

5.4 Superconductors

There exists a class of materials whose resistance decreases to zero below a specific temperature of T_c, known as the critical temperature. These materials are known as *superconductors*. If we would plot the resistance as a function of temperature for a superconductor, it follows that a superconductor behaves like a standard metal for $T > T_c$, and for $T \leq T_c$ its resistivity suddenly becomes zero. That was discovered by the Dutch physicist Heike Kamerlingh-Onnes (1853–1926), in 1911, working with mercury (a superconductor material below 4.2 K). Recently, it has been shown that the resistivity of superconductors for $T < T_c$ is less than 4×10^{-25} $\Omega \cdot$ m; that is, around 10^{17} times lower than the resistivity of copper metal, which can practically be considered zero. There are thousands of superconductors with critical temperatures that are substantially higher than initially thought possible. Because of low values of resistivity of superconductors, once a current set up in a superconductor wire, it will persist without any applied potential difference. Note that, already, steady currents are observed to persist in superconducting loops for several years with no apparent decay.

5.5 Electric Energy and Power

In Fig. 5.8, we show a simple circuit consisting of a battery and a resistor. Consider a positive amount of charge ΔQ moves clockwise around the circuit. That is, the charge passes from point a through the battery, then resistor and back to point a. Points a and d are grounded; that is, we take the electric potential at these two points to be zero. Electric potential energy U increases as the charge moves from point a to b through the battery as

Electric potential energy

$$\Delta U = \Delta Q \Delta V = \Delta Q(\phi_b - \phi_a) \tag{5.36}$$

Fig. 5.8 A simple circuit consisting of a battery whose terminals are connected to a resistor

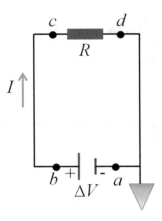

where ϕ_a and ϕ_b are electric potentials at the points a and b, respectively.

In contrast, as the charge moves through the resistor, from c to d, it loses the same amount of electric potential energy due to collisions with atoms of the resistor. That potential energy transfers into internal energy in the resistor. Note that we have neglected the resistance of the connecting wires; that is, electric potential energy faces no loss of energy in the paths bc and da. Therefore, when the charge arrives back at point a, it has lost all the electric potential energy, and hence its potential energy is zero.

Electric Power

The rate at which the charge ΔQ loses potential energy in going through the resistor is

$$\frac{\Delta U}{\Delta t} = \frac{\Delta Q}{\Delta t} \Delta V = I \Delta V \tag{5.37}$$

where I is the current in circuit. The charge loses energy equals the electric power:

$$\mathcal{P} = I \Delta V \tag{5.38}$$

Using Ohm's Law: $\Delta V = I R$, we get

$$\mathcal{P} = I^2 R = \frac{(\Delta V)^2}{R} \tag{5.39}$$

When I expresses in amperes, ΔV in volts and R in ohms, the SI unit of power is the watt.

5.6 Electromotive Force

A battery is a device that supplies electrical energy. Often, it is called either a *source of electromotive force* or *emf source*. In general, the internal resistance of the battery is neglected, and the potential difference between points a and b, see Fig. 5.8, is equal to the emf ϵ of the battery:

$$\Delta V = \phi_b - \phi_a = \epsilon \tag{5.40}$$

Therefore, the current in the circuit, based on Ohm's law, is

$$I = \frac{\Delta V}{R} = \frac{\epsilon}{R} \tag{5.41}$$

Because $\Delta V = \epsilon$, the power supplied by the emf source can be expressed as

$$\mathcal{P} = I\epsilon \tag{5.42}$$

5.7 Exercises

Exercise 5.1 Consider 1 mol, or 63.5 g, copper wire in a residential building. The wire has a cross-sectional area of 3.31×10^{-6} m^2, and a current of 10.0 A, determine the drift speed of the electrons. Assume that each copper atom contributes one free electron to the current. The density of copper is 8.95 g/cm^3.

Solution 5.1 The volume is

$$V = \frac{m}{\rho} = \frac{63.5 \text{ g}}{8.95 \text{ g/cm}^3} = 7.09 \text{ cm}^3 \tag{5.43}$$

Since in each mol of a substance there are $N_A = 6.02 \times 10^{23}$ atoms, and since each atom contributes with one electron, the total number of free electrons is

$$n = \frac{6.02 \times 10^{23}}{7.09 \text{ cm}^3} = 8.49 \times 10^{28} \text{ electrons/m}^3 \tag{5.44}$$

Using the equation of current $I = nq A v_d$, with $q = 1.6 \times 10^{-19}$ C being the absolute value of charge of electron, we get

$$v_d = \frac{I}{nq A} = 2.22 \times 10^{-4} \text{ m/s} \tag{5.45}$$

Exercise 5.2 Find the resistance of an 10.0 cm long aluminum cylinder with a cross-sectional area of 2.00×10^{-4} m^2. Compare the calculations for a cylinder of the same dimensions made of glass having a resistivity of 3.0×10^{10} $\Omega \cdot$ m.

Solution 5.2 The resistance for aluminum with $\rho = 2.82 \times 10^{-8}$ $\Omega \cdot$ m is calculated as

$$R_{Al} = \rho \frac{\ell}{A} \tag{5.46}$$

$$= (2.82 \times 10^{-8} \ \Omega \cdot m) \left(\frac{0.100 \ m}{2.00 \times 10^{-4} \ m^2} \right)$$

$$= 1.41 \times 10^{-5} \ \Omega$$

For the glass, we get

$$R_{gl} = \rho \frac{\ell}{A} \tag{5.47}$$

$$= (3.0 \times 10^{10} \ \Omega \cdot m) \left(\frac{0.100 \ m}{2.00 \times 10^{-4} \ m^2} \right)$$

$$= 1.5 \times 10^{13} \ \Omega$$

Exercise 5.3 Cable television uses a coaxial cable, which consists of two cylindrical conductors (see Fig. 5.9). The empty space between the conductors is completely filled with silicon material. Assume that the current leakage through the silicon is unwanted; that is, the cable is designed to conduct current along its length. The radii of the inner and outer conductors are $a = 0.500$ cm and $b = 1.75$ cm, respectively. The length of the cable is $L = 15.0$ cm. What is the resistance of the silicon between the two conductors?

Solution 5.3 We divide the object whose resistance we are calculating into concentric elements of infinitesimal thickness dr.

Then, equation $R = \rho \ell / A$ is written as

$$dR = \rho \frac{dr}{A} \tag{5.48}$$

dR is the resistance of a small element of silicon that has a thickness dr and surface area A: $A = 2\pi r L$. A current that passes from the inner to outer conductor must

Fig. 5.9 A coaxial cable consists of two cylindrical conductors with radius a and b, respectively

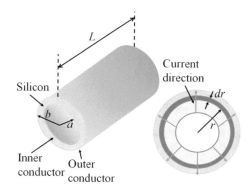

pass radially through this concentric small element, and the area through which this current passes is A. Therefore, we can write the resistance of that hollow silicon cylinder as

$$dR = \rho \frac{dr}{2\pi r L} \tag{5.49}$$

Total resistance is

$$R = \int_a^b \rho \frac{dr}{2\pi r L} = \frac{\rho}{2\pi L} \ln\left(\frac{b}{a}\right) \tag{5.50}$$

Given $\rho = 640 \, \Omega \cdot m$ for silicon, we obtain

$$R = \frac{640 \, \Omega \cdot m}{2\pi 0.150 \, m} \ln\left(\frac{1.75 \, cm}{0.500 \, cm}\right) \tag{5.51}$$

$$= 851 \, \Omega$$

Exercise 5.4 Consider again an 63.5 g (1 mol) copper wire with a cross-sectional area of $3.31 \times 10^{-6} \, m^2$, carrying a current of 10.0 A. Assume that each copper atom contributes to the current with one free electron. The density of copper is $8.95 \, g/cm^3$. What is: (a) the average time between collisions for electrons? (b) the mean free path for electrons in copper? Assume that the average speed for free electrons in copper is $1.6 \times 10^6 \, m/s$.

Solution 5.4 (a) The average collision time is

$$\tau = \frac{m_e}{nq^2\rho} \tag{5.52}$$

where $\rho = 1.7 \times 10^{-8}\ \Omega \cdot m$ for the cooper and charge carrier density is $n = 8.49 \times 10^{28}$ electrons/m^3, thus

$$\tau = \frac{9.11 \times 10^{-31}\ \text{kg}}{(8.49 \times 10^{28}\ \text{m}^{-3})(1.6 \times 10^{-19}\ \text{C})^2(1.7 \times 10^{-8}\ \Omega \cdot \text{m})} \tag{5.53}$$
$$= 2.5 \times 10^{-14}\ \text{s}$$

(b) Mean free path is

$$\ell = \tau \bar{v} = (2.5 \times 10^{-14}\ \text{s})(1.6 \times 10^6\ \text{m/s}) \tag{5.54}$$
$$= 4.0 \times 10^{-8}\ \text{m} = 40\ \text{nm}$$

Knowing that the atomic spacing is about 0.2 nm, that result indicates that the electron in the wire travels about 200 atomic spacing until the next collision.

Exercise 5.5 A resistance thermometer measures the temperature by measuring the change in resistance of a conductor. It is made from platinum and has a resistance of 50.0 Ω at 20.0 °C. The resistance increases to 76.8 Ω when immersed in a vessel containing melting indium. What is the melting point of the indium?

Solution 5.5 Using $R = R_0\,(1 + \alpha(T - T_0))$, we get

$$\Delta T = \frac{R - R_0}{\alpha R_0} \tag{5.55}$$

where $T_0 = 20.0\,°C$, $R = 76.8\ \Omega$ and $R_0 = 50.0\ \Omega$, then

$$\Delta T = 137\,°C \tag{5.56}$$

Then

$$T = T_0 + \Delta T = 157\,°C \tag{5.57}$$

Exercise 5.6 Consider an electric heater where we apply a potential difference of 120 V to a Nichrome wire with a total resistance of 8.00 Ω. What are the current in the wire and the power rating of the heater?

Solution 5.6 Using Ohm's law:

$$I = \frac{\Delta V}{R} \tag{5.58}$$

$$= \frac{120 \text{ V}}{8.00 \ \Omega} = 15.0 \text{ A}$$

The power is

$$\mathcal{P} = I \Delta V \tag{5.59}$$
$$= (15.0 \text{ A})(120.0 \text{ V})$$
$$= 1800 \text{ W} = 1.80 \text{ kW}$$

Exercise 5.7 (a) How much energy will it take to cook a turkey in an oven that works at 20.0 A and 240 V for 4 h? (b) If the energy is purchased at an estimated price of 8.00 cent per kilowatt hour, what is the cost?

Solution 5.7 (a) The power is

$$\mathcal{P} = I \Delta V = (20.0 \text{ A})(240.0 \text{ V}) = 4800 \text{ W} = 4.80 \text{ kW} \tag{5.60}$$

Energy is

$$Energy = \mathcal{P}t = (4.80 \text{ kW})(4 \text{ h}) = 19.2 \text{ kWh} \tag{5.61}$$

(b) The cost is

$$Cost = (19.2 \text{ kWh})(\$0.080/\text{kWh}) = \$1.54 \tag{5.62}$$

Exercise 5.8 Find the cost to operate a 100 W light-bulb for 24 h if the power company charges \$0.08/kWh.

Solution 5.8 Energy is

$$Energy = \mathcal{P}t = (0.10 \text{ kW})(24 \text{ h}) = 2.4 \text{ kWh} \tag{5.63}$$

The cost is

$$Cost = (2.4 \text{ kWh})(\$0.08/\text{kWh}) = \$0.19 \tag{5.64}$$

Exercise 5.9 Electrons that emerge in a particle accelerator have energy of 40.0 MeV (1 MeV = 1.60×10^{-13} J). Assume that the electrons emerge in pulses at the rate of 250 pulses/s, corresponding to a time between pulses of 4.00 ms. Duration of the pulse is 200 ns, and the electrons in the pulse constitute a current of 250 mA. The current is zero between pulses. (a) How many electrons are delivered by the accelerator per pulse? (b) What is the average current per pulse delivered by the accelerator? (c) What is the maximum power delivered by the electron beam?

Solution 5.9 (a) We know that

$$I = \frac{dQ}{dt} \tag{5.65}$$

or,

$$dQ = I\,dt \tag{5.66}$$

then

$$Q_{pulse} = I \int dt \tag{5.67}$$
$$= I\Delta t = (250 \times 10^{-3}\ \text{A})(200 \times 10^{-9}\ \text{s})$$
$$= 5.00 \times 10^{-8}\ \text{C}$$

Dividing this quantity of charge per pulse by the electronic charge gives the number of electrons per pulse:

$$\text{Electrons per pulse} = \frac{5.00 \times 10^{-8}\ \text{C/pulse}}{1.60 \times 10^{-19}\ \text{C/electron}} \tag{5.68}$$
$$= 3.13 \times 10^{11}\ \text{electrons/pulse}$$

(b) Average current is given as $I_{av} = \Delta Q / \Delta t$. Because the time interval between pulses is 4.00 ms, and because we know the charge per pulse from part (a), we obtain

$$I_{av} = \frac{Q_{pulse}}{t} \tag{5.69}$$
$$= \frac{5.00 \times 10^{-8}\ \text{C}}{4.00 \times 10^{-3}\ \text{s}}$$
$$= 12.5\ \mu\text{A}$$

(c) By definition, power is energy delivered per unit time. Thus, the maximum power is equal to the energy delivered by a pulse divided by the pulse duration (Fig. 5.10):

Fig. 5.10 Train of pulses of electrons in a particle accelerator

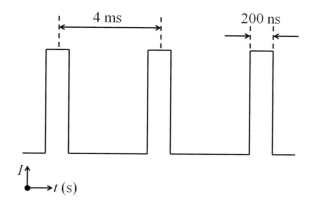

$$\mathcal{P} = \frac{E}{\Delta t} \tag{5.70}$$
$$= \frac{(3.13 \times 10^{11} \text{ electrons/pulse})(40.0 \text{ MeV/electron})}{2.00 \times 10^{-7} \text{ s/pulse}}$$
$$= (6.26 \times 10^{19} \text{ MeV/s})(1.60 \times 10^{13} \text{ J/MeV})$$
$$= 1.00 \times 10^{7} \text{ W} = 10.0 \text{ MW}$$

Exercise 5.10 In a particular cathode ray tube, the measured beam current is 30.0 μA. How many electrons strike the tube screen every 40.0 s?

Solution 5.10 From the definition

$$I = \frac{Q}{\Delta t} = \frac{N_e e}{\Delta t} \tag{5.71}$$

where e is the magnitude of the electron charge 1.6×10^{-19} C, and N_e number of electrons, then

$$N_e = \frac{I \Delta t}{e} \tag{5.72}$$
$$= \frac{(30.0 \times 10^{-6} \text{ A})(40.0 \text{ s})}{1.6 \times 10^{-19} \text{ C}}$$
$$= 75 \times 10^{14}$$

Exercise 5.11 If the drift velocity of free electrons in a copper wire is 7.84×10^{-4} m/s, what is the electric field in the conductor?

Solution 5.11 For cooper $\rho = 1.7 \times 10^{-8}$ $\Omega \cdot$ m and charge carrier density is $n = 8.49 \times 10^{28}$ electrons/m^3, therefore, the average collision time is:

$$\tau = \frac{m_e}{nq^2\rho} \tag{5.73}$$

or

$$\tau = \frac{9.11 \times 10^{-31} \text{ kg}}{(8.49 \times 10^{28} \text{ m}^{-3})(1.6 \times 10^{-19} \text{ C})^2 (1.7 \times 10^{-8} \Omega \cdot \text{m})} \tag{5.74}$$
$$= 2.5 \times 10^{-14} \text{ s}$$

Then, using the equation

$$v_d = q \frac{E}{m_e} \tau \tag{5.75}$$

we get

$$E = \frac{m_e v_d}{q\tau} \tag{5.76}$$
$$= \frac{(9.11 \times 10^{-31} \text{ kg})(7.84 \times 10^{-4} \text{ m/s})}{(1.6 \times 10^{-19} \text{ C})(2.5 \times 10^{-14} \text{ s})}$$
$$= 0.18 \text{ N/C}$$

Exercise 5.12 Use data from the previous exercise to calculate the collision mean free path of electrons in copper, assuming that the average thermal speed of conduction electrons is 8.60×10^5 m/s.

Solution 5.12 The collision mean free path is

$$\ell = v_d \tau \tag{5.77}$$

For cooper $\rho = 1.7 \times 10^{-8}$ $\Omega \cdot$ m and charge carrier density is $n = 8.49 \times 10^{28}$ electrons/m^3, therefore, the average collision time is

$$\tau = \frac{m_e}{nq^2\rho} \tag{5.78}$$

or

$$\tau = \frac{9.11 \times 10^{-31} \text{ kg}}{(8.49 \times 10^{28} \text{ m}^{-3})(1.6 \times 10^{-19} \text{ C})^2(1.7 \times 10^{-8} \ \Omega \cdot \text{m})} \tag{5.79}$$
$$= 2.5 \times 10^{-14} \text{ s}$$

then

$$\ell = (8.60 \times 10^5 \text{ m/s})(2.5 \times 10^{-14} \text{ s}) \tag{5.80}$$
$$= 2.15 \times 10^{-8} \text{ m}$$
$$= 21.5 \text{ nm}$$

Reference

Holliday D, Resnick R, Walker J (2011) Fundamentals of physics, John Wiley and Sons, New York

Chapter 6
Magnetic Field

This chapter aims to introduce the magnetic field and the motion of charges in a magnetic field.

In this chapter, we will introduce the magnetic field. We described the interactions between charged objects in terms of electric fields. An electric field associates with any stationary or moving electric charge. In addition to an electric field, the region of space surrounding any moving electric charge also contains a *magnetic field*. A magnetic field also encompasses any magnetic substance. As extra reading material, the reader can also consider other literature (Holliday et al. 2011).

6.1 Magnetic Field

The magnetic field is usually represented by symbol **B**. To determine the direction of the magnetic field **B** at some location the compass needle is used, which points along **B** at that location. The magnetic field lines outside a magnet go from north poles to south poles, as shown in Fig. 6.1. It is common to use small iron filings to display magnetic field line patterns of a bar magnet. To define the magnetic field **B** at any location in space, the magnetic force $\mathbf{F}_B$ that the field exerts on a test object can be used. For that, we can use a charged particle moving with some velocity of **v**. Furthermore, we ignore the presence of the gravitational field or an electric field at the position of the test object.

© The Author(s), under exclusive license to Springer Nature Switzerland AG 2022
H. Kamberaj, *Electromagnetism*, Undergraduate Texts in Physics,
https://doi.org/10.1007/978-3-030-96780-2_6

Fig. 6.1 A diagram of
magnetic field lines going
from north pole to south pole

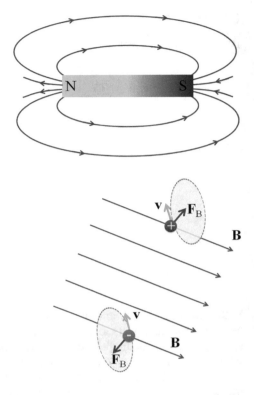

Fig. 6.2 Direction of the
magnetic force acting on a
charge particle q (positive or
negative) moving with
velocity **v**

From the experimental results of the charged particles moving in a magnetic
field (see also Fig. 6.2), the following statements can be written

1. The magnitude of the magnetic force, F_B, acting on a charged particle is
   proportional to the charge q and the speed v of that particle.
2. The magnitude and direction of the magnetic force $\mathbf{F}_B$ depend on the velocity of the particle and the magnitude and direction of the magnetic field **B**.
3. A charged particle moving parallel to the magnetic field vector **B** experiences no magnetic force, $\mathbf{F}_B = 0$.
4. For a charged particle moving with velocity **v** making an angle $\theta \neq 0$ with
   the magnetic field **B**, the magnetic force $\mathbf{F}_B$ exerted on that particle is
   perpendicular to both **v** and **B**. Thus, $\mathbf{F}_B$ is normal to the plane formed by
   v and **B**.
5. Furthermore, the direction of the magnetic force $\mathbf{F}_B$ acting on a positive
   charge moving with velocity **v** is opposite to the direction of the magnetic
   force acting on a negative charge moving with the same velocity **v**.

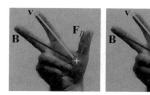

Fig. 6.3 The right-hand rule demonstration for finding the direction of the magnetic force acting on a charge particle q (positive or negative) moving with velocity $\mathbf{v}$

6. Moreover, the magnitude of the magnetic force $\mathbf{F}_B$ acting on the moving charged particle is proportional to $\sin \theta$, where θ is the angle between the particle velocity $\mathbf{v}$ and the direction of the magnetic field $\mathbf{B}$.

Mathematically, those observations can be formulated in the following form:

$$\mathbf{F}_B = q\mathbf{v} \times \mathbf{B} \tag{6.1}$$

Equation (6.1) indicates that the direction of $\mathbf{F}_B$ applied on a positive charge particle q is in the direction of $\mathbf{v} \times \mathbf{B}$, and hence, by definition of the cross product, it is perpendicular to both $\mathbf{v}$ and $\mathbf{B}$ (see Fig. 6.3). Furthermore, if q is a negative charge, then $\mathbf{F}_B$ is opposite to the direction of $\mathbf{v} \times \mathbf{B}$. Moreover, Eq. (6.1) implies that the magnitude of the magnetic force F_B is

$$F_B = \mid q \mid vB \sin \theta \tag{6.2}$$

Here, θ is the smaller angle between $\mathbf{v}$ and $\mathbf{B}$. Equation (6.2) implies that $F_B = 0$ if $\mathbf{v}$ is parallel or antiparallel to $\mathbf{B}$ (that is, $\theta = 0$ or $180°$) and maximum ($F_{B,max} = qvB$) if $\mathbf{v}$ is perpendicular to $\mathbf{B}$ (that is, $\theta = 90°$). Equation (6.2) defines the operational function of the magnetic field $\mathbf{B}$ at some point in space; that is, if a moving charged particle is placed at that location in space, the magnetic field is defined in terms of the force acting on that charge.

It is important to summarize the differences between electric and magnetic forces as follows:
- The electric field applies a force on a charged particle that acts along the direction of the field. In contrast, the magnetic force is perpendicular to the direction of the magnetic field.
- The electric force acts on both moving and charged particle at rest, whereas the magnetic force acts only on a moving charged particle.

- The electric force does work in moving a charged particle in space; in contrast, the magnetic force associated with a uniform magnetic field does not do any work on a moving particle.
- Therefore, using the work-kinetic energy theorem, we can say that the kinetic energy of a charged particle moving through only a magnetic field remains constant.
- That is, when on a moving charged particle with a velocity **v** acts a magnetic field, the direction of the velocity may change due to the field, but its speed or kinetic energy remains constant.

From Eq. (6.1), the SI unit of the magnetic field is newtons per coulomb-meter per second, which is called the *tesla* (T):

$$1\,T = 1\frac{N}{C \cdot m/s} \tag{6.3}$$

Since we defined a coulomb per second as an ampere, we write that

$$1\,T = 1\frac{N}{A \cdot m} \tag{6.4}$$

Note that gauss (G) is another unit used for the magnetic field, which relates to the SI unit, tesla, as follows:

$$1T = 10^4\,G \tag{6.5}$$

6.2   Magnetic Force Acting on a Current-Carrying Conductor

Consider the magnetic force exerted on a single charged particle that moves through a magnetic field given by Eq. (6.1). If we suppose having a conducting wire in which a current is maintained, for example, by utilizing a battery, then it should experience a force when placed in a magnetic field. That is because the current is a stream of mobile charged particles. Thus, the net force acting by the field on the wire is the directorial sum of the individual forces exerted on all the charged particles making up that current. When the mobile charged particles collide with atoms of the conducting wire, then those magnetic forces used on the charged particles transmit to the wire.

We will consider a straight segment of wire with length L and cross-sectional area A, carrying a steady current I, to quantify the magnetic force applied by the magnetic force on a current-carrying wire. Furthermore, suppose that the wire is placed in a uniform magnetic field **B**, as shown in Fig. 6.4. The magnetic force exerted on a charge q moving with a drift velocity $\mathbf{v}_d$ is given by Eq. (6.1) as

Fig. 6.4 Magnetic force acting on a charge particle q moving with a drift velocity $\mathbf{v}_d$

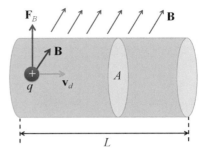

$$\mathbf{f}_B = q\mathbf{v}_d \times \mathbf{B} \tag{6.6}$$

We denote by N_q the total number of charges in that segment, given as: $N_q = nAL$, where n is the number of charges per unit volume and AL gives the volume of the segment. Then, to find the total force exerted on the wire, we multiply the force $\mathbf{f}_B$ exerted on one charge by N_q. Hence, the total magnetic force on the wire of length L is

$$\mathbf{F}_B = N_q\mathbf{f}_B = nALq\mathbf{v}_d \times \mathbf{B} \tag{6.7}$$

Since the current is $I = nqv_dA$, then

$$\mathbf{F}_B = I\,(\mathbf{L} \times \mathbf{B}) \tag{6.8}$$

In Eq. (6.8), $\mathbf{L}$ is a vector pointing in the same direction as the current I and has a magnitude equal to the length L of the segment.

Note that the expression in Eq. (6.8) applies only to a straight segment of wire in a uniform magnetic field. To generalize it for any arbitrary shaped wire, consider the wire segment of uniform cross-section in a magnetic field, as shown in Fig. 6.5. We partition the segment into small linear segments of length ds, where the vector $d\mathbf{s}$ is in the direction of the current flow in the wire. Then, the magnetic force exerted on $d\mathbf{s}$ in a field $\mathbf{B}$ is

$$d\mathbf{F}_B = I\,(d\mathbf{s} \times \mathbf{B}) \tag{6.9}$$

$d\mathbf{F}_B$, in Eq. (6.9), is directed out of the page, if we assume that the magnetic field is pointing along y-axis. The total force $\mathbf{F}_B$ acting on the wire is

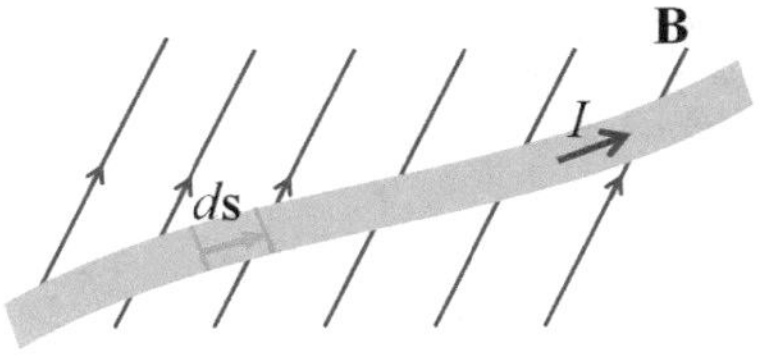

Fig. 6.5 An arbitrarily shaped wire segment of uniform cross-section in which is passing a current I in a magnetic field

Fig. 6.6 An arbitrarily
shaped wire segment of
uniform cross-section in
which is passing a current I
in a uniform magnetic field

$$\mathbf{F}_B = I \int_a^b d\mathbf{s} \times \mathbf{B} \qquad (6.10)$$

a and b, in Eq. (6.10), represent the end points of the wire. Note that the magnitude of the magnetic field and the direction the field makes with the vector $d\mathbf{s}$ may differ at different points.

To evaluate the formula given by Eq. (6.10), we can consider the following situations. First, we consider that the magnetic field is uniform; that is, $\mathbf{B}$ is constant, as shown in Fig. 6.6, then

$$\mathbf{F}_B = I \int_a^b d\mathbf{s} \times \mathbf{B} \qquad (6.11)$$
$$= I \left(\int_a^b d\mathbf{s} \right) \times \mathbf{B}$$
$$= I \left(\mathbf{L}' \times \mathbf{B} \right)$$

Using the rule of sum of vectors, $\mathbf{L}'$ in Eq. (6.11) denotes the vector pointing from a to b.

In a second situation, we consider an arbitrarily shaped closed loop carrying the steady current I, which is placed in a uniform magnetic field (see Fig. 6.7). We can again express the force acting on the loop in the form:

$$\mathbf{F}_B = I \oint d\mathbf{s} \times \mathbf{B} \qquad (6.12)$$
$$= I \left(\oint d\mathbf{s} \right) \times \mathbf{B}$$

Fig. 6.7 A closed shaped
wire segment of uniform
cross-section in which is
passing a current I in a
uniform magnetic field

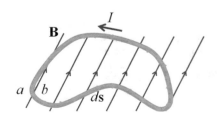

Because the set of length elements $d\mathbf{s}$ forms a closed polygon, the vector sum must be zero:

$$\oint d\mathbf{s} = 0 \tag{6.13}$$

Hence,

$$\mathbf{F}_B = 0 \tag{6.14}$$

Therefore, the net magnetic force acting on any closed current loop in a uniform magnetic field is zero.

6.3   Torque on a Current Loop in a Uniform Magnetic Field

We showed that a force acts on a current-carrying conductor placed in a magnetic field, given mathematically by Eq. (6.10). Now, consider a current loop placed in a magnetic field, as shown in Fig. 6.8. Equation (6.10) implied that the net magnetic force acting on the loop is zero (see also Eq. (6.14)). However, we will prove in the following that the magnetic field applies a torque on the current loop. For that, consider a rectangular loop carrying a current I in the presence of a uniform magnetic field. First, we assume that the magnetic field is directed parallel to the plane of the loop, as shown in Fig. 6.8.

The magnetic forces exerted on sides 12 and 34 are both zeroes because these wire segments are parallel to the field $\mathbf{B}$; hence, $\mathbf{L} \times \mathbf{B} = 0$ for those two segments. However, the magnetic field is normal to the portions 23 and 34; therefore, the magnetic forces on Sects. 23 and 41 are different from zero.

We calculate the magnitude of the forces acting on segments 23 and 41 carrying the current I and length $L = a$ from Eq. (6.8), as

$$F_{23} = F_{41} =\mid I\,(\mathbf{L} \times \mathbf{B} \mid= I a B \tag{6.15}$$

Fig. 6.8 A current loop in a uniform magnetic field directed parallel to the plane of loop

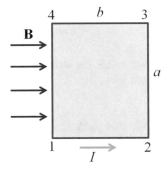

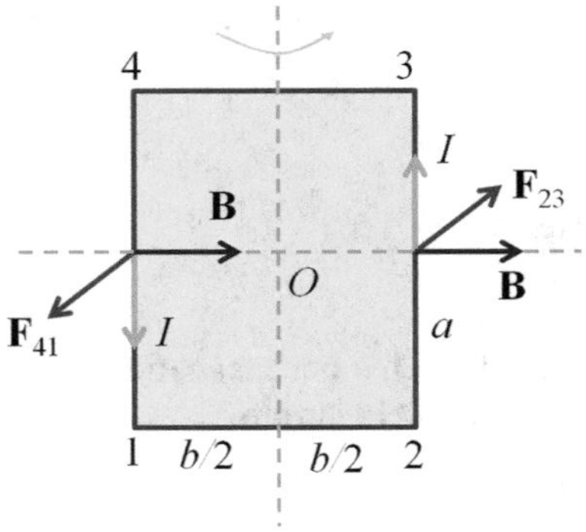

Fig. 6.9 A diagram of forces
acting on the current loop

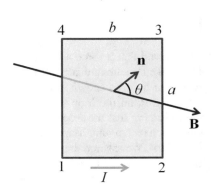

Fig. 6.10 A current loop in
a uniform magnetic field
forming an angle θ to the
plane of loop

However, the direction of $\mathbf{F}_{41}$, the force exerted on segment 41, is out of the page,
and that of $\mathbf{F}_{23}$, the force exerted on segment 23, is into the page, as indicated in
Fig. 6.9. Therefore, those two forces point in opposite directions, and they are not
directed along the same line of action; that is, the distance between their lines of
action is b. Thus, those two forces exert a torque on the loop, such that the loop
rotates counterclockwise about the axis passing through the point O, as shown in
Fig. 6.9. The magnitude of that torque τ is

$$\tau = F_{23}\frac{b}{2} + F_{41}\frac{b}{2} = (IaB)\frac{b}{2} + (IaB)\frac{b}{2} = IabB \qquad (6.16)$$

where $b/2$ is the moment arm about O of each force. Denoting $A = ab$ the area of
closed loop, then

$$\tau = IAB \qquad (6.17)$$

We can generalize that result by considering the same current-carrying closed
loop, but in a uniform magnetic field making an angle $\theta < 90°$ with the normal
vector to the plane of the loop, as indicated in Fig. 6.10. We also assume that $\mathbf{B}$ is
perpendicular to the segment lines 12 and 34.

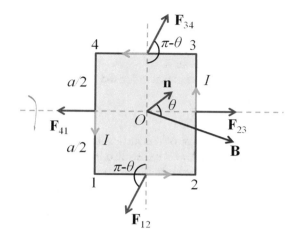

Fig. 6.11 Diagram of magnetic forces acting on a current loop in a uniform magnetic field forming an angle θ to the plane of loop

Magnetic forces $\mathbf{F}_{23}$ and $\mathbf{F}_{41}$ exerted on segments 23 and 41 are perpendicular to those segment lines, and they have the same line of action, and they point in the opposite direction to one another (see also Fig. 6.11); thus, they do not produce any torque. On the other hand, $\mathbf{F}_{12}$ and $\mathbf{F}_{34}$ acting on segment lines 12 and 34 have different lines of action, they point perpendicular to the plane of the loop, and they have opposite direction. Hence, they produce a torque about the horizontal axis passing through point O, as indicated in Fig. 6.11.

The moment arm of $\mathbf{F}_{12}$ about the point O is equal to $(a/2) \sin \theta$. Similarly, the moment arm of $\mathbf{F}_3$ about O is also $(a/2) \sin \theta$. Because $F_{12} = F_{34} = IbB$, the net torque about O has the magnitude

$$\tau = F_1 \frac{a}{2} \sin \theta + F_3 \frac{a}{2} \sin \theta \qquad (6.18)$$
$$= IbB \left(\frac{a}{2} \sin \theta \right) + IbB \left(\frac{a}{2} \sin \theta \right)$$
$$= IabB \sin \theta = IAB \sin \theta$$

where $A = ab$ is the area of the loop.

Equation (6.18) indicates that the torque is maximum when the magnetic field is perpendicular to the normal vector of the plane of the loop ($\theta = 90°$):

$$\tau_{max} = IAB \qquad (6.19)$$

That is when the field $\mathbf{B}$ is parallel with the plane of the loop. Furthermore, it is zero when the field $\mathbf{B}$ is parallel to the normal vector of the plane of the loop ($\theta = 0°$):

$$\tau = 0 \qquad (6.20)$$

Note that due to the magnetic field, the loop rotates in the direction of decreasing the angle θ. That is, the loop rotates in the direction such that the normal vector **n** to the surface of the loop (or equivalently, surface area vector $\mathbf{A} = A\mathbf{n}$) aligns with the magnetic field vector **B**.

Torque vector

A convenient expression for the torque applied on a loop in a uniform magnetic field **B** is

$$\tau = I\,(\mathbf{A} \times \mathbf{B}) \tag{6.21}$$

where **A**, the vector of the surface area of the loop, is perpendicular to the plane of the loop, that is, $\mathbf{A} = A\mathbf{n}$ (see also Fig. 6.11). Direction of **A** is determined using the right-hand rule shown in Fig. 6.10.

Magnetic dipole

The product $I\mathbf{A}$ is defined to be the magnetic dipole moment μ (or the "magnetic moment") of the loop:

$$\mu = I\mathbf{A} \tag{6.22}$$

The SI unit of magnetic dipole moment is ampere-meter^2 ($A \cdot m^2$). Using this definition, we can express the torque exerted on a current-carrying loop in a magnetic field **B** as

$$\tau = \mu \times \mathbf{B} \tag{6.23}$$

The direction of the magnetic moment is the same as the direction of **A** (Fig. 6.12).

Fig. 6.12 Right-hand rule demonstration in determining the direction of magnetic dipole

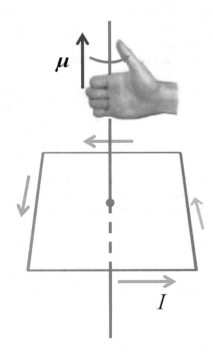

6.4   Motion of a Charged Particle in a Uniform Magnetic Field

From Eq. (6.1), the force exerted on a charged particle moving in a magnetic field is normal to the velocity of the particle, and hence, perpendicular to its displacement. Therefore, the work done on the particle by the magnetic force is zero because the force is perpendicular to the displacement vector of the particle.

As a particular case, we consider a positively charged particle moving in a uniform magnetic field with an initial velocity vector of the particle perpendicular to the field (see Fig. 6.13). The direction of the magnetic field is pointing into the page. Using the right-hand rule in Eq. (6.1), we find that the direction of the magnetic force is pointing toward a single point at the center of a circle. Therefore, the particle is going to move in a circle in a plane perpendicular to the magnetic field.

The particle moves in this way because the magnetic force $\mathbf{F}_B$ is perpendicular to both $\mathbf{v}$ and $\mathbf{B}$ and has a constant magnitude of qvB. As the force deflects the particle, the directions of both $\mathbf{v}$ and $\mathbf{F}_B$ change continuously, and $\mathbf{F}_B$ points toward the center of the circle at each position of the particle. Thus, force changes only the direction of $\mathbf{v}$, but it does change its magnitude. The rotation is counterclockwise for that positive charge, and if q is negative, the rotation would be clockwise.

Using the second law of Newton for circular motion, we get

$$\sum_i F_{ir} = ma_r \qquad (6.24)$$

or

$$F_B = qvB = m\frac{v^2}{r} \qquad (6.25)$$

The radius of the circle is

$$r = \frac{mv^2}{qvB} = \frac{mv}{qB} \qquad (6.26)$$

Fig. 6.13 A positive charge particle moving in a uniform magnetic field with initial velocity perpendicular to the field

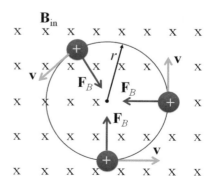

That is, the radius of the path is proportional to the linear momentum mv of the particle and inversely proportional to the magnitude of the charge on the particle and to the magnitude of the magnetic field. The angular speed of the particle is

$$\omega = \frac{v}{r} = \frac{qB}{m} \tag{6.27}$$

The period of the motion (the time that the particle takes to complete one revolution) is equal to the circumference of the circle divided by the linear speed of the particle:

$$T = \frac{2\pi r}{v} = \frac{2\pi}{\omega} = \frac{2\pi m}{qB} \tag{6.28}$$

6.5 Exercises

Exercise 6.1 Consider an electron in a television picture tube, which moves toward the front of the tube (that is, toward us) with a speed of 8.0×10^6 m/s along the x axis (see Fig. 6.14). The faces of the tube are coils of wire that create a magnetic field of magnitude 0.025 T, at an angle of $60°$ to the x axis and lying in the xy plane. Determine: (a) the magnetic force exerted on the electron; (b) its acceleration.

Solution 6.1 From Eq. (6.2), we obtain

$$F_B = (1.6 \times 10^{-19} \text{ C})(8.0 \times 10^6 \text{ m/s})(0.025 \text{ T})(\sin 60°) \tag{6.29}$$
$$= 2.8 \times 10^{-14} \text{ N}$$

Fig. 6.14 An electron in a television picture tube moves toward the front of the tube with a constant speed

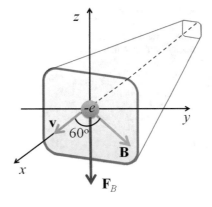

In addition, the vector $\mathbf{v} \times \mathbf{B}$ is along the positive z direction (from the right-hand rule) and the charge is negative, and hence, $\mathbf{F}_B$ is along the negative z direction. Considering that the electron mass is $m_e = 9.11 \times 10^{-31}$ kg, the electron acceleration is

$$a = \frac{F_B}{m_e} \tag{6.30}$$
$$= \frac{2.8 \times 10^{-14} \text{ N}}{9.11 \times 10^{-31} \text{ kg}}$$
$$= 3.1 \times 10^{16} \text{ m/s}^2$$

In addition, the vector $\mathbf{a}$ is along the direction negative z axis.

Exercise 6.2 Consider a conducting wire loop in the form of a semicircle of radius R carrying a current I, as shown in Fig. 6.15. Suppose a uniform magnetic field is applied in the direction along the positive vertical direction. Determine the magnitude and direction of the magnetic force acting on the straight and the curved portions of the wire.

Solution 6.2 The two forces acting on the wire are $\mathbf{F}_1$ applied on the straight portion and $\mathbf{F}_2$ applied on the semicircle portion.

The magnitude of $\mathbf{F}_1$ is

$$F_1 = ILB = 2IRB \tag{6.31}$$

because the straight portion of the wire is perpendicular to $\mathbf{B}$. The direction of $\mathbf{F}_1$ is out of the page, since $\mathbf{L} \times \mathbf{B}$ is along the positive z axis.

We first write an expression for the force $d\mathbf{F}_2$ on a small element of length $d\mathbf{s}$:

$$dF_2 = I \mid d\mathbf{s} \times \mathbf{B} \mid = IB \sin\theta \, ds \tag{6.32}$$

Since $s = R\theta$, then $ds = Rd\theta$, then $dF_2 = IRB \sin\theta \, d\theta$. Hence, the total force is

Fig. 6.15 A wire bent into a semicircle of radius R forms a closed circuit and carries a current I

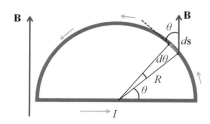

$$F_2 = \int_0^\pi IRB \sin\theta \, d\theta \qquad (6.33)$$
$$= IRB \, [-\cos\theta]_0^\pi$$
$$= 2IRB$$

Because $\mathbf{F}_2$, with a magnitude of $2IRB$, is directed into the page and because $\mathbf{F}_1$, with a magnitude of $2IRB$, is directed out of the page, the net force on the closed loop is zero.

Exercise 6.3 A rectangular coil with dimensions of 5.40 cm × 8.50 cm and 25 turns of wire. It carries a steady current of 15.0 mA. In addition, the magnetic force of 0.350 T is applied parallel to the plane of the loop. Determine: (a) the magnitude of the magnetic dipole moment; (b) the magnitude of the torque acting on the loop.

Solution 6.3 (a) For a coil with $N = 25$ turns, the magnetic dipole is

$$\mu_{coil} = NIA = 25(15.0 \times 10^{-3} \text{ A})(0.0540 \times 0.0850 \text{ m}^2) \qquad (6.34)$$
$$= 1.72 \times 10^{-3} \text{ A} \cdot \text{m}^2$$

(b) Since $\mathbf{B}$ is perpendicular to μ_{coil}, then

$$\tau = \mu_{coil} B = (1.72 \times 10^{-3} \text{ A} \cdot \text{m}^2)(0.350 \text{ T}) \qquad (6.35)$$
$$= 6.02 \times 10^{-4} \text{ N} \cdot \text{m}$$

Exercise 6.4 A satellite uses a device made up of coils to adjust the orientation by interacting with the Earth's magnetic field. A torque is created on the satellite in the x, y, or z direction. This control system is used solar-generated electricity. If we assume that the device has a magnetic dipole moment of 250 A · m^2, find the maximum torque applied to the satellite when it is at a location where the magnitude of the Earth's magnetic field is 3.0×10^{-5} T.

Solution 6.4 The torque is given as

$$\boldsymbol{\tau} = \boldsymbol{\mu} \times \mathbf{B} \tag{6.36}$$

and the magnitude is

$$\tau = \mu B \sin \theta \tag{6.37}$$

where θ is angle between $\boldsymbol{\mu}$ and $\mathbf{B}$. The maximum torque is obtained when

$$\tau_{max} = \mu B = (250 \text{ A} \cdot \text{m}^2)(3.0 \times 10^{-5} \text{ T}) \tag{6.38}$$

$$= 750 \times 10^{-5} \left(\text{A} \cdot \text{m}^2 \cdot \frac{\text{N}}{\text{A} \cdot \text{m}} \right)$$

$$= 750 \times 10^{-5} \text{ N} \cdot \text{m}$$

Exercise 6.5 A proton of mass 1.67×10^{-27} kg moves in a circular orbit of radius 14 cm. Assume a uniform magnetic field of magnitude 0.35 T is perpendicular to plane of the orbit. What is the linear speed of the proton?

Solution 6.5 The linear speed is of proton is

$$v = \frac{q B r}{m_p} \tag{6.39}$$

$$= \frac{(1.60 \times 10^{-19} \text{ C})(0.35 \text{ T})(14 \times 10^{-2} \text{ m})}{1.67 \times 10^{-27} \text{ kg}}$$

$$= 4.7 \times 10^6 \text{ m/s}$$

Exercise 6.6 An electron of charge $q = -1.60 \times 10^{-19}$ C and mass 9.11×10^{-31} kg moves in a direction perpendicular to a uniform magnetic field of magnitude 0.35 T with a linear speed 4.7×10^6 m/s. Find the radius of its circular orbit.

Solution 6.6 The radius of the circle is given by

$$r = \frac{m_e v}{q B} \tag{6.40}$$

Then,

$$r = \frac{(9.11 \times 10^{-31} \text{ kg})(4.7 \times 10^6 \text{ m/s})}{(1.60 \times 10^{-19} \text{ C})(0.35 \text{ T})}$$

$$= 7.6 \times 10^{-5} \text{ m} \tag{6.41}$$

Exercise 6.7 To measure the magnitude of a uniform magnetic field, electrons are accelerated from rest in a potential difference of 350 V. They travel along a curved path because of the magnetic force exerted on electrons. Assume that the radius of the path is 7.5 cm and the magnetic field is perpendicular to the beam of electrons. Determine: (a) the magnitude of the field; (b) the angular speed of the electrons; (c) the period of revolution of the electrons.

Solution 6.7 (a) The change in the electric potential energy is

$$\Delta U = (U_f - U_i) = - \mid e \mid \Delta V \tag{6.42}$$

ΔV is the potential difference. Using the conservation law of energy: $\Delta K = -\Delta U$. Since

$$K_i = 0; \quad K_f = m_e v^2/2 \tag{6.43}$$

Thus, we get

$$m_e \frac{v^2}{2} = \mid e \mid \Delta V \tag{6.44}$$

Or,

$$v = \sqrt{\frac{2 \mid e \mid \Delta V}{m_e}} \tag{6.45}$$

Replacing the numerical values:

$$v = \sqrt{\frac{2(1.60 \times 10^{-19} \text{ C})(350 \text{ V})}{9.11 \times 10^{-31} \text{ kg}}}$$

$$= 1.11 \times 10^7 \text{ m/s} \tag{6.46}$$

and the magnetic field is

$$B = \frac{m_e v}{|e|r} \tag{6.47}$$

$$= \frac{(9.11 \times 10^{-31} \text{ kg})(1.11 \times 10^7 \text{ m/s})}{(1.60 \times 10^{-19} \text{ C})(0.075 \text{ m})}$$

$$= 8.4 \times 10^{-4} \text{ T}$$

(b) From the equation

$$\omega = \frac{v}{r} \tag{6.48}$$

we get

$$\omega = \frac{1.11 \times 10^7 \text{ m/s}}{0.075 \text{ m}} = 1.5 \times 10^8 \text{ rad/s} \tag{6.49}$$

(c) For the period

$$T = \frac{2\pi}{\omega} = 4.1888 \times 10^{-8} \text{ s} \approx 42 \text{ ns} \tag{6.50}$$

Exercise 6.8 Suppose a proton of charge 1.6×10^{-19} C moves with a velocity of $\mathbf{v} = (2\mathbf{i} - 4\mathbf{j} + \mathbf{k})$ m/s in a region in which the magnetic field is $\mathbf{B} = (\mathbf{i} + 2\mathbf{j} - 3\mathbf{k})$ T. What is the magnitude of the magnetic force this charge experiences?

Solution 6.8 The magnetic force is

$$\mathbf{F}_B = q\mathbf{v} \times \mathbf{B} \tag{6.51}$$

where q is the charge of proton $q = 1.6 \times 10^{-19}$ C. The components are

$$F_{Bx} = q(\mathbf{v} \times \mathbf{B})_x = q(v_y B_z - v_z B_y) \tag{6.52}$$
$$F_{By} = q(\mathbf{v} \times \mathbf{B})_y = -q(v_x B_z - v_z B_x)$$
$$F_{Bz} = q(\mathbf{v} \times \mathbf{B})_z = q(v_x B_y - v_y B_x)$$

or

$$F_{Bx} = (1.6 \times 10^{-19} \text{ C})((-4.0)(-3.0) - (1.0)(2.0)) \text{ T} \cdot \text{m/s} \tag{6.53}$$
$$= 16 \times 10^{-19} \text{ N}$$
$$F_{By} = -(1.6 \times 10^{-19} \text{ C})((2.0)(-3.0) - (1.0)(1.0)) \text{ T} \cdot \text{m/s} \tag{6.54}$$
$$= 11.2 \times 10^{-19} \text{ N}$$

$$F_{Bz} = (1.6 \times 10^{-19} \text{ C})((2.0)(2.0) - (-4.0)(1.0)) \text{ T} \cdot \text{m/s} \tag{6.55}$$
$$= 12.8 \times 10^{-19} \text{ N}$$

The magnitude of the force is

$$F_B = \sqrt{F_{Bx}^2 + F_{By}^2 + F_{Bz}^2} \tag{6.56}$$
$$= \sqrt{256 + 125.44 + 163.84} \times 10^{-19} \text{ N}$$
$$\approx 23.4 \times 10^{-19} \text{ N}$$

Exercise 6.9 An electron of charge -1.6×10^{-19} C is moving in a uniform magnetic field of $\mathbf{B} = (1.40\mathbf{i} + 2.10\mathbf{j})$ T. Find the magnetic force on the electron if its velocity is $\mathbf{v} = 3.70 \times 10^5\mathbf{j}$ m/s.

Solution 6.9 The magnetic force given by Eq. (6.1), where q is the charge of electron $q = -1.6 \times 10^{-19}$ C. The components are

$$F_{Bx} = q(\mathbf{v} \times \mathbf{B})_x = q(v_y B_z - v_z B_y) \tag{6.57}$$
$$F_{By} = q(\mathbf{v} \times \mathbf{B})_y = -q(v_x B_z - v_z B_x)$$
$$F_{Bz} = q(\mathbf{v} \times \mathbf{B})_z = q(v_x B_y - v_y B_x)$$

or

$$F_{Bx} = (-1.6 \times 10^{-19} \text{ C})((3.70)(0.0) - (0.0)(2.10)) \times 10^5 \text{ T} \cdot \text{m/s} \tag{6.58}$$
$$= 0.0 \text{ N}$$
$$F_{By} = -(-1.6 \times 10^{-19} \text{ C})((0.0)(0.0) - (0.0)(1.40)) \text{ T} \cdot \text{m/s}$$
$$= 0.0 \text{ N}$$
$$F_{Bz} = (-1.6 \times 10^{-19} \text{ C})((0.0)(2.10) - (3.70)(1.40)) \times 10^5 \text{ T} \cdot \text{m/s}$$
$$= 8.3 \times 10^{-14} \text{ N}$$

The magnetic force is
$$\mathbf{F}_B = 8.3 \times 10^{-14}\mathbf{k} \text{ N} \tag{6.59}$$

Exercise 6.10 Consider a wire with a mass per unit length of 0.500 g/cm. A current of 2.00 A flows horizontally to the east. Determine the direction and magnitude of the minimum magnetic field needed to lift this wire vertically upward.

Solution 6.10 The magnetic force on the current-carrying wire is

$$F_B = | I\mathbf{L} \times \mathbf{B} | \qquad (6.60)$$
$$= ILB \sin \theta$$

where θ is the angle between the direction of the current and magnetic field. The force per unit length is then

$$f_B = \frac{F_B}{L} = IB \sin \theta \qquad (6.61)$$

In order to lift the wire vertically up, it should be larger than the force of gravity $f_g = mg/L$ per unit length and directed upwards to lift the wire, so $\theta = 90°$: $f_B = IB$. Hence the minimum magnetic field is given by

$$f_B = f_g \rightarrow B = \frac{(m/L)g}{I} \qquad (6.62)$$

where $m/L = 0.05$ kg/m. For the force to point upward, the field, by the right-hand rule, should point east. Its magnitude, from the above formula

$$B = 0.05 \times 9.8/2 = 0.25 \text{ T} \qquad (6.63)$$

Exercise 6.11 Consider a wire loop in the form of a square of side length 10.0 cm, which carries a steady current of 10 A. The loop is placed in a uniform magnetic field of magnitude 4.0 T. The magnetic field vector makes an angle of 30° with the normal vector to the plane of the loop. Determine the magnitude of the torque acting on the loop.

Solution 6.11 The torque is given by

$$\tau = I\mathbf{A} \times \mathbf{B} \qquad (6.64)$$

and its magnitude is

$$\tau = IAB \sin \theta \qquad (6.65)$$

where θ is the angle between cross-sectional area vector **A** of the plane and **B**: $\theta = 30°$, hence

$$\tau = (10 \text{ A})(0.10 \times 0.10 \text{ m}^2)(4.0 \text{ T}) \sin(30°) \tag{6.66}$$
$$= 0.20 \text{ N} \cdot \text{m}$$

Exercise 6.12 Consider an electron that undergoes a circular motion with a radius of 23 mm, and its kinetic energy is 6.4×10^{-17} J in a uniform magnetic field. Find the magnetic field.

Solution 6.12 The radius is given as

$$r = \frac{mv}{qB} \rightarrow B = \frac{mv}{qr} \tag{6.67}$$

where mv is the momentum of the particle, i.e.,

$$mv = \sqrt{2mK} \tag{6.68}$$

where K is the kinetic energy. Using this expression, we get

$$B = \frac{\sqrt{2mK}}{qr} = 2.93 \text{ mT} \tag{6.69}$$

Exercise 6.13 Consider a steady current of 2.40 A in a wire with a straight section of 0.750 m long placed along the x-axis in a uniform magnetic field of magnitude $B = 1.60$ T applied in the positive z direction. Suppose the current is in the $+x$ direction. Find the magnetic force on the section of wire.

Solution 6.13 The magnetic force is

$$\mathbf{F}_B = I\mathbf{L} \times \mathbf{B} \tag{6.70}$$

where **L** is pointing in the direction of I, so

$$\mathbf{L} = 0.750\mathbf{i} \ m \tag{6.71}$$

and

$$\mathbf{B} = 1.60\mathbf{k} \text{ T} \qquad\qquad (6.72)$$

Therefore,

$$F_{Bx} = I(L_y B_z - L_z B_y) = 0 \qquad\qquad (6.73)$$
$$F_{By} = -I(L_x B_z - L_z B_z) = -(2.40 \text{ A})(0.750 \text{ m})(1.60 \text{ T}) = -2.88 \text{ N}$$
$$F_{Bz} = I(L_x B_y - L_y B_x) = 0$$

Hence, $\mathbf{F}_B$ is in the negative y direction, and its magnitude is 2.88 N.

Reference

Holliday D, Resnick R, Walker J (2011) Fundamentals of physics. John Wiley and Sons

Chapter 7
Sources of Magnetic Field

The chapter aims to introduce sources of magnetic field, Gauss's and Maxwell-Ampére's laws.

In this chapter, we introduce the sources of the magnetic field. In particular, we introduce Biot-Savart law and the magnetic forces between the current-carrying parallel conductors. Also, we introduce Ampére's law and the magnetic field flux. Then, Gauss's law and Maxwell-Ampére's law will be presented. As extra reading material, the reader can also consider other literature Holliday et al. (2011), Jackson (1999), Griffiths (1999).

7.1 Biot-Savart Law

Biot-Savart law gives a mathematical formula about the magnetic field at a location in space in terms of the current that produces the field. According to Biot-Savart law, the magnetic field $d\mathbf{B}$ produced by a length element $d\mathbf{s}$ of a wire carrying a steady current I at any point P (see also Fig. 7.1):

$$d\mathbf{B} = \frac{\mu_0}{4\pi} \frac{I\,d\mathbf{s} \times \hat{\mathbf{r}}}{r^2} \qquad (7.1)$$

where μ_0 is a constant called the *permeability of free space*:

$$\mu_0 = 4\pi \times 10^{-7} \text{ T} \cdot \text{m/A} \qquad (7.2)$$

In Eq. (7.1), r gives the distance from the position of element $d\mathbf{s}$ to the point P, and $\hat{\mathbf{r}}$ is a unit vector along that direction, as indicated in Fig. 7.1.

© The Author(s), under exclusive license to Springer Nature Switzerland AG 2022
H. Kamberaj, *Electromagnetism*, Undergraduate Texts in Physics,
https://doi.org/10.1007/978-3-030-96780-2_7

Fig. 7.1 The magnetic field $d\mathbf{B}$ created by a small segment of current-carrying wire at the point P and P': $d\mathbf{B}$ is proportional to the vector $d\mathbf{s} \times \hat{\mathbf{r}}$, and hence, at point P it points out of the viewing plane and at the point P' it points in the viewing plane

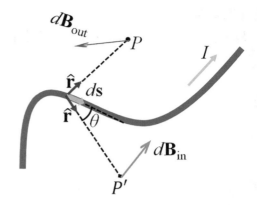

To find the total magnetic field **B** created at the point P by a current-carrying wire of finite size and arbitrary shape, we sum up those contributions from all current elements $I\,d\mathbf{s}$ that make up the current (see Eq. (7.1)) to obtain

$$\mathbf{B} = \frac{\mu_0 I}{4\pi} \int_{\mathcal{L}} \frac{d\mathbf{s} \times \hat{\mathbf{r}}}{r^2} \tag{7.3}$$

where $\mathcal{L}$ is the path along the wire.

In Eq. (7.3), the integral is about the entire current distribution over the wire. Furthermore, notice that the unit vector $\hat{\mathbf{r}}$ is along the line that joins the position of the length element $d\mathbf{s}$ along the direction of the current in the wire and the point P, and hence, it is part of the integral.

Biot-Savart law also applies to a current consisting of charges flowing through space, such as the electron beam in a television tube, or any other beam of some charges (positive or negative). In that case, $d\mathbf{s}$ denotes the length of a small segment vector of space along the direction of charge flow.

Similarities between Biot-Savart law for magnetism and Coulomb's law for electrostatics exist as follows.

First, the current element produces a magnetic field, whereas a fixed point charge in space produces an electric field. Hence, the current is a source of the magnetic field in space, and the charge at rest is a source of the electrostatic field in that space.

Furthermore, both the magnitude of the magnetic field and electrostatic field vary as the inverse square of the distance from the source (current element or static charge) of that field to any location P in space. However, the directions of the two fields are substantially different. The electric field created by a point charge is radial, that is, it is either parallel (if the charge is positive) or antiparallel (if the charge is negative) to the unit vector $\hat{\mathbf{r}}$. But, the magnetic field created by a current element is perpendicular to both the small portion element $d\mathbf{s}$ and the unit vector $\hat{\mathbf{r}}$.

Moreover, the electric field is due to an isolated electric charge at some point, which exists either an isolated single point-like charge or as a portion of an extended distribution of charges. Biot-Savart law, on the other hand, relates the magnetic field of an isolated current element at some point that exists only as a portion of an extended current distribution in a complete circuit where the charges flow. Therefore, Biot-Savart law is only the first step in the determination of a magnetic field; mathematically, it is the integrand of an integral over the current distribution.

The magnetic field produced by a current-carrying long and straight wire often occurs, as shown in Fig. 7.2 where the magnetic field surrounding a long, straight current-carrying wire is presented. The magnetic field lines are circles concentric with the wire, due to the symmetry, laying in planes perpendicular to the wire. Furthermore, the magnitude of **B** is constant on any point of a circle of radius a given by (see solved exercises)

Fig. 7.2 Right-hand rule demonstration

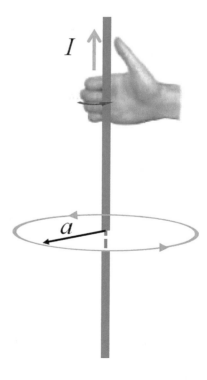

$$B = \frac{\mu_0 I}{2\pi a} \tag{7.4}$$

The right-hand rule is a convenient way for determining the direction of **B**; that is, we grasp the wire with the right hand, positioning the thumb along the direction of the current, then the four fingers curl in the direction of the magnetic field, as shown in Fig. 7.2.

7.2 Magnetic Force Between Two Parallel Conductors

Suppose we have two long, straight, parallel current-carrying wires (named 1 and 2, respectively) at a distance a from one another. Let I_1 and I_2 be the currents in each wire that flow in the same direction, as shown in Fig. 7.3. Based on Biot-Savart law, wire 2, carrying the current I_2, creates a magnetic field $\mathbf{B}_2$ at the location of wire 1. Using the right-hand rule, the direction of $\mathbf{B}_2$ is perpendicular to wire 1 at its location, and the magnetic force on a length ℓ of wire 1 is given by Eq. (6.8) (Chap. 6):

$$\mathbf{F}_1 = I_1\ell \times \mathbf{B}_2 \tag{7.5}$$

Direction of the force $\mathbf{F}_1$ is shown in Fig. 7.3(A).

Because ℓ is perpendicular to $\mathbf{B}_2$, the magnitude of force in Eq. (7.5) is

$$F_1 = I_1\ell B_2 \tag{7.6}$$

Using Eq. (7.4) for the magnitude of $\mathbf{B}_2$, we have

$$B_2 = \frac{\mu_0 I_2}{2\pi a} \tag{7.7}$$

Substituting Eq. (7.7) into Eq. (7.6), we obtain

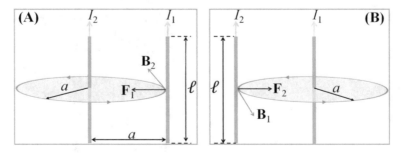

Fig. 7.3 Demonstration of the interaction between two long and straight current-carrying wires I_1 and I_2

$$F_1 = I_1 \ell B_2 = \ell \frac{\mu_0 I_1 I_2}{2\pi a} \tag{7.8}$$

Similarly, we can now calculate the force $\mathbf{F}_2$ acting on wire 2 that is placed in the field set up by wire 1, as shown in Fig. 7.3b. Based on third law of Newton, $\mathbf{F}_2$ is equal in magnitude and opposite in direction to $\mathbf{F}_1$:

$$\mathbf{F}_2 = I_2 \ell \times \mathbf{B}_1 = -\mathbf{F}_1 \tag{7.9}$$

$$F_2 = I_2 \ell B_1 = \ell \frac{\mu_0 I_1 I_2}{2\pi a} = F_1 \tag{7.10}$$

For the currents flowing in opposite directions (that is when one of the currents is reversed), the forces exerted on each current-carrying wire are reversed, and the wires repel each other. Hence, we find that when parallel conductors carry currents in the same direction, they attract each other, and when they take currents in opposite directions repel each other. Because the magnitudes of the forces are the same on both wires, we denote the magnitude of the magnetic force between the wires as simply F_B.

We can rewrite this magnitude in terms of the force per unit length:

$$\frac{F_B}{\ell} = \frac{\mu_0 I_1 I_2}{2\pi a} \tag{7.11}$$

Following the result of the force per unit length, given by Eq. (7.11), between any two parallel wires, Ampére defined the unit of ampere as 1 ampere (denoted 1 A) is the current carried in each of two long, parallel identical wires at distance of 1 m from one another, which interact between them with a force of magnitude per unit length 2×10^{-7} N/m:

Definition of Ampere

$$2 \times 10^{-7} \text{ N/m} = \frac{(4\pi \times 10^{-7} \text{ T} \cdot \text{m/A})(1 \text{ A})(1 \text{ A})}{2\pi(1 \text{ m})} \tag{7.12}$$

7.3   Ampére's Law

The following experiment by Oersted, in 1819, about deflection of the compass needles demonstrated that a current-carrying conductor produces a magnetic field, and it also lead to Ampére's law.

Fig. 7.4 Compass needles
are placed in a horizontal
plane nearby a long vertical
wire

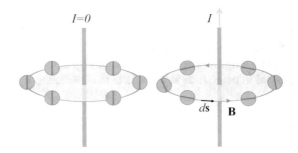

In this experiment, several needles of a compass are distributed in a horizontal
plane nearby a long vertical wire, as shown in Fig. 7.4. If no current flows in the
wire, all the needles point in the same direction along the Earth's magnetic field
because the magnetic field of Earth exerts a torque on the magnetic moment of
each needle. If a steady current flows in each wire, all needles deflect and align
tangent to the circle along the direction of the magnetic field produced by the
long current-carrying wire. The observation demonstrated that the direction of
the magnetic field produced by the current in the wire is consistent with the
right-hand rule. Furthermore, if the current changes in the opposite direction,
the needles reverse.

The experiment demonstrated that the compass needles point in the direction of
B, and the magnetic field of **B** forms circles around the wire. The magnitude of **B**
is the same at every point on a circular path centered on the wire, by symmetry,
and it is lying in the circle plane perpendicular to the wire. This is illustrated in
Fig. 7.5. Moreover, varying the current and distance a from the wire, **B** varies, and
it is proportional to the current and inversely proportional to the distance a from the
wire as in Eq. (7.4).

We can evaluate the product $\mathbf{B} \cdot d\mathbf{s}$ for a small element $d\mathbf{s}$ on the circular path $\mathcal{L}$,
as shown in Fig. 7.5. Note that the vector **B** is tangent to the circular path at the tail
of vector $d\mathbf{s}$, and hence, they are approximately parallel. Summing up the products
for all small elements over the closed circular path, we obtain

$$\mathbf{B} \cdot d\mathbf{s} = B\,ds \qquad (7.13)$$

Furthermore, the magnitude of B is constant along the path $\mathcal{L}$ at every point, and
hence, the sum of the products $B\,ds$ over the closed line becomes the line integral of
$\mathbf{B} \cdot d\mathbf{s}$:

Fig. 7.5 The magnetic field
at every point on a circular
path $\mathcal{L}$ centered at a long
vertical current-carrying
wire, I. $d\mathbf{s}$ is a small potion
of the path $\mathcal{L}$ such that $\mathbf{B}$ is
approximately parallel to $d\mathbf{s}$

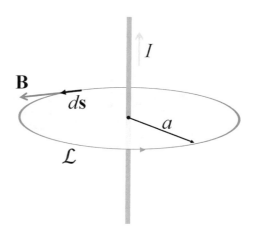

$$\oint_{\mathcal{L}} \mathbf{B} \cdot d\mathbf{s} = B \oint_{\mathcal{L}} ds \qquad (7.14)$$

$$= \frac{\mu_0 I}{2\pi r}(2\pi r)$$

$$= \mu_0 I$$

Note that Eq. (7.14) is derived for the special case of a circular path surrounding
a wire; it, however, holds for any arbitrary shaped closed path surrounding a current-
carrying wire that is part of a complete circuit. In the general case of an arbitrary
closed path, Ampére's law states: The closed path line integral of $\mathbf{B} \cdot d\mathbf{s}$ is $\mu_0 I$,
where I is the total continuous current passing through any surface supported by the
closed path $\mathcal{L}$. Mathematically, we write

Ampére's law

$$\oint_{\mathcal{L}} \mathbf{B} \cdot d\mathbf{s} = \mu_0 I \qquad (7.15)$$

7.4 Magnetic Flux

The flux related to a magnetic field is defined similarly to electric flux. For that,
we consider a small surface element of area dA on an arbitrarily shaped surface, as
indicated in Fig. 7.6. Let $\mathbf{B}$ be the magnetic field vector at this small surface element
vector $d\mathbf{A}$, the magnetic flux through the surface element is $\mathbf{B} \cdot d\mathbf{A}$. Here, $d\mathbf{A}$ defines
a vector that is perpendicular to the surface and has a magnitude equal to the area

Fig. 7.6 Magnetic flux
through an arbitrary shaped
surface

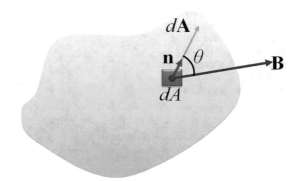

dA: $d\mathbf{A} = dA\mathbf{n}$, where $\mathbf{n}$ is a unit normal vector at the center of surface element. Hence, the total magnetic flux Φ_B through the surface is

$$\Phi_B = \int_A \mathbf{B} \cdot d\mathbf{A} \tag{7.16}$$

If we consider a planar surface of area A and a uniform field $\mathbf{B}$ that makes an angle θ with $d\mathbf{A}$ at every point on the surface. Then, the magnetic flux through the plane is

$$\Phi_B = BA\cos\theta \tag{7.17}$$

From Eq. (7.17), when the magnetic field $\mathbf{B}$ is parallel to the plane, $\theta = 90°$ (that is, $\cos\theta = 0$) and the flux is zero: $\Phi_B = 0$. If the magnetic field $\mathbf{B}$ is perpendicular to the plane of the surface, then $\theta = 0$ (that is, $\cos\theta = 1$) and the flux is maximum $\Phi_B = BA$.

The SI unit of flux is the $T \cdot m^2$, defined as the weber (Wb):

$$1\ Wb = 1\ T \cdot m^2 \tag{7.18}$$

7.5   Gauss's Law in Magnetism

In Chap. 2, we expressed the electric flux through a closed surface surrounding a net charge is proportional to that charge inside the volume, as Gauss's law. That is, the net number of electric field lines leaving the surface depends only on the net charge within it. That is related to the fact that electric field lines originate from some positive electrical charges and terminate on some other electric negative charges; that is, often, there is a non-zero net difference between the number of electric field lines leaving and entering any closed surface surrounding the distribution of charges.

However, that is substantially different for magnetic field lines, which are continuous closed loops. Hence, magnetic field lines do not begin at some point and end at any other location. Furthermore, the number of lines entering any closed surface is equal to the number of lines leaving that surface, and thus, the net magnetic flux must be zero. In contrast, the net electric flux through a closed surface enclosing only one of the charges of an electric dipole moment is not zero.

In magnetism, Gauss's law defines the net magnetic flux through any closed surface to be always zero:

$$\Phi_B = \oint_A \mathbf{B} \cdot d\mathbf{A} = 0 \tag{7.19}$$

Gauss's law in magnetism

The experimental facts support Gauss's law in magnetism that isolated magnetic poles (monopoles) have not been detected yet, and they perhaps do not exist. The research for identifying magnetic monopoles continues because specific theories suggest the possibility of finding monopoles experimentally.

7.6 Displacement Current

We showed that mobile charges produce magnetic fields, such as the free charge carrier of a current-carrying conductor, as indicated in Ampére's law. Furthermore, if a current-carrying conductor has high symmetry, we can use Ampére's law to calculate the produced magnetic field, such as a long, straight current-carrying wire. In Eq. (7.15), the line integral is over any closed path $\mathcal{L}$ through which the conduction current passes, defined as

$$I = \frac{dq}{dt} \tag{7.20}$$

We now suppose that an electric field that varies with time is present, and show that Ampére's law in the form given by Eq. (7.15) is valid only if that electric field is constant in time, which was proved by Maxwell. For that, let us consider a capacitor that is being charged, as shown in Fig. 7.7. In the presence of a conduction current, the charge on the positive plate changes, let say in the time interval dt it changes by an amount $dq = I dt$. However, the conduction current between the plates is zero, $I = 0$.

Let us take two surfaces S_1 and S_2, bounded by the same closed line path P, as indicated in Fig. 7.7. Note that the choice of the shape of these two surfaces is completely arbitrary. In Fig. 7.7, S_1 is one of the faces of a cube, and the other five faces of the same cube construct the open surface S_2.

Based on Ampére's law, Eq. (7.15), when the closed line path $\mathcal{L}$ is the path P bounding the surface S_1, then

$$\oint_P \mathbf{B} \cdot d\mathbf{s} = \mu_0 I \tag{7.21}$$

Fig. 7.7 A capacitor formed
by two charged plates at the
charge Q

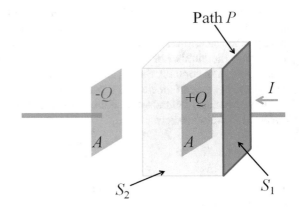

where I is the conduction current passing through S_1. In contrast, when the closed
line path is considered as bounding S_2, then the right-hand side of Eq. (7.15) is zero
because no conduction current passes through S_2:

$$\oint_P \mathbf{B} \cdot d\mathbf{s} = 0 \tag{7.22}$$

Equations (7.21) and (7.22) show a disagreement by applying Ampére's law as
in Eq. (7.15), which arises from the discontinuity of the current, as discovered by
Maxwell. To solve this contradiction, Maxwell postulated to add another term on the
right-hand side of Eq. (7.15) that included a factor called the displacement current
I_d, as

$$I_d = \epsilon_0 \frac{d\Phi_E}{dt} \tag{7.23}$$

where ϵ_0 is the permittivity of free space and $\Phi_E = \int_A \mathbf{E} \cdot d\mathbf{A}$ is the electric flux.

Physical basis of that term can be explained as follows. When the capacitor is
being charged (or discharged), the change on electric field between the plates may
be considered equivalent to adding a current (that is, the displacement current) flowing only inside the capacitor that makes the conduction current in the wire to be
continuous. That current, I_d, can be added to the right of the expression in Eq. (7.15)
as

$$\oint_{\mathcal{L}} \mathbf{B} \cdot d\mathbf{s} = \mu_0(I + I_d) = \mu_0 I + \mu_0 \epsilon_0 \frac{d\Phi_E}{dt} \tag{7.24}$$

Now, it does matter which surface bounded by the closed path (path P in Fig. 7.7) is
chosen, either conduction current (I) for the surface S_1 or displacement current (I_d)
for the surface S_2 passes through it. Equation (7.24) is a general form of Ampére's
law, and it is called the Ampére-Maxwell law.

The electric flux through surface S_2 is

$$\Phi_E = \int_{S_2} \mathbf{E} \cdot d\mathbf{A} = EA \tag{7.25}$$

where A is the surface area of the capacitor plates and E is the magnitude of the uniform electric field between the plates. If Q is the charge on the plates at any instant, then

$$E = \frac{Q}{\epsilon_0 A} \tag{7.26}$$

Therefore, the electric flux through S_2 is simply

$$\Phi_E = EA = \frac{Q}{\epsilon_0} \tag{7.27}$$

Thus,

$$I_d = \epsilon_0 \frac{d\Phi_E}{dt} \tag{7.28}$$
$$= \frac{dQ}{dt}$$

Equation (7.28) implies that the displacement current through open surface S_2 is precisely equal to the conduction current I through S_1. Thus, the displacement current is the source of the magnetic field on the surface boundary of S_2. Moreover,

Fig. 7.8 A capacitor formed by two charged plates at the charge Q.Electric field inside the capacitor

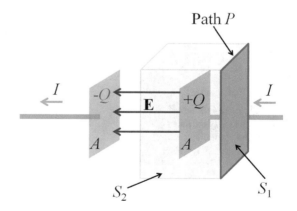

the physical origin of the displacement current is the time-varying electric field (see Fig. 7.8). The central point of this formalism, then, is that magnetic fields are produced both by conduction currents and by time-varying electric fields.

7.7 Exercises

Exercise 7.1 Consider a thin, straight wire carrying a constant current I and placed along the x axis (Fig. 7.9). Find (a) the magnitude and (b) direction of the magnetic field at point P due to this current.

Solution 7.1 Using Biot-Savart law:

$$d\mathbf{B} = \frac{\mu_0 I}{4\pi} \frac{d\mathbf{s} \times \hat{\mathbf{r}}}{r^2} \tag{7.29}$$

where $d\mathbf{s} \times \hat{\mathbf{r}}$ has the direction along positive z axis, thus

$$d\mathbf{B} = \frac{\mu_0 I}{4\pi} \frac{dx \sin \theta}{r^2} \mathbf{k} \tag{7.30}$$

where $\mid d\mathbf{s} \mid = dx$. We can get

$$dx = -a\, d \left(\frac{1}{\tan \theta} \right) = a \frac{d\theta}{\sin^2 \theta} \tag{7.31}$$

Then,

Fig. 7.9 A straight wire carrying a steady current I and placed along the x axis

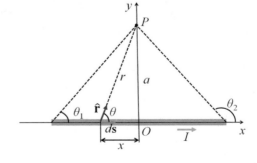

$$dB = \frac{\mu_0 I}{4\pi} \frac{a d\theta}{\sin \theta r^2} \mathbf{k} \qquad (7.32)$$

$$= \frac{\mu_0 I}{4\pi} \frac{\sin \theta \, d\theta}{a} \mathbf{k}$$

Then, for the total magnitude of magnetic field we get:

$$B = \frac{\mu_0 I}{4a\pi} \int_{\theta_1}^{\theta_2} \sin \theta \, d\theta \qquad (7.33)$$

$$= \frac{\mu_0 I}{4a\pi} (\cos \theta_1 - \cos \theta_2)$$

Consider the special case of an infinitely long, straight wire: $\theta_1 = 0$ and $\theta_2 = \pi$, thus

$$B = \frac{\mu_0 I}{2a\pi} \qquad (7.34)$$

Exercise 7.2 What is the magnetic field at point O for the current-carrying wire segment in Fig. 7.10? The wire consists of two straight segments and a circular arc of radius R, which subtends an angle θ. The arrowheads on each portion of the wire indicate the direction of the current (see also Fig. 7.10).

Solution 7.2 The magnetic field at O due to the current in the straight segments AA' and CC' is zero because $d\mathbf{s}$ is parallel to $\hat{\mathbf{r}}$ along these paths; this means that

$$d\mathbf{s} \times \hat{\mathbf{r}} = 0 \qquad (7.35)$$

Fig. 7.10 A wire consists of two straight portions and a circular arc of radius R, which subtends an angle θ. The arrowheads on the wire indicate the direction of the current

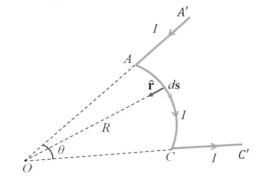

Each length element ds along path AC is at the same distance R from O, and the current in each contributes a field element $d\mathbf{B}$ directed into the page at O. Furthermore, at every point on AC, $d\mathbf{s}$ is perpendicular to $\hat{\mathbf{r}}$; hence,

$$| \, d\mathbf{s} \times \hat{\mathbf{r}} \, | = ds \tag{7.36}$$

The magnitude of the field at O due to the current in an element of length ds is:

$$dB = \frac{\mu_0 I}{4\pi} \frac{ds}{R^2} \tag{7.37}$$

Therefore, the total magnetic field magnitude is

$$\begin{aligned} B &= \frac{\mu_0 I}{4\pi} \int \frac{ds}{R^2} \\ &= \frac{\mu_0 I}{4\pi} \frac{s}{R^2} \\ &= \frac{\mu_0 I}{4\pi} \frac{\theta}{R} \end{aligned} \tag{7.38}$$

Exercise 7.3 Consider a circular wire loop of radius R located in the yz plane and carrying a steady current I (see Fig. 7.11). Calculate the magnetic field at an axial point P a distance x from the center of the loop.

Solution 7.3 Every length element $d\mathbf{s}$ is perpendicular to vector $\hat{\mathbf{r}}$ at the location of the element. Thus,

$$| \, d\mathbf{s} \times \hat{\mathbf{r}} \, | = ds \tag{7.39}$$

Furthermore, all small segment elements around the loop are at the same distance r from P, where $r^2 = x^2 + R^2$. Hence, the magnitude of $d\mathbf{B}$ due to the current in any length element $d\mathbf{s}$ is

Fig. 7.11 A circular wire loop of radius R located in the yz-plane and carrying a steady current I

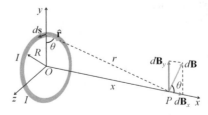

$$dB = \frac{\mu_0 I}{4\pi} \frac{|ds \times \hat{\mathbf{r}}|}{r^2} = \frac{\mu_0 I}{4\pi} \frac{ds}{x^2 + R^2} \tag{7.40}$$

We can write

$$d\mathbf{B} = dB_x \mathbf{i} + dB_y \mathbf{j} \tag{7.41}$$

and

$$\mathbf{B} = \oint dB_x \mathbf{i} + \oint dB_y \mathbf{j} = \oint dB_x \mathbf{i} = B_x \mathbf{i} \tag{7.42}$$

where $dB_x = dB \cos \theta$, hence

$$B_x = \oint \frac{\mu_0 I}{4\pi} \frac{ds \cos \theta}{x^2 + R^2} \tag{7.43}$$

where

$$\cos \theta = \frac{R}{\sqrt{x^2 + R^2}} \tag{7.44}$$

Thus, we obtain

$$B_x = \oint \frac{\mu_0 I}{4\pi} \frac{R ds}{(x^2 + R^2)^{3/2}} \tag{7.45}$$
$$= \frac{\mu_0 I}{4\pi} \frac{R 2\pi R}{(x^2 + R^2)^{3/2}}$$
$$= \frac{\mu_0 I R^2}{2(x^2 + R^2)^{3/2}}$$

Thus,

$$\mathbf{B} = \frac{\mu_0 I R^2}{2(x^2 + R^2)^{3/2}} \mathbf{i} \tag{7.46}$$

Exercise 7.4 A long, straight cylindrical wire with a radius R carries a steady current I_0, which is uniformly distributed through its cross-section, as shown in Fig. 7.12. Calculate the magnetic field at the region 1 ($r \geq R$) and 2 ($r < R$).

Solution 7.4 (a) Let us choose for our path of integration a circle, labeled 1 in Fig. 7.12. From symmetry, $\mathbf{B}$ must be constant in magnitude and approximately parallel to ds at every point of the circle 1. Because the total current passing through the plane of the circle is I_0, Ampére's law gives

$$\oint \mathbf{B} \cdot ds = B \oint ds = B(2\pi r) = \mu_0 I_0 \tag{7.47}$$

Fig. 7.12 A long, straight cylindrical wire of radius R carrying a steady current I_0 uniformly distributed through its cross section

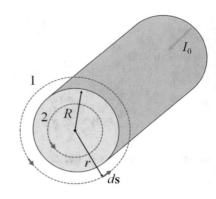

or (for $r \geq R$) we get

$$B = \frac{\mu_0 I_0}{2\pi r} \tag{7.48}$$

(b) Let us choose, for our path of integration, the circle 2. The same symmetry applies in this case. Using Ampére's law:

$$\oint \mathbf{B} \cdot d\mathbf{s} = B \oint ds = B(2\pi r) = \mu_0 I \tag{7.49}$$

Then, the magnetic field magnitude is

$$B = \frac{\mu_0 I}{2\pi r} \tag{7.50}$$

where the current I is determined according to

$$\frac{I_0}{\pi R^2} = \frac{I}{\pi r^2} \tag{7.51}$$

From here, the current is

$$I = I_0 \frac{r^2}{R^2} \tag{7.52}$$

Therefore, we obtain (for $r < R$)

$$B = \frac{\mu_0 I_0}{2\pi R^2} r \tag{7.53}$$

Fig. 7.13 A toroid with
inner radius a and outer
radius b consists of a
conducting wire wrapped
around a ring (a torus) made
of a nonconducting material

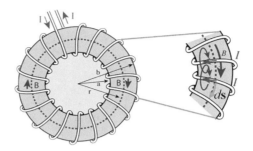

Exercise 7.5 A device called a toroid is often used to create an almost uniform magnetic field in some enclosed area (see also Fig. 7.13). The device consists of a conducting wire wrapped around a ring (a torus) made of a nonconducting material. For a toroid having N closely spaced turns of wire, calculate the magnetic field in the region occupied by the torus at a distance r from the center.

Solution 7.5 Using Ampére's law:

$$\oint \mathbf{B} \cdot d\mathbf{s} = \mu_0 N I \tag{7.54}$$

where N is the number of wires. By symmetry, we see that the magnitude of the field is constant on this circle and tangent to it, so $\mathbf{B} \cdot d\mathbf{s} = B ds$. Hence,

$$B \oint ds = B(2\pi r) = \mu_0 N I \tag{7.55}$$

Therefore, B is

$$B = \frac{\mu_0 N I}{2\pi r} \tag{7.56}$$

Exercise 7.6 Let us now consider an extended object carrying a current. A thin, infinitely large sheet lying in the yz plane carries a current of linear current density $\mathbf{J}_s$, as shown in Fig. 7.14. The current is in the y direction, and J_s represents the current per unit length measured along the z axis. Find the magnetic field near the sheet.

Fig. 7.14 A thin, infinitely
large sheet lying in the
yz-plane carries a current of
linear current density $\mathbf{J}_s$

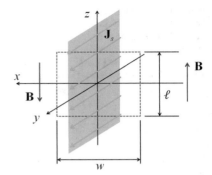

Solution 7.6 Using Ampére's law:

$$\oint \mathbf{B} \cdot d\mathbf{s} = \mu_0 I = \mu_0 J_s \ell \qquad (7.57)$$

where the rectangle has dimensions ℓ and w, with the sides of length ℓ parallel to
the sheet surface. Or,

$$2B\ell = \mu_0 J_s \ell \qquad (7.58)$$

Thus,

$$B = \frac{\mu_0 J_s}{2} \qquad (7.59)$$

Exercise 7.7 A current-carrying wire (labeled 1) is oriented along the y axis
in which the current is I_1. A rectangular loop is located to the right of the wire
1 and in the xy plane. Wire 2 carries a current I_2 (see also Fig. 7.15). What is
(a) the magnetic force applied on the top wire of length b in the loop, labeled
Wire 2 in Fig. 7.15 and (b) the force exerted on the bottom wire of length b?

Solution 7.7 Let consider the force exerted by wire 1 on a small segment $d\mathbf{s}$ of wire
2 by using:

$$d\mathbf{F}_B = I_2 d\mathbf{s} \times \mathbf{B} \qquad (7.60)$$

From Ampére's law, the field at a distance x from wire 1 is

$$\mathbf{B} = \frac{\mu_0 I_1}{2\pi x}(-\mathbf{k}) \qquad (7.61)$$

where $\mathbf{k}$ is a unit vector along z axis. Here, $d\mathbf{s} = dx\mathbf{i}$, then we obtain

Fig. 7.15 A wire 1 oriented along the y axis and carrying a steady current I_1, and another rectangular loop located to the right of the wire and in the xy-plane carries a current I_2

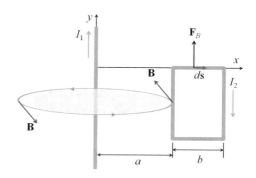

$$dF_B = I_2 dx\mathbf{i} \times \frac{\mu_0 I_1}{2\pi x}(-\mathbf{k}) \tag{7.62}$$
$$= \frac{\mu_0 I_1 I_2}{2\pi}\frac{dx}{x}\mathbf{j}$$

Total force is

$$\mathbf{F}_B = \int_a^{a+b} \frac{\mu_0 I_1 I_2}{2\pi}\frac{dx}{x}\mathbf{j} \tag{7.63}$$
$$= \frac{\mu_0 I_1 I_2}{2\pi}\mathbf{j}\,[\ln x]_a^{a+b}$$
$$= \frac{\mu_0 I_1 I_2}{2\pi}\mathbf{j}\ln\left(1 + \frac{b}{a}\right)$$

(b) Force exerted on the bottom wire of length b is

$$\mathbf{F}_B = -\frac{\mu_0 I_1 I_2}{2\pi}\mathbf{j}\ln\left(1 + \frac{b}{a}\right) \tag{7.64}$$

Exercise 7.8 Consider a rectangular loop of length b and width a located close to a long wire carrying a current I, as shown in Fig. 7.16. Assume that a distance c between the wire and the nearest side of the rectangle, and that the wire is parallel to the long side of the rectangle. Find the total magnetic flux through the loop due to the current in the wire.

Solution 7.8 The magnitude of the magnetic field created by the wire at a distance r from the wire is

$$B = \frac{\mu_0 I}{2\pi r} \tag{7.65}$$

Fig. 7.16 A rectangular
loop of width a and length b
is located near a long wire
carrying a current I

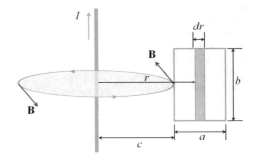

The factor $1/r$ indicates that the field varies over the loop, and the field is directed into
the page. Because **B** and $d\mathbf{A}$ are parallel at any point within the loop, the magnetic
flux through a small surface element dA is

$$\Phi_B = \int B \, dA \tag{7.66}$$

$$= \int \frac{\mu_0 I}{2\pi r} dA$$

$$= \int \frac{\mu_0 I}{2\pi r} b \, dr$$

Performing the integration

$$\Phi_B = \int_c^{a+c} \frac{\mu_0 I}{2\pi r} b \, dr = \frac{\mu_0 I b}{2\pi} \ln\left(1 + \frac{a}{c}\right) \tag{7.67}$$

Apply the series expansion formula for $\ln(1 + x) \approx x$ (for $x \ll 1$) to this equation
to show that it gives a reasonable result when the loop is far from the wire relative
to the loop dimensions, $c \gg a$:

$$\Phi_B \approx 0 \tag{7.68}$$

Exercise 7.9 A sinusoidally varying voltage is applied across an 8.00 μF
capacitor. The angular frequency of the voltage is $\omega = 2.00 \times 10^4$ rad/s, and
the voltage amplitude is 30.0 V. Find the displacement current between the
plates of the capacitor.

Solution 7.9 The angular frequency of the source is

$$\omega = 2\pi f = 2\pi(3.00 \times 10^3 \text{ Hz}) = 1.88 \times 10^4 \text{ rad/s} \tag{7.69}$$

Hence the voltage across the capacitor in terms of t is

$$\Delta V = \Delta V_{max} \sin \omega t = (30.0 \text{ V}) \sin(1.88 \times 10^4 t) \tag{7.70}$$

More, we know $Q = C\Delta V$. Thus,

$$I_d = \frac{dQ}{dt} = \frac{d}{dt}(C\Delta V) = C\frac{d}{dt}(\Delta V) \tag{7.71}$$

$$= (8.00 \times 10^{-6} \text{ F})\frac{d}{dt}\left[(30.0 \text{ V}) \sin(1.88 \times 10^4 \ t)\right]$$

$$= (4.52 \text{ A}) \cos(1.88 \times 10^4 \ t)$$

The displacement current varies sinusoidally with time and has a maximum value of 4.52 A.

Exercise 7.10 In Niels Bohr's 1913 model of the hydrogen atom, an electron circles the proton at a distance of 5.29×10^{-11} m with a speed of 2.19×10^6 m/s. Compute the magnitude of the magnetic field that this motion produces at the location of the proton.

Solution 7.10 The current is

$$I = \frac{|e| v_d}{2\pi r} \tag{7.72}$$

where $r = 5.29 \times 10^{-11}$ m, $v_d = 2.19 \times 10^6$ m/s, and $| e | = 1.6 \times 10^{-19}$ C. Using the Biot-Savart law:

$$d\mathbf{B} = \frac{\mu_0 I}{4\pi} \frac{(d\mathbf{s} \times \hat{\mathbf{r}})}{r^2} \tag{7.73}$$

Therefore, we obtain

$$dB = \frac{\mu_0 I}{4\pi} \frac{ds \sin \theta}{r^2} \tag{7.74}$$

$$= \frac{\mu_0 I}{4\pi} \frac{r \sin \theta d\theta}{r^2}$$

Total magnetic field is

$$B = \int_0^\pi \frac{\mu_0 I}{4\pi} \frac{\sin \theta d\theta}{r} \tag{7.75}$$

$$= \frac{\mu_0 I}{2\pi r} = \frac{\mu_0 \ | e | \ v_d}{(2\pi r)^2}$$

$$= \frac{(4\pi \times 10^{-7} \text{ T} \cdot \text{m/A})(1.6 \times 10^{-19} \text{ C})(2.19 \times 10^{6} \text{ m/s})}{(2\pi 5.29 \times 10^{-11} \text{ m})^2}$$

$$= 3.99 \ T$$

References

Holliday D, Resnick R, Walker J (2011) Fundamentals of physics. John Wiley and Sons
Jackson JD (1999) Classical electrodynamics, 3rd edn. John Wiley and Sons
Griffiths DJ (1999) Introduction to electrodynamics, 3rd edn. Prentice Hall

Chapter 8
Magnetism in Matter

The chapter aims to introduce Maxwell equations of electromagnetism in free space and the medium.

In this chapter, we introduce magnetism in the matter. We first discuss the magnetic moments of an atom of matter. Then, we will discuss the magnetic substances, such as ferromagnetic, paramagnetic, and diamagnetic substances. Next, we will introduce Faraday's law of induction. Finally, we will introduce the famous Maxwell equations of electromagnetism in free space and the medium. As extra reading material, the reader can also consider other literature Jackson (1999), Landau and Lifshitz (1971), Sykja (2006), Griffiths (1999).

8.1 Magnetic Moments of Atoms

The magnetic field created by a current-carrying coil of wire can explain what makes some materials to have strong magnetic properties. In general, based on Biot–Savart law (Eq. (7.1), Chap. 7), any current loop produces a magnetic field, as shown in Fig. 8.1. Thus, it has a magnetic dipole moment, including the atomic-level current loops described in some models of the atom. Those atomic-level current loops may define the magnetic moments in a magnetized substance. In the Bohr model of the atom, the current loops are associated with the circular motion of electrons around the nucleus. Besides, another magnetic moment, which is intrinsic to electrons, protons, neutrons, and other particles, arises from a property called *spin*.

Now, let us consider the classical model of the atom in which electrons move in circular orbits around the nucleus. Each orbiting electron creates a current loop because it is a moving charge e. Therefore, there exists a magnetic moment of the

© The Author(s), under exclusive license to Springer Nature Switzerland AG 2022
H. Kamberaj, *Electromagnetism*, Undergraduate Texts in Physics,
https://doi.org/10.1007/978-3-030-96780-2_8

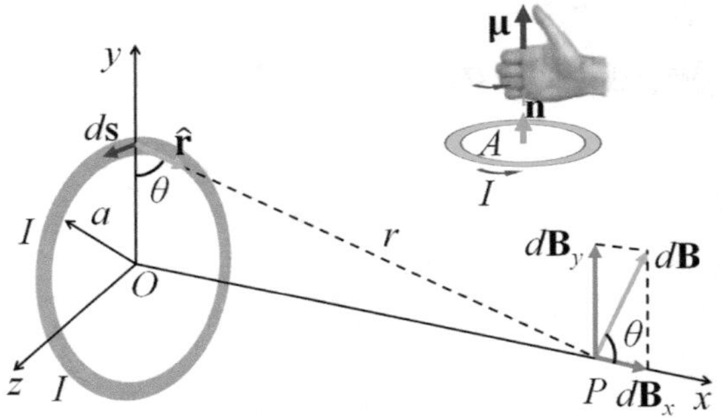

Fig. 8.1 Magnetic field and magnetic dipole moment created by a current loop. $A = \pi a^2$ is the surface area of the loop and **n** is unit normal vector to the surface of the current loop

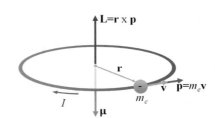

Fig. 8.2 Angular momentum $\mathbf{L} = \mathbf{r} \times \mathbf{p}$ created by an orbiting electron and the magnetic dipole moment $\boldsymbol{\mu}$ of the current loop I

electron associated with its orbital motion (see also Fig. 8.2). Suppose electron is moving with constant speed v in a circular orbit of radius r about the nucleus counterclockwise, as shown in Fig. 8.2. During a full period T, the electron travels the length $2\pi r$, which is the circumference of the circle, and hence its speed is

$$v = \frac{2\pi r}{T} \tag{8.1}$$

The current I associated with this orbiting electron with a charge e is given as

$$I = \frac{e}{T} \tag{8.2}$$

Using the following relations:

$$T = \frac{2\pi}{\omega} \tag{8.3}$$

$$\omega = \frac{v}{r}$$

we get

$$I = \frac{e\omega}{2\pi} = \frac{ev}{2\pi r} \tag{8.4}$$

Fig. 8.3 The orbital magnetic moments μ produced by two current orbits of two electrons moving in opposite direction

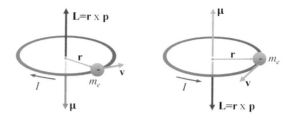

The magnitude of magnetic moment associated with this current loop is $\mu = IA$ with a direction of the vector shown in Fig. 8.2, where $A = \pi r^2$ is the surface area enclosed by the orbit. Therefore, we obtain the magnetic moment

$$\mu = IA = \frac{ev}{2\pi r}\pi r^2 = \frac{1}{2}evr \tag{8.5}$$

Knowing that the orbital angular momentum of electron is $\mathbf{L} = \mathbf{r} \times \mathbf{p}$ and its magnitude

$$L = m_e v r \sin 90° = m_e v r \tag{8.6}$$

the magnetic dipole moment can be written as

$$\mu = \frac{eL}{2m_e} \tag{8.7}$$

Equation (8.7) implies that the magnetic moment of the electron is proportional to its orbital angular momentum. Note that because the electron is negatively charged, the vectors μ and $\mathbf{L}$ point in opposite directions, as shown in Fig. 8.2. Both vectors are perpendicular to the plane of the orbit.

Based on the quantum physics, the orbital angular momentum is quantized and it is equal to multiples of $\hbar$, which relates to Planck's constant h:

Quantization

$$\hbar = \frac{h}{2\pi} = 1.05 \times 10^{-34} \text{ J} \cdot \text{s} \tag{8.8}$$

The smallest non-zero value of the electron's magnetic moment resulting from its orbital motion is

$$\mu = \sqrt{2}\frac{e}{2m_e}\hbar \tag{8.9}$$

Although the orbital motion of electrons in atoms of every substance produces magnetic dipole moments, not all materials are magnetic. That is because the orbital magnetic moment created by an electron in an atom may be canceled by the orbital magnetic moment of an electron in another atom that is moving in the opposite direction in orbit, as illustrated in Fig. 8.3. Therefore, the net magnetic effect produced by the orbital motion of the electrons in atoms is approximately zero in some materials.

Fig. 8.4 Classical
description of the electron
viewed as spinning about its
axis while it orbits the
nucleus

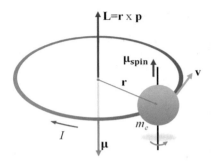

In addition to the orbital magnetic moment, an electron has a *spin* that may also contribute to its net magnetic moment. In classical description, the electron spins about its axis while it moves around the nucleus in an orbit, as shown in Fig. 8.4. The magnitudes of the angular momentum **S** of the spinning motion and of the angular momentum **L** due to the orbital motion are of the orders given in the following. Based on the quantum physics, the magnitude of the spin angular momentum is

$$S = \frac{\sqrt{3}}{2}\hbar \tag{8.10}$$

Therefore, the spin magnetic moment of an electron is

$$\mu_{spin} = \frac{e\hbar}{2m_e} \tag{8.11}$$

That is also called the *Bohr magneton*:

$$\mu_B = \frac{e\hbar}{2m_e} = 9.27 \times 10^{-24} \text{ J/T} = 9.27 \times 10^{-24} \text{ A} \cdot \text{m}^2 \tag{8.12}$$

Thus, atomic magnetic moments can be expressed as multiples of the Bohr magneton.

Usually, in an atom with many electrons, the electrons pair up with their spins opposite to one another, giving a net spin magnetic moment of zero. However, in an atom with an odd number of electrons, there exists at least one unpaired electron, and hence there is a spin magnetic moment different from zero.

The total magnetic moment of an atom is the sum of the orbital and spin magnetic moment vectors. The nucleus of an atom also contributes with a net magnetic moment that originates from protons and neutrons. However, in practice, the magnetic moments of the protons and neutrons are much smaller than that of the electrons and can usually be omitted. For instance, the magnetic moment associated with a proton or neutron is given as in Eq. (8.12), by replacing the mass of the electron with the mass of a proton or neutron:

$$\mu_B^{p,n} = \frac{e\hbar}{2m_{p,n}} \tag{8.13}$$

Since the masses of the proton (m_p) and neutron (m_n) are much greater than the mass of an electron (m_e), we get

$$\frac{\mu_B^e}{\mu_B^{p,n}} = \frac{m_{p,n}}{m_e} \sim 10^3 \tag{8.14}$$

Equation (8.14) indicates that the magnetic moments of the proton and neutron are about 1000 times smaller than that of the electron.

8.2   Magnetization Vector and Magnetic Field Strength

To describe the magnetic properties of a substance, the *magnetization vector* **M** is introduced. In a substance of volume V with a large number of molecules or atoms per unit of volume, where each has a magnetic moment $\boldsymbol{\mu}_i$, the average macroscopic magnetization vector is

$$\mathbf{M}(\mathbf{r}) = \sum_i n_i \langle \boldsymbol{\mu}_i \rangle \tag{8.15}$$

where n_i is the average number density of molecules or atoms of type i and $\langle \boldsymbol{\mu}_i \rangle$ is the average magnetic moment in a macroscopically small volume element centered at position $\mathbf{r}$ of type i. The magnetization contributes an effective current density:

$$\mathbf{J}_M = \nabla \times \mathbf{M} \tag{8.16}$$

In the microscopic description, Ampére's law for static electric fields can be rewritten as

$$\oint_{\mathcal{L}} \mathbf{B}_{micro} \cdot d\mathbf{s} = \mu_0 \int_A \mathbf{J}_{micro} \cdot d\mathbf{A} \tag{8.17}$$

where A is any surface enclosed by the path $\mathcal{L}$. Using Stokes' formula, we obtain

$$\int_A (\nabla \times \mathbf{B}_{micro}) \cdot d\mathbf{A} = \mu_0 \int_A \mathbf{J}_{micro} \cdot d\mathbf{A} \tag{8.18}$$

The macroscopic equivalent of Eq. (8.18) is given by

$$\int_A (\nabla \times \mathbf{B}) \cdot d\mathbf{A} = \mu_0 \int_A \mathbf{J}' \cdot d\mathbf{A} \tag{8.19}$$

where the macroscopic surface current density, $\mathbf{J}'$, is

$$\mathbf{J}' = \mathbf{J} + \mathbf{J}_M \tag{8.20}$$

where $\mathbf{J}$ represents the surface charge density from the flow of free charge in the medium. Substituting Eq. (8.20) into Eq. (8.19), we obtain

$$\int_A (\nabla \times \mathbf{B}) \cdot d\mathbf{A} = \mu_0 \int_A (\mathbf{J} + \mathbf{J}_M) \cdot d\mathbf{A} \tag{8.21}$$

Comparing both sides in Eq. (8.21), we have

$$\nabla \times \mathbf{B} = \mu_0 (\mathbf{J} + \nabla \times \mathbf{M}) \tag{8.22}$$

where Eq. (8.16) is used. Equation (8.22) can be re-arranged in a convenient form as

$$\nabla \times (\mathbf{B} - \mu_0 \mathbf{M}) = \mu_0 \mathbf{J} \tag{8.23}$$

The magne-tization vector The contribution of the magnetization of the medium is expressed in terms of the magnetization vector as

$$\mathbf{B}_m = \mu_0 \mathbf{M} \tag{8.24}$$

The total magnetic field $\mathbf{B}$ at a point within a substance will be sum of both the applied (external) field $\mathbf{B}_0$ and the magnetization of the substance, $\mathbf{B}_m$:

$$\mathbf{B} = \mathbf{B}_0 + \mathbf{B}_m \tag{8.25}$$

Therefore, the total magnetic field in the region where both the external magnetic field and magnetization of the substance are present becomes

$$\mathbf{B} = \mathbf{B}_0 + \mu_0 \mathbf{M} \tag{8.26}$$

To analyze the magnetic fields produced from the magnetization, we introduce another field quantity, namely, the *magnetic field strength* $\mathbf{H}$ within the substance. The magnetic field strength characterizes the effect of the conduction currents in wires on a material. To indicate the differences between the magnetic field strength $\mathbf{H}$ and the magnetic field $\mathbf{B}$, often, $\mathbf{B}$ is called the *magnetic flux density* or the *magnetic induction*.

The mag-netic field strength The magnetic field strength $\mathbf{H}$ is a vector quantity defined as

$$\mathbf{H} = \frac{\mathbf{B}_0}{\mu_0} = \frac{\mathbf{B}}{\mu_0} - \mathbf{M} \tag{8.27}$$

Or,

$$\mathbf{B} = \mu_0 (\mathbf{H} + \mathbf{M}) \tag{8.28}$$

Both **H** and **M** have the same units. In SI units, **M**, as a magnetic moment per unit volume, has the units of (ampere)(meter)^2 /(meter)^3 or equivalently amperes per meter. Then, Eq. (8.23) can be written in terms of **H** as

$$\nabla \times \mathbf{H} = \mathbf{J} \tag{8.29}$$

which represents the differential form of Ampére's law for magnetism in a medium with a static electric field.

8.3 Classification of Magnetic Substances

Materials can be classified based on their magnetic properties as

1. Paramagnetic;
2. Diamagnetic;
3. Ferromagnetic.

Diamagnetic materials are made up of atoms that do not have permanent magnetic moments. In contrast, paramagnetic and ferromagnetic materials are made up of atoms that have permanent magnetic moments. For a paramagnetic and diamagnetic material, the magnetization vector **M** depends linearly on the external magnetic field strength **H** as

$$\mathbf{M} = \chi \mathbf{H} \tag{8.30}$$

In Eq. (8.30), χ (Greek letter chi) is a dimensionless proportionality constant, called the *magnetic susceptibility*. For a paramagnetic material, χ is positive, and hence **M** is parallel with **H**. On the other hand, for a diamagnetic material, χ is negative, and **M** points the opposite direction to **H**. The linearity of Eq. (8.30) between **M** and **H** does not obey for ferromagnetic materials.

Using the above equation:

$$\mathbf{B} = \mu_0(\mathbf{H} + \mathbf{M}) \tag{8.31}$$

we obtain

$$\mathbf{B} = \mu_0(\mathbf{H} + \chi\mathbf{H}) = \mu_0(1 + \chi)\mathbf{H} \tag{8.32}$$

Or,

$$\mathbf{B} = \mu_m \mathbf{H} \tag{8.33}$$

where

$$\mu_m = \mu_0(1 + \chi) \tag{8.34}$$

is called *magnetic permeability* of the material.

By comparing the magnetic permeability, μ_m, with the permeability of free space, μ_0, materials classify as follows:

1. Paramagnetic: $\mu_m > \mu_0$;
2. Diamagnetic: $\mu_m < \mu_0$.

Since χ is very small for paramagnetic and diamagnetic materials, $\mu_m \approx \mu_0$ for those materials. However, for a ferromagnetic substance, μ_m is up to several thousand times greater than μ_0, and hence χ is very high for ferromagnetic materials.

8.3.1 Ferromagnetism

In Fig. 8.5, we show a substance in which the atoms have permanent magnetic moments. That material exhibits strong magnetic effects called *ferromagnetism*. Such ferromagnetic substances include iron, cobalt, and nickel. In a ferromagnetic material, there exist atomic magnetic moments (see Fig. 8.5) that when we put them in an external magnetic field, the atomic magnetic moments align parallel to one another and external magnetic field. Interestingly, the magnetic moments remain aligned after the external field is removed due to strong interactions between neighboring moments. Therefore, we say that the material remains magnetized.

Ferromagnetic materials are made up of microscopic domains with aligned magnetic moments of about 10^{17} to 10^{21} atoms. The volume of each of those domains is about 10^{-12} to 10^{-8} m^3. There exist walls between the domains. When the material is not magnetized, the net microscopic magnetic moments of domains are randomly oriented, and hence the net macroscopic magnetic moment vanishes. When an external magnetic field is applied across the sample, the microscopic magnetic moments of the domains tend to align with the field, yielding a non-zero net macroscopic magnetization of the sample. When the external magnetic field reduces slowly to zero, the sample may have a net magnetization in the direction of the applied external field. Usually, at standard temperatures, thermal fluctuations are not significantly large to cause any changes to that preferred orientation of domain magnetic moments.

Fig. 8.5 a Orientation of permanent magnetic moments of a ferromagnetic material in absence of external magnetic field and **b** in presence of an external magnetic field

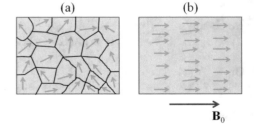

8.3.2 *Paramagnetism*

Paramagnetic materials are characterized by a small positive magnetic susceptibility $0 < \chi \ll 1$ that is due to the existence of the permanent magnetic moments of atoms or ions. The dipole moments interact very weakly with one another, and hence they are randomly oriented when there is no external magnetic field. On the other hand, atomic magnetic moments of the paramagnetic line up with the field when an external magnetic field is present. The alignment process, however, is opposed by the thermal motion, which tends to randomize the magnetic moment orientations.

It was found experimentally by Pierre Curie and others that, under a wide range of conditions, the magnetization of a paramagnetic is proportional to the applied magnetic field B_0 and inversely proportional to the temperature T (in Kelvin):

$$M = C\frac{B_0}{T} \tag{8.35}$$

Equation (8.51) is known as Curie's law, and the constant C is called Curie's constant. The law implies that when $B_0 = 0$, the magnetization is zero, corresponding to a random orientation of magnetic moments. With increasing the ratio of the magnetic field to temperature, the magnetization approaches a maximum value, which corresponds to the complete alignment of its moments, and Eq. (8.51) becomes invalid.

When the temperature of a ferromagnetic substance is greater or equal to the critical Curie temperature, T_C, the substance loses its residual magnetization and becomes paramagnetic. Below T_C, the magnetic moments are aligned and the substance is ferromagnetic. For $T > T_C$, the thermal fluctuations are high such that they cause a random orientation of the magnetic moments, and the substance becomes paramagnetic, as shown in Fig. 8.6.

Fig. 8.6 Magnetization versus temperature for a ferromagnetic and paramagnetic

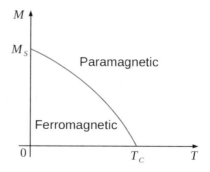

8.3.3 Diamagnetism

When an external magnetic field $\mathbf{B}_0$ is applied, a weak magnetic moment is induced opposite to the applied field direction, and a magnet only weakly repels the diamagnetic substances. Usually, the ferromagnetic and paramagnetic effects are much more significant than diamagnetic effects in all materials. Therefore, those effects are dominant only when the ferromagnetic and paramagnetic effects do not exist. To understand the diamagnetism, we may consider a classical model of two atomic electrons orbiting around the nucleus with the same speed in opposite directions. The electrons continue their circular orbits around the nucleus because of the attractive electrostatic force exerted by the positive charge of the nucleus. Since the magnetic moments of the two electrons are equal in magnitude but have opposite directions, their net magnetic moment in the atom is zero. However, when an external magnetic field is applied, it exerts an additional force on the electrons, $q\mathbf{v} \times \mathbf{B}$. That magnetic force combines with the electrostatic force increasing the orbital motion linear speed of the electron with a magnetic moment opposite to the field and in a decrease of that speed when the magnetic moment of the electron is parallel to the field. Therefore, the two magnetic moments of the electrons no longer cancel, and the substance gains a net magnetic moment opposite of the applied field.

As we mentioned, the superconductor is a material with zero electrical resistance below some critical temperature. There exist some types of superconductors that exhibit perfect diamagnetism in the superconducting state. Hence, the applied magnetic field is excluded by the superconductor such that the field is zero in its interior, known as the *Meissner effect*. When a permanent magnet is nearby a superconductor, they repel one another.

8.4   The Magnetic Field of the Earth

Usually, we say about a compass magnet having a "north-seeking" pole and a "south-seeking" pole. That is, one pole of the magnet seeks or points to the north geographic pole of the Earth.

Since the north pole of a magnet is attracted by the north geographic pole of the Earth, we have that the Earth's south magnetic pole is located near the north geographic pole, and the Earth's north magnetic pole is located near the south geographic pole. The configuration of the Earth's magnetic field is very much like the one that would be achieved by burying a gigantic bar magnet deep in the interior of the Earth.

8.5 Faraday's Law of Induction

In 1831, Faraday was the first who observed quantitatively the phenomena related to time-dependent electric and magnetic fields. In particular, the behavior of currents in circuits was observed when placed in a time-varying magnetic field. Faraday found that a transient current is induced in a loop if the steady current flow in an adjacent circuit is turned on or off. Also, he observed that when the circuit moves relative to the circuit in which a constant current is flowing, then a current is induced in the moving circuit. Moreover, Faraday observed that when a magnet is approaching or moving away from a circuit, then a current is produced in the circuit. Similarly, he found that no current is induced when the current flow on the second circuit was not changing or when either the second circuit or magnet was not moving relative to the first circuit.

Faraday explained the observation of the induced current with the change of the magnetic flux linked by the circuit. That is, the change in the magnetic flux induces an electric field around the circuit; the line integral of the induced electric field yields a potential difference, called *electromotive force*, ε. Then, the electromotive force produces a current flow based on Ohm's law.

Fara-day's Law

To obtain a mathematical formulation of Faraday's law, we consider the circuit $\mathcal{L}$ bounded by an open surface A with unit normal vector $\mathbf{n}$, as shown in Fig. 8.7. Furthermore, the magnetic field $\mathbf{B}$ near the circuit is shown. The magnetic flux through the surface of circuit is given by

$$\Phi_B = \int_A \mathbf{B} \cdot d\mathbf{A} \tag{8.36}$$

where $d\mathbf{A} = \mathbf{n}\,dA$ with dA being a small surface element (see Fig. 8.7). The electromotive force around the circuit or induced voltage is given as

$$\varepsilon = -\left(-\oint_{\mathcal{L}} \mathbf{E}_{ind} \cdot d\mathbf{s}\right) = \oint_{\mathcal{L}} \mathbf{E}_{ind} \cdot d\mathbf{s} \tag{8.37}$$

Lenz's Law

The first minus sign in Eq. (8.37) indicates that the polarity of induced electromotive force, ε, is opposing the change on the magnetic flux, Φ_B. That is, induced electromotive force produces a current in the circuit, which creates a magnetic field, based on Biot–Savart's law, to oppose the change in the magnetic flux Φ_B through the area enclosed by the current circuit. That is known as Lenz's law.

In Eq. (8.37), $\mathbf{E}_{ind}$ is the induced electric field at $d\mathbf{s}$. Faraday's law, mathematically, is formulated as[1]:

$$\varepsilon = -\frac{d\Phi_B}{dt} \tag{8.38}$$

---

[1] In general, Faraday's law is written as $\varepsilon = -k\dfrac{d\Phi_B}{dt}$, where the constant k is used to adjust the units. For SI units, $k = 1$, and for Gaussian units, $k = 1/c$, where c is the speed of light. Here, we use SI units, and thus $k = 1$.

Fig. 8.7 Magnetic field lines
passing through an arbitrary
surface A supported by the
contour $\mathcal{L}$

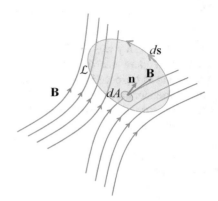

where the magnetic flux is given by Eq. (8.36).

In particular, in the following, we consider a uniform magnetic field **B**. Then, from Eq. (8.36), the magnetic flux is

$$\Phi_B = \mathbf{B} \cdot \mathbf{A} = BA \cos\theta \tag{8.39}$$

θ is the angle formed by the magnetic field vector and unit normal vector **n**. Equation (8.39) implies that $\Phi_B = 0$, when $\theta = 90°$, that is, when the magnetic field **B** and the open surface are parallel. Furthermore, $\Phi_{B,max} = BA$, maximum value of the magnetic flux is obtained for $\theta = 0°$, that is, when magnetic field **B** is perpendicular to the open surface. In addition, Φ_B is between zero and its maximum value, $\Phi_{B,max}$, for any other angle θ.

Moreover, Eq. (8.39) indicates that to change the magnetic flux Φ_B value, and hence to cause induction of an electromotive force, ε, we can perform it in different ways:

1. By changing the magnitude of the magnetic field B and keeping constant A and θ.
2. By fixing the magnitude of magnetic field B and surface area of the circuit A and changing its orientation with respect to **B**, that is, the angle θ.
3. By changing the surface area A of the circuit and fixing the magnitude B and the orientation of the surface relative to **B**, that is, the angle θ.
4. In addition, since the magnetic flux is equal to the net magnetic field lines passing through the circuit surface, then we can cause induced electromotive force by keeping B, A, and θ constant, but moving the circuit (either moving it closer to the magnet or moving it away) to change the number of magnetic field lines passing through the surface (either increasing or decreasing it).

To explain the origin of the electromotive force ε, consider a conductor moving to the right with constant velocity $\mathbf{v}$ in a uniform magnetic field $\mathbf{B}$, which has a direction in the viewing plane, as shown in Fig. 8.8. Knowing that in a conductor there are two kind of charges, namely, the free electrons (negative charge) and positively charged atoms, which move along with the conductor, then there is a Lorentz force acting on both electrons and positive ions: $\mathbf{F}_B = q(\mathbf{v} \times \mathbf{B})$, where $q = -e$ for electrons and $q = +e$ for positive ions. Here, e is the magnitude of the charge of an electron. The direction of the force acting on the electron and positive charge is indicated in Fig. 8.8; they cause the electron to move down and positively charge atoms up. Therefore, after some time, there will be a collection of negative charges (electrons) at the lower end and positive charges at the upper end, as shown in Fig. 8.8. That causes a potential difference between the upper and lower sides of the conductor, $\phi_+ - \phi_-$, which by definition is the electromotive force (or induced potential difference), ε:

$$\varepsilon = \phi_+ - \phi_- \tag{8.40}$$

That is equivalent to the appearance of an induced current I_{ind}, which is opposite to the direction of the movement of electrons, that is, it flows upward (see Fig. 8.8). Thus, in addition to magnetic force, on the charges is acting the electric force: $\mathbf{F}_e = q\mathbf{E}_{ind}$, which has the opposite direction to the magnetic force. The charges will continue accumulating at each end of the conductor until the two forces balance each other. Therefore, based on Newton's third law:

$$\mathbf{F}_B = -\mathbf{F}_e \tag{8.41}$$

or

$$qvB = qE_{ind} \tag{8.42}$$

From Eq. (8.42), we obtain induced electric field:

$$E_{ind} = vB \tag{8.43}$$

Note that after the electrons stop moving, the induced electric field becomes constant and produces a potential difference between the upper end and the lower end, which is the induced electromotive force, given as

$$\varepsilon = LE_{ind} = LvB \tag{8.44}$$

which is positive, and hence $\phi_+ > \phi_-$. Note that this polarity will change if the direction of the movement of the conductor reverses, that is, if the conductor moves to the left.

Note that this is in agreement with Faraday's viewpoint. That is, the effect of the electromotive force is to induce a current in the conductor to oppose the changes, that is, to oppose the movement of the conductor to the right. Indeed, since in the conductor the current I_{ind} is induced, then the magnetic field exerts a force on the

Fig. 8.8 A conductor in a
uniform magnetic field **B**
point in the viewing plane.
Conductor is moving to the
right with constant velocity **v**

Fig. 8.9 The magnetic force
F$_B$ on the conductor moving
to the right with constant
velocity **v** in which is
induced a current I_{ind}

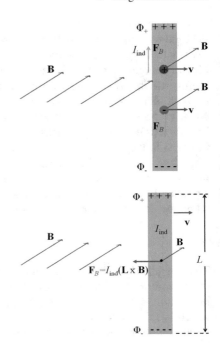

current conductor, based on previous sections, which is given as (see also Fig. 8.9)

$$\mathbf{F}_B = I_{ind}(\mathbf{L} \times \mathbf{B}) \tag{8.45}$$

As seen in Fig. 8.9, the direction of this force opposes the movement of the conductor
to the right.

Combining Eq. (8.37) and Eq. (8.38), we obtain Faraday's law in a general form:

$$\oint_{\mathcal{L}} \mathbf{E}_{ind} \cdot d\mathbf{s} = -\frac{d\Phi_B}{dt} = -\frac{d}{dt}\int_A \mathbf{B} \cdot d\mathbf{A} \tag{8.46}$$

8.6 Rowland Ring Apparatus

Consider a ferromagnetic material, which consists of a torus made up of some material
(for example, iron) within N turns of wire, as shown in Fig. 8.10. Another coil is
connected to a galvanometer (G), which is used to measure the total magnetic flux
through the torus.

To measure the magnetic field **B** in the torus, the current in the toroid increases
from zero to some maximum value I:

Fig. 8.10 Rowland ring
apparatus

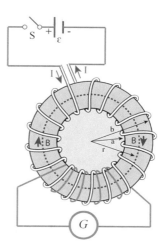

$$B = \mu_0 N I / (2\pi r) \qquad (8.47)$$

$$\oint \mathbf{B} \cdot d\mathbf{s} = \mu_0 N I$$

As the current increases, the magnetic flux through the second coil connected to G
changes by

$$\Phi_B = B A \qquad (8.48)$$

In Eq. (8.48), A is the cross-sectional surface area of the toroid. Based on Faraday's
law, the change on flux induces an electromotive force in the secondary coil:

$$|\varepsilon| = \frac{d\Phi_B}{dt} \qquad (8.49)$$

If the galvanometer is calibrated in advance, the value of the current in the primary
coil corresponds to a value of **B**. First, the magnetic field **B** is measured in the absence
of the torus, and then in the presence of the torus. In that way, the magnetic properties
of the torus material can be obtained by comparing the magnetic field in the two
measurements.

Consider a torus made of iron with no magnetization. When the current in the
primary coil increases from zero to its maximum value I, the magnitude of the
magnetic field strength H increases according to

$$H = n I \qquad (8.50)$$

Fig. 8.11 shows that the magnitude of the total field **B** also increases with I, along
the curve from point O to point a. At O, the domains in the iron are randomly
oriented, and hence $B_m = 0$. With increasing further the current in the primary coil,

Fig. 8.11 Magnetization
curve

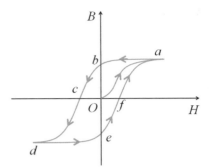

the external field $\mathbf{B}_0$ increases, and the domains become nearly aligned at point a. At point a, the iron core is approaching saturation in which all domains in the iron are aligned. If the current decreases slowly to zero, the external field $\mathbf{B}_0$ also decreases to zero, following the path ab, as shown in Fig. 8.11. At the point, b, $B \neq 0$, only the external field is $\mathbf{B}_0 = 0$ because the iron is magnetized due to the alignment of a large number of its domains (that is, $\mathbf{B} = \mathbf{B}_m$). Therefore, the iron has a *remanent* magnetization. The curve giving B as a function of H is called the *magnetization curve*.

Now, if the current in the primary coil is reversed, then the external magnetic field $\mathbf{B}_0$ direction is also changed. Therefore, the domains experience an opposite direction torque to reorient; when the point c is reached, the sample is again unmagnetized $\mathbf{B} = 0$. A further increase of the current in the reverse direction produces an opposite direction magnetization of the iron compared with that at point b. The saturation point is reached at d. The same sequence of events occurs when the current decreases to zero and then increases back in the original direction, shown by the path def in Fig. 8.11.

Further increase in the current returns the magnetization curve to point a, with a maximum magnetization of the sample. That effect is called *magnetic hysteresis*, and the closed loop is known as the *hysteresis loop*. That is, the magnetization of a ferromagnetic substance depends on both the history of magnetization and on the magnitude of the applied external field. Moreover, a ferromagnetic substance has a memory that makes the material to remain magnetized after the external field is removed.

The shape and size of the hysteresis loop depend on the ferromagnetic properties and in addition to the strength of the applied magnetic field. For *hard ferromagnetic* materials it is wide, that is, a large remanent magnetization (see Fig. 8.12). Those materials are more difficult to be demagnetized by an external field. For *soft ferromagnetic* materials, such as iron, the hysteresis loop is very narrow, and hence it has a small remanent magnetization. Those materials are easily magnetized and demagnetized.

Ideally, a soft ferromagnet may not exhibit a hysteresis, and hence it would have no remanent magnetization. Furthermore, a ferromagnetic sample can be demagnetized

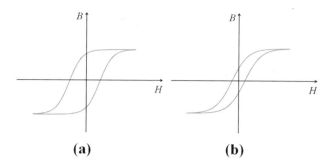

Fig. 8.12 Magnetic hysteresis loop

Fig. 8.13 Illustration of the demagnetization of a ferromagnetic material using successive hysteresis loops

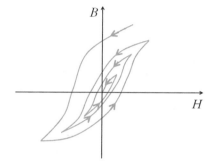

using successive hysteresis loops, by decreasing applied magnetic field, as shown in Fig. 8.13. Moreover, the magnitude of the surface area enclosed by the magnetization curve corresponds to the work required to perform one hysteresis loop. That energy transferred to the material during a magnetization process comes from the source of the external field, such as the *emf* in the circuit of the toroidal coil, that is, the electrical energy transfers into magnetic energy.

When a magnetization cycle repeats, magnetic energy transfers into internal energy due to the dissipation processes within the material as the domains try to re-align. That will increase the temperature of the substance. Therefore, electric devices, such as alternating current adapters for cell phones, or power tools, subject to alternating fields, use soft ferromagnetic substance cores, having narrow hysteresis loops, and hence little energy loss per cycle.

Magnetic computer disks store information by alternating the direction of **B** in a thin layer of ferromagnetic material. Old floppy disks are constructed of a layer on a circular plastic sheet. Hard drives are made up of several rigid platters, each with a magnetic coating on each side. Also, audiotapes and videotapes are similar to floppy disks with a ferromagnetic material on a very long strip of plastic.

8.7   Maxwell's Equations of Magnetism

Now, we present Maxwell's equations that characterize the magnetic phenomena. These laws include all the laws of the magnetism discussed in this chapter. Thus, we assume here that $\mathbf{E} = 0$.

First, we start with Maxwell's equations in free space. In the following, we write these equations in the integral form:

$$\oint_{\mathcal{L}} \mathbf{B} \cdot d\mathbf{s} = \mu_0 I \tag{8.51}$$

$$\oint_A \mathbf{B} \cdot d\mathbf{A} = 0$$

In Eq. (8.51), the first Maxwell's equation is Ampére's law, where the first term on the right-hand side gives the net current through the open surface enclosed by the contour $\mathcal{L}$. The second Maxwell's equation in Eq. (8.51) implies that magnetic field flux through a closed surface is equal to zero. Alternatively, the net number of magnetic field lines passing through a closed surface is zero, that is, so many magnetic lines are leaving the closed surface as entering it. That indicates that there do not exist free magnetic poles.

It is important to note that in Eq. (8.51) we have assumed that $\mathbf{B}$ is only a function of the position $\mathbf{r}$. In the next chapter (Chap. 9), we will discuss the magnetic and electrostatic fields that depend on both position $\mathbf{r}$ and time t.

Equation (8.51) can also be written in a differential form. For instance, the differential form of the first Maxwell's equation is defined using Stokes' formula and current density vector $\mathbf{J}$ as follows:

$$\oint_{\mathcal{L}} \mathbf{B} \cdot d\mathbf{s} = \int_A (\nabla \times \mathbf{B}) \cdot d\mathbf{A} \tag{8.52}$$

$$= \mu_0 \int_A \mathbf{J} \cdot d\mathbf{A}$$

Comparing both sides in Eq. (8.52), we obtain the first Maxwell's equation of magnetism in the following differential form:

$$\nabla \times \mathbf{B} = \mu_0 \mathbf{J} \tag{8.53}$$

To derive the differential form of the second Maxwell's equation of magnetism, we use Gauss's formula in the last expression of Eq. (8.51), as

$$\oint_A \mathbf{B} \cdot d\mathbf{A} = \int_V \nabla \cdot \mathbf{B} \, dV = 0 \tag{8.54}$$

Here, V is the volume enclosed by the surface A. Therefore, we obtain

$$\nabla \cdot \mathbf{B} = 0 \tag{8.55}$$

Finally, we can summarize Maxwell's equations of magnetism in the following differential form:

$$\nabla \times \mathbf{B} = \mu_0 \mathbf{J} \tag{8.56}$$
$$\nabla \cdot \mathbf{B} = 0 \tag{8.57}$$

Next, we will consider Maxwell's equations of magnetism in a medium, which are written in terms of the magnetic field strength $\mathbf{H}$:

$$\mathbf{B} = \mu_0 (\mathbf{H} + \mathbf{M}) \tag{8.58}$$

where $\mathbf{M}$ is magnetization vector.

To obtain the first Maxwell's equation of magnetism in the media, we start with differential form of Ampére's law :

$$\begin{aligned}\nabla \times \mathbf{B} &= \mu_0 \nabla \times (\mathbf{H} + \mathbf{M}) \\ &= \mu_0 \left(\nabla \times \mathbf{H}\right) + \mu_0 \left(\nabla \times \mathbf{M}\right) \\ &= \mu_0 \mathbf{J}'\end{aligned} \tag{8.59}$$

where $\mathbf{J}'$ is the equivalent macroscopic current, which is sum of the microscopic current $\mathbf{J}$ and an effective current density due to magnetization:

$$\mathbf{J}_M = \nabla \times \mathbf{M} \tag{8.60}$$

Combining Eqs. (8.59) and (8.60), we obtain

$$\mu_0 \left(\nabla \times \mathbf{H}\right) = \mu_0 \mathbf{J} \tag{8.61}$$

or

$$\nabla \times \mathbf{H} = \mathbf{J} \tag{8.62}$$

The second Maxwell's equation of magnetism in a medium remains the same expression as in Eq. (8.57).

8.8 Vector Potential

We return to Biot–Savart law, and rewrite it as follows (refer also to Fig. 8.14):

Fig. 8.14 Biot–Savart law

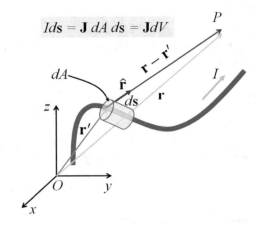

$$\mathbf{B} = \frac{\mu_0}{4\pi} \nabla \times \int_V \frac{\mathbf{J}(\mathbf{r}')}{|\mathbf{r} - \mathbf{r}'|} d\mathbf{r}' \tag{8.63}$$

where we have used that

$$\nabla \left(\frac{1}{|\mathbf{r} - \mathbf{r}'|} \right) = -\frac{\mathbf{r} - \mathbf{r}'}{|\mathbf{r} - \mathbf{r}'|^3} \tag{8.64}$$

and

$$\frac{I d\mathbf{s} \times \hat{\mathbf{r}}}{|\mathbf{r} - \mathbf{r}'|^2} = -d\mathbf{r}' \frac{\left((\mathbf{r} - \mathbf{r}') \times \mathbf{J}(\mathbf{r}') \right)}{|\mathbf{r} - \mathbf{r}'|^3} \tag{8.65}$$

where $dV = d\mathbf{r}'$ is a small volume element, as indicated in Fig. 8.14. We can now introduce a *vector potential* of the magnetic field as

$$\mathbf{A}(\mathbf{r}) = \frac{\mu_0}{4\pi} \int_V \frac{\mathbf{J}(\mathbf{r}')}{|\mathbf{r} - \mathbf{r}'|} d\mathbf{r}' \tag{8.66}$$

and the magnetic field can be written as

$$\mathbf{B} = \nabla \times \mathbf{A}(\mathbf{r}) \tag{8.67}$$

In general, the vector potential is defined up to the gradient of an arbitrary scalar function $\Psi(\mathbf{r})$, that is,

$$\mathbf{A}(\mathbf{r}) = \frac{\mu_0}{4\pi} \int_V \frac{\mathbf{J}(\mathbf{r}')}{|\mathbf{r} - \mathbf{r}'|} d\mathbf{r}' + \nabla \Psi(\mathbf{r}) \tag{8.68}$$

Equation (8.68) indicates that vector potential can be transformed as

$$\mathbf{A}(\mathbf{r}) \rightarrow \mathbf{A}(\mathbf{r}) + \nabla \Psi(\mathbf{r}) \tag{8.69}$$

which is known as *gauge transformation*. Transformation of the form given by Eq. (8.69) is allowed, since they do not change the magnetic field. For instance, using transformation in Eq. (8.69) and definition of the magnetic field in Eq. (8.67), we get

$$\mathbf{B} = \nabla \times (\mathbf{A}(\mathbf{r}) + \nabla \Psi(\mathbf{r})) = \nabla \times \mathbf{A}(\mathbf{r}) \tag{8.70}$$

since $\nabla \times \nabla \Psi(\mathbf{r}) = 0$.

Consider the differential form of Maxwell's equation of magnetism given by Eq. (8.53):

$$\nabla \times \mathbf{B} = \mu_0 \mathbf{J} \tag{8.71}$$

Substituting Eq. (8.67) into Eq. (8.71), we get

$$\nabla \times (\nabla \times \mathbf{A}) = \mu_0 \mathbf{J} \tag{8.72}$$

Simplifying the left-hand side of Eq. (8.72), we obtain

$$\nabla(\nabla \cdot \mathbf{A}) - \nabla^2 \mathbf{A} = \mu_0 \mathbf{J} \tag{8.73}$$

The freedom of gauge transformations allows to make another convenient choice of gauge as $\nabla \cdot \mathbf{A} = 0$, then the vector potential satisfies Poisson equation:

$$\nabla^2 \mathbf{A} = -\mu_0 \mathbf{J} \tag{8.74}$$

Note that the choice $\nabla \cdot \mathbf{A} = 0$, using Eq. (8.69), implies that $\nabla^2 \Psi = 0$. Therefore, if $\nabla^2 \Psi = 0$ in all space, then Ψ is a constant assuming there are no currents at the infinity.

8.9 Multipole Expansion

Equation (8.66) gives the vector potential of the magnetic field in terms of the current density $\mathbf{J}(\mathbf{r}')$ in a localized finite volume V. Furthermore, $\mathbf{J}(\mathbf{r}')$ is zero outside the volume. Suppose that we are interested on finding A outside that volume. For that, similar to scalar potential in electrostatics, we expand the term $1/\mid \mathbf{r} - \mathbf{r}' \mid$ around $\mathbf{r}' = 0$ using Taylor expansion, as given by Eq. (3.50) (Chap. 3).

Assuming that $\mid \mathbf{r} \mid \gg \mid \mathbf{r}' \mid$, we can rewrite Eq. (8.66) as follows:

$$A(r) = \frac{\mu_0}{4\pi} \int_V J(r') \left(\frac{1}{r} + \frac{r' \cdot r}{r^3} + \frac{1}{2} \sum_{i,j=1}^{3} \frac{3x_i' x_j' - \delta_{ij}(r')^2}{r^5} x_i x_j + \cdots \right) dr'$$

$$(8.75)$$

Equation (8.75) can be considered sum of three contributions, namely, A_0, A_1 and A_3, if we neglect higher order term in the expansion.

The first term, which corresponds to the monopole term in the electrostatic expansion, is

$$A_0(r) = \frac{\mu_0}{4\pi r} \int_V J(r') dr' = \frac{\mu_0}{4\pi r} \int_A (\nabla \cdot J(r')) \, dA \qquad (8.76)$$

where the integration is over the surface enclosing the volume V and Stokes' formula is used. Using the continuity equation of the current density ($\rho = 0$):

$$\nabla \cdot J = 0 \qquad (8.77)$$

we obtain

$$A_0(r) = 0 \qquad (8.78)$$

which indicates that there are no isolated magnetic monopoles (or magnetic charges).

To calculate the second term of the expansion, we use the following mathematical relation:

$$c \times (a \times b) = (b \cdot c)a - b(c \cdot a) + a(c \cdot b) - (a \cdot c)b \qquad (8.79)$$

with $r \equiv c$, $r' \equiv a$ and $J \equiv b$, we obtain

$$r \times (r' \times J) = (J \cdot r)r' - J(r \cdot r') + r'(r \cdot J) - (r' \cdot r)J \qquad (8.80)$$
$$= 2(J \cdot r)r' - 2J(r' \cdot r)$$

Or,

$$J(r' \cdot r) = (J \cdot r)r' - \frac{1}{2}r \times (r' \times J) \qquad (8.81)$$

Substituting Eq. (8.81) into the second term of Eq. (8.75), we obtain

$$A_1(r) = \frac{\mu_0}{4\pi} \int_V J \frac{r' \cdot r}{r^3} dr' \qquad (8.82)$$

$$= \frac{\mu_0}{4\pi} \int_V (J \cdot r) \frac{r'}{r^3} dr' - \frac{\mu_0}{4\pi} r \times \left(\frac{1}{2} \int_V \frac{r' \times J}{r^3} dr' \right)$$

The first integral in Eq. (8.82) vanishes since the integrand is an odd function of r'. Furthermore, we define the magnetic moment density or magnetization as

$$M(\mathbf{r}) = \frac{1}{2}(\mathbf{r} \times \mathbf{J}(\mathbf{r})) \tag{8.83}$$

and its integral over the volume gives the magnetic dipole moment:

$$\mathbf{m} = \frac{1}{2} \int_V (\mathbf{r}' \times \mathbf{J}) \, d\mathbf{r}' \tag{8.84}$$

Therefore, Eq. (8.82) can be written as

$$\mathbf{A}_1(\mathbf{r}) = \frac{\mu_0}{4\pi} \frac{\mathbf{m} \times \mathbf{r}}{r^3} \tag{8.85}$$

which is the lowest non-zero term in the expansion of $\mathbf{A}$ for a steady current distribution in a localized volume V. To evaluate the third term, we introduce the magnetic quadrupole moment components as follows:

$$Q_{ijk} = \int_V J_k(\mathbf{r}') \left(3x_i' x_j' - \delta_{ij}(r')^2\right) d\mathbf{r}' \tag{8.86}$$

where the third index k expresses the anisotropic distribution of the current along x-, y-, and z-axes. Then, the third term of the expansion becomes

$$\mathbf{A}_2(\mathbf{r}) = \frac{\mu_0}{8\pi} \sum_{i,j,k=1}^{3} Q_{ijk} x_i x_j \hat{\mathbf{x}}_k \tag{8.87}$$

where $\hat{\mathbf{x}}_k$ (for $k = 1, 2, 3$) are unit vectors along x-, y-, and z-axes.

To calculate the expansion terms of the magnetic field, we use Eq. (8.67) with vector potential expanded according to Eq. (8.75). We consider only the first non-vanishing term of the expansion, and write the first term of expansion of $\mathbf{B}$ as

$$\mathbf{B}_1(\mathbf{r}) = \nabla \times \mathbf{A}_1 = \frac{\mu_0}{4\pi} \left[\nabla \times \mathbf{m} \times \left(\frac{\mathbf{r}}{r^3}\right) \right] \tag{8.88}$$

Using Eq. (8.79), where $\nabla \equiv \mathbf{c}$, $\mathbf{m} \equiv \mathbf{a}$ and $\dfrac{\mathbf{r}}{r^3} \equiv \mathbf{b}$, we can write

$$\nabla \times \left(\mathbf{m} \times \frac{\mathbf{r}}{r^3}\right) = \left(\frac{\mathbf{r}}{r^3} \cdot \nabla\right) \mathbf{m} - \frac{\mathbf{r}}{r^3} (\nabla \cdot \mathbf{m}) + \mathbf{m} \left(\nabla \cdot \frac{\mathbf{r}}{r^3}\right) - (\mathbf{m} \cdot \nabla) \frac{\mathbf{r}}{r^3} \tag{8.89}$$

where

$$(\mathbf{r} \cdot \nabla) \mathbf{m} = \sum_{i,j=1}^{3} x_i \left(\nabla_i m_j \right) \hat{\mathbf{x}}_j = 0 \tag{8.90}$$

$$\mathbf{r} \left(\nabla \cdot \mathbf{m} \right) = \sum_{i,j=1}^{3} x_j \left(\nabla_i m_i \right) \hat{\mathbf{x}}_j = 0 \tag{8.91}$$

because $\mathbf{m}$ is independent on $\mathbf{r}$. Furthermore,

$$\begin{aligned}
\mathbf{m} \left(\nabla \cdot \frac{\mathbf{r}}{r^3} \right) &= \mathbf{m} \left(\frac{\partial}{\partial x} \frac{x}{r^3} + \frac{\partial}{\partial y} \frac{y}{r^3} + \frac{\partial}{\partial z} \frac{z}{r^3} \right) \\
&= \mathbf{m} \left(\frac{r^2 - 3x^2}{r^5} + \frac{r^2 - 3y^2}{r^5} + \frac{r^2 - 3z^2}{r^5} \right) \\
&= 0
\end{aligned} \tag{8.92}$$

and

$$\begin{aligned}
(\mathbf{m} \cdot \nabla) \frac{\mathbf{r}}{r^3} &= \left(m_x \frac{\partial}{\partial x} + m_y \frac{\partial}{\partial y} + m_z \frac{\partial}{\partial z} \right) \frac{x\mathbf{i} + y\mathbf{j} + z\mathbf{k}}{r^3} \\
&= m_x \left(\frac{r^2 - 3x^2}{r^5}\mathbf{i} - \frac{3xy}{r^5}\mathbf{j} - \frac{3xz}{r^5}\mathbf{k} \right) \\
&\quad + m_y \left(-\frac{3xy}{r^5}\mathbf{i} + \frac{r^2 - 3y^2}{r^5}\mathbf{j} - \frac{3yz}{r^5}\mathbf{k} \right) \\
&\quad + m_z \left(-\frac{3xz}{r^5}\mathbf{i} - \frac{3yz}{r^5}\mathbf{j} + \frac{r^2 - 3z^2}{r^5}\mathbf{k} \right) \\
&= \frac{\mathbf{m}}{r^3} - 3\frac{\mathbf{m} \cdot \mathbf{r}}{r^5}x\mathbf{i} - 3\frac{\mathbf{m} \cdot \mathbf{r}}{r^5}y\mathbf{j} - 3\frac{\mathbf{m} \cdot \mathbf{r}}{r^5}z\mathbf{k} \\
&= \frac{\mathbf{m}}{r^3} - 3\frac{(\mathbf{m} \cdot \hat{\mathbf{r}})\hat{\mathbf{r}}}{r^3}
\end{aligned} \tag{8.93}$$

where $\hat{\mathbf{r}}$ is a unit vector along $\mathbf{r}$, that is, $\mathbf{r} = r\hat{\mathbf{r}}$. Combining Eqs. (8.88), (8.89), (8.90), (8.92), and (8.93), we obtain

$$\mathbf{B}_1(\mathbf{r}) = \nabla \times \mathbf{A}_1 = \frac{\mu_0}{4\pi} \frac{3(\mathbf{m} \cdot \hat{\mathbf{r}})\hat{\mathbf{r}} - \mathbf{m}}{r^3} \tag{8.94}$$

Equation (8.94) gives the magnetic field created by a magnetic dipole moment, which has the same form as the electric field of an electric dipole (see also the second term in Eq. (3.64), Chap. 3).

8.10 Energy of the Magnetic Field

Consider a steady current-carrying single circuit. When the magnetic flux through the circuit changes, an electromotive force ε is induced around it, based on Faraday's law. To maintain a constant current in the circuit, the external sources, such as the battery, must do work. The rate change of the work is

$$\frac{dW}{dt} = -I\varepsilon = I\frac{d\Phi_B}{dt} \tag{8.95}$$

where Φ_B is the magnetic flux through the circuit and the negative sign is due to Lenz's law. Form Eq. (8.95), the work done by the sources to keep current constant for a change of the magnetic flux with $d\Phi_B$ is

$$\delta W = I\delta\Phi_B \tag{8.96}$$

Consider an element of the circuit with cross-sectional surface area ΔS perpendicular to the direction of the current flow, then $I = J\Delta S$. Then, Eq. (8.96) can be written as

$$\Delta(\delta W) = J\Delta S \int_A \delta\mathbf{B} \cdot d\mathbf{A} \tag{8.97}$$

where A is the surface area of circuit, and $d\mathbf{A} = \mathbf{n}dA$ is an small surface element with $\mathbf{n}$ a unit normal vector to the surface of the circuit. Using Eq. (8.67), we obtain

$$\Delta(\delta W) = J\Delta S \int_A (\nabla \times \delta\mathbf{A}) \cdot \mathbf{n}dA \tag{8.98}$$

Applying Stokes' theorem, we write Eq. (8.98) in the following form:

$$\Delta(\delta W) = J\Delta S \oint_{\mathcal{L}} \delta\mathbf{A} \cdot d\mathbf{s} \tag{8.99}$$

where $\mathcal{L}$ is a closed contour line of the portion of circuit and $d\mathbf{s}$ is a small element in this contour parallel to the current density vector $\mathbf{J}$. Therefore, $J\Delta S d\mathbf{s} = \mathbf{J}dV$, where $dV = d\mathbf{r}$ is a volume element. Summing up all those closed path portion of the circuit, we obtain the total increment of work done by the external sources due to a change of magnetic field $\delta\mathbf{A}$:

$$\delta W = \int_V \delta\mathbf{A}(\mathbf{r}) \cdot \mathbf{J}\,d\mathbf{r} \tag{8.100}$$

Using Ampére's law, Eq. (8.62) for the static electric field, we write Eq. (8.100) as

$$\delta W = \int_V \delta \mathbf{A}(\mathbf{r}) \cdot (\nabla \times \mathbf{H}) \, d\mathbf{r} \tag{8.101}$$

Using the following vector identity:

$$\nabla \cdot (\mathbf{a} \times \mathbf{b}) = \mathbf{b} \cdot (\nabla \times \mathbf{a}) - \mathbf{a} \cdot (\nabla \times \mathbf{b}) \tag{8.102}$$

Equation (8.101) reduces to

$$\delta W = \int_V (\mathbf{H} \cdot (\nabla \times \delta \mathbf{A}) + \nabla \cdot (\mathbf{H} \times \delta \mathbf{A})) \, d\mathbf{r} \tag{8.103}$$

If we assume that the magnetic field distribution is localized, the second term is zero. Therefore, using $\delta \mathbf{B} = \nabla \times \delta \mathbf{A}$, we obtain

$$\delta W = \int_V \mathbf{H} \cdot \delta \mathbf{B} \, d\mathbf{r} \tag{8.104}$$

For paramagnetic or diamagnetic medium, $\mathbf{H}$ and $\mathbf{B}$ are linearly dependent, and hence

$$\mathbf{H} \cdot \delta \mathbf{B} = \frac{1}{2} \delta (\mathbf{H} \cdot \mathbf{B}) \tag{8.105}$$

Substituting Eq. (8.105) into Eq. (8.104), we get

$$\delta W = \frac{1}{2} \int_V \delta (\mathbf{H} \cdot \mathbf{B}) \, d\mathbf{r} \tag{8.106}$$

If we consider a change of the fields from zero to their final values, we obtain the total magnetic energy:

$$W = \frac{1}{2} \int_V \mathbf{H} \cdot \mathbf{B} \, d\mathbf{r} \tag{8.107}$$

Equation (8.107) gives the magnetic analogue of the electrostatic energy.

We can obtain another expression for the magnetic energy analogue to the electrostatic energy given in terms of the charge and electric potential. For that, we can use Eq. (8.100), and assuming a linear relation between $\mathbf{J}$ and $\mathbf{A}$:

$$\delta \mathbf{A} \cdot \mathbf{J} = \frac{1}{2} \delta (\mathbf{J} \cdot \mathbf{A}) \tag{8.108}$$

Substituting Eq. (8.108) into Eq. (8.100), we get

$$\delta W = \frac{1}{2} \int_V \delta(\mathbf{J} \cdot \mathbf{A}) \, d\mathbf{r} \tag{8.109}$$

Summing up all small changes of the field, we obtain the total magnetic energy in another analogue form:

$$W = \frac{1}{2} \int_V \mathbf{J} \cdot \mathbf{A} \, d\mathbf{r} \tag{8.110}$$

8.11 Exercises

Exercise 8.1 A toroid wound with 60.0 turns/m of wire carries a current of 5.00 A. The torus is iron, which has a magnetic permeability of $\mu_m = 5000\mu_0$ under the given conditions. Find **H** and **B** inside the iron.

Solution 8.1 Consider the torus region of a toroid that carries a current I. If this region is a vacuum, $\mathbf{M} = 0$ (because no magnetic material is present), the total magnetic field is that arising from the current alone, and $\mathbf{B} = \mathbf{B_0} = \mu_0 \mathbf{H}$. Because $B_0 = \mu_0 n I$ in the torus region, where n is the number of turns per unit length of the toroid,

$$H = B_0/\mu_0 = \mu_0 n I / \mu_0 \tag{8.111}$$

or

$$H = nI = (60.0 \; turns/m)(5.00 \text{ A}) = 300 \; \frac{A \cdot turns}{m} \tag{8.112}$$

and

$$B = \mu_m H = 5000\mu_0 H \tag{8.113}$$

$$= (5000)(4\pi \times 10^{-7} \frac{T \cdot m}{A})(300 \; \frac{A \cdot turns}{m}) = 1.88 \text{ T}$$

This value of B is 5000 times the value in the absence of iron.

Exercise 8.2 Estimate the saturation magnetization in a long cylinder of iron, assuming one unpaired electron spin per atom.

Solution 8.2 The saturation magnetization is obtained when all the magnetic moments in the sample are aligned. If the sample contains n atoms per unit volume, then the saturation magnetization is

$$M_S = n\mu \tag{8.114}$$

where μ is the magnetic moment of atom. Because the molar mass of iron is 55 g/mol and its density is 7.9 g/cm^3, the value of n for iron is 8.6×10^{28} atoms/m^3. Assuming that each atom contributes one Bohr magneton (due to one unpaired spin) to the magnetic moment, we obtain

$$M_S = (8.6 \times 10^{28} \frac{atoms}{m^3})(9.27 \times 10^{-24} \frac{A \cdot m^2}{atom}) \tag{8.115}$$
$$= 8.0 \times 10^5 \text{ A/m}$$

Exercise 8.3 The current in the long, straight wire is $I_1 = 5.00$ A, and the wire lies in the plane of the rectangular loop, which carries 10.0 A, as shown in Fig. 8.15. The dimensions are $c = 0.100$ m, $a = 0.150$ m, and $\ell = 0.450$ m. Find the magnitude and direction of the net force exerted on the loop by the magnetic field created by the wire.

Solution 8.3 The magnitude of magnetic field created by the current I_1 at any distance r is

$$B_1 = \frac{\mu_0 I_1}{2\pi r} \tag{8.116}$$

and the direction is tangent to the circle in the place perpendicular to the direction of the current I_1. The force exerted by $\mathbf{B}_1$ on the I_2 is given as $\mathbf{F} = I_2\mathbf{L} \times \mathbf{B}_1$ where $\mathbf{L}$ is along the direction of I_2. Thus, the forces on the horizontal directions of the loop where I_2 passes (i.e., 12 and 34) are zero, and it is different from zero only on the vertical directions (41 and 23). The directions of these two forces are shown in the figure. Their magnitudes are

$$F_{41} = I_2\ell B_1 \sin 90° = \frac{\mu_0 I_1 I_2}{2\pi c}\ell \tag{8.117}$$
$$F_{23} = I_2\ell B_1 \sin 90° = \frac{\mu_0 I_1 I_2}{2\pi (a + c)}\ell$$

Replacing the numerical values, we get

Fig. 8.15 The current in the long, straight wire is $I_1 = 5.00$ A, and the wire lies in the plane of the rectangular loop, which carries 10.0 A

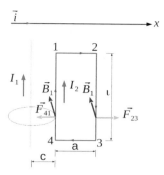

$$F_{41} = \frac{(4\pi \times 10^{-7}\ \text{T} \cdot \text{m/A})(5.00\ \text{A})(10.0\ \text{A})}{2\pi\,(0.100\ \text{m})}(0.450\ \text{m}) \qquad (8.118)$$

$$= 0.450 \times 10^{-4}\ \text{N}$$

$$F_{23} = \frac{(4\pi \times 10^{-7}\ \text{T} \cdot \text{m/A})(5.00\ \text{A})(10.0\ \text{A})}{2\pi\,(0.250\ \text{m})}(0.450\ \text{m})$$

$$= 1.80 \times 10^{-5}\ \text{N}$$

The resultant force is then given as

$$\mathbf{F} = -4.50 \times 10^{-5}\mathbf{i}\ \text{N} + 1.80 \times 10^{-5}\mathbf{i}\ \text{N} \qquad (8.119)$$

$$= -2.70 \times 10^{-5}\mathbf{i}\ \text{N}$$

Sign $(-)$ indicates that force has opposite direction to positive x-axis.

Exercise 8.4 Two long, parallel conductors separated by 10.0 cm carry currents in the same direction (see Fig. 8.16). The first wire carries current $I_1 = 5.00$ A, and the second carries $I_2 = 8.00$ A. (a) What is the magnitude of the magnetic field created by I_1 and acting on I_2? (b) What is the force per unit length exerted on I_2 by I_1? (c) What is the magnitude of the magnetic field created by I_2 at the location of I_1? (d) What is the force per unit length exerted by I_2 on I_1?

Solution 8.4 In Fig. 8.16, we have shown direction of the magnetic fields created by the two currents and the forces they exert on each other.
   (a) The magnitude of $\mathbf{B}_1$ is

Fig. 8.16 Two long, parallel
conductors separated by
10.0 cm carry currents in the
same direction

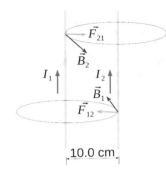

$$B_1 = \frac{\mu_0 I_1}{2\pi r} \tag{8.120}$$

$$= \frac{(4\pi \times 10^{-7} \text{ T} \cdot \text{m/A})(5.00 \text{ A})}{2\pi(0.100 \text{ m})}$$

$$= 1.00 \times 10^{-5} \text{ T}$$

(b) The magnetic force is

$$\mathbf{F}_{12} = I_2 \mathbf{L} \times \mathbf{B}_1 \tag{8.121}$$

and the magnitude:

$$F_{12} = I_2 L B_1 \tag{8.122}$$

and its magnitude per unit length is

$$f_{12} = \frac{F_{12}}{L} = I_2 B_1 \tag{8.123}$$

$$= \frac{\mu_0 I_1 I_2}{2\pi r}$$

$$= \frac{(4\pi \times 10^{-7} \text{ T} \cdot \text{m/A})(5.00 \text{ A})(8.00 \text{ A})}{2\pi(0.100 \text{ m})}$$

$$= 8.00 \times 10^{-5} \text{ N}$$

(c) The magnitude of $\mathbf{B}_2$ is

$$B_2 = \frac{\mu_0 I_2}{2\pi r} \tag{8.124}$$

$$= \frac{(4\pi \times 10^{-7} \text{ T} \cdot \text{m/A})(8.00 \text{ A})}{2\pi(0.100 \text{ m})}$$

$$= 1.60 \times 10^{-5} \text{ T}$$

(d) The magnetic force is

$$\mathbf{F}_{21} = I_1 \mathbf{L} \times \mathbf{B}_2 \tag{8.125}$$

and the magnitude:

$$F_{21} = I_1 L B_2 \tag{8.126}$$

and its magnitude per unit length is

$$f_{21} = \frac{F_{21}}{L} = I_1 B_2 \tag{8.127}$$
$$= \frac{\mu_0 I_1 I_2}{2\pi r}$$
$$= \frac{(4\pi \times 10^{-7} \text{ T} \cdot \text{m/A})(5.00 \text{ A})(8.00 \text{ A})}{2\pi (0.100 \text{ m})}$$
$$= 8.00 \times 10^{-5} \text{ N}$$

Note that the forces have the same magnitude, but have opposite direction, in agreement with the third law of Newton.

Exercise 8.5 A circular wire loop of radius r carries a current I, as shown in Fig. 8.17. What is the magnitude of the magnetic field at its center?

Solution 8.5 Using Biot–Savart law:

$$d\mathbf{B} = \frac{\mu_0 I}{4\pi} \frac{d\mathbf{s} \times \hat{\mathbf{r}}}{r^2} \tag{8.128}$$

For the magnitude, we obtain

$$dB = \frac{\mu_0 I}{4\pi} \frac{ds}{r^2} \tag{8.129}$$

Total magnetic field is

$$B = \frac{\mu_0 I}{4\pi} \int \frac{ds}{r^2} \tag{8.130}$$
$$= \frac{\mu_0 I}{4\pi r^2} (2\pi r)$$
$$= \frac{\mu_0 I}{2r}$$

Fig. 8.17 A circular wire
loop of radius r carries a
current I

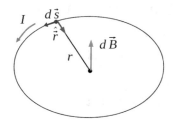

References

Fermi E (1930) Über die magnetischen Momente der Atomkerne. Z Phys 60:5–6
Griffiths DJ (1999) Introduction to electrodynamics, 3rd edn. Prentice Hall
Jackson JD (1999) Classical electrodynamics, 3rd edn. John Wiley and Sons
Landau LD, Lifshitz EM (1971) The classical theory of fields. Pergamon Press
Sykja H (2006) Bazat e Elektrodinamikës. SHBUT

Chapter 9
Maxwell's Equations
of Electromagnetism

This chapter aims to introduce Maxwell's equations of electromagnetism for the free space and the medium.

In this chapter, we discuss Maxwell's equations of electromagnetism. Magnetic and electrostatic fields exist together, and hence it makes sense to always discuss the two fields simultaneously under electromagnetism. That is, often, the electromagnetism is represented by electromagnetic field, which is the pair of vectors $(\mathbf{E}, \mathbf{B})$ for the free space or $(\mathbf{D}, \mathbf{H})$ for the medium. Also, in this chapter, we will present the energy of the electromagnetic field and the motion of the charged particles in an electromagnetic field. For further reading, one can also consider other available literature Jackson (1999), Landau and Lifshitz (1971), Sykja (2006), Griffiths (1999).

9.1 Maxwell's Equations of Electromagnetism

First, we start with Maxwell's equations in free space. In the following, we write these equations in the integral form:

$$\oint_A \mathbf{E} \cdot d\mathbf{A} = \frac{Q}{\epsilon_0} \tag{9.1}$$

$$\oint_{\mathcal{L}} \mathbf{E} \cdot d\mathbf{s} = -\frac{\partial}{\partial t} \int_A \mathbf{B} \cdot d\mathbf{A}$$

$$\oint_{\mathcal{L}} \mathbf{B} \cdot d\mathbf{s} = \mu_0 \left(I + \epsilon_0 \frac{\partial}{\partial t} \int_A \mathbf{E} \cdot d\mathbf{A} \right)$$

$$\oint_A \mathbf{B} \cdot d\mathbf{A} = 0$$

© The Author(s), under exclusive license to Springer Nature Switzerland AG 2022
H. Kamberaj, *Electromagnetism*, Undergraduate Texts in Physics,
https://doi.org/10.1007/978-3-030-96780-2_9

The first Maxwell's equation in Eq. (9.1) is simply Gauss's law of the electric field flux through a surface A enclosing a volume V with a net charge inside that volume Q. Note that here electric field $\mathbf{E}$ is created by all the charges in free space (including the charge Q inside the volume). This equation indicates the existence of the free electric poles of the electric field (charges). The second Maxwell's equation in Eq. (9.1) derives from Faraday's law; the left-hand side is the induced electromotive force, and the right-hand side gives the rate change with time of the magnetic field flux through the surface enclosed by contour $\mathcal{L}$.

In Eq. (9.1), the third Maxwell's equation is Ampére's law, where the first term on the right-hand side gives the net current through the open surface enclosed by the contour $\mathcal{L}$, and the second term determines the displacement current.

The fourth Maxwell's equation in Eq. (9.1) implies that magnetic field flux through a closed surface is equal to zero. Alternatively, the net number of magnetic field lines passing through a closed surface is zero; that is, so many magnetic lines are leaving the closed surface as entering it. That indicates that there do not exist free magnetic poles.

It is important to note that in Eq. (9.1) we have assumed that $\mathbf{B}$ and $\mathbf{E}$ are functions of both position $\mathbf{r}$ and time t. Therefore, the formulae consider partial derivatives.

Equation (9.1) can also be written in a differential form. For instance, the differential form of the first Maxwell's equation is given by expression in Eq. (4.69) (Chap. 4). To derive the second Maxwell's equation in a differential form, we use Stokes' formula in Eq. (9.1):

$$\int_A \nabla \times \mathbf{E}\, d\mathbf{A} = -\frac{\partial}{\partial t} \int_A \mathbf{B} \cdot d\mathbf{A} \tag{9.2}$$

Comparing both sides of Eq. (9.2), we get the second Maxwell's equation in a differential form:

$$\nabla \times \mathbf{E} = -\frac{\partial \mathbf{B}}{\partial t} \tag{9.3}$$

Similarly, using Stokes' formula and current density vector $\mathbf{J}$, we can rewrite the third Maxwell's equation as follows:

$$\oint_{\mathcal{L}} \mathbf{B} \cdot d\mathbf{s} = \int_A (\nabla \times \mathbf{B}) \cdot d\mathbf{A} \tag{9.4}$$

$$= \mu_0 \left(\int_A \mathbf{J} \cdot d\mathbf{A} + \epsilon_0 \frac{\partial}{\partial t} \int_A \mathbf{E} \cdot d\mathbf{A} \right)$$

Comparing both sides in Eq. (9.4), we obtain the third Maxwell's equation in the following differential form:

$$\nabla \times \mathbf{B} = \mu_0 \left(\mathbf{J} + \epsilon_0 \frac{\partial \mathbf{E}}{\partial t} \right) \tag{9.5}$$

To derive the differential form of the fourth Maxwell's equation, we use Gauss's formula in the last expression of Eq. (9.1), as

$$\oint_A \mathbf{B} \cdot d\mathbf{A} = \int_V \nabla \cdot \mathbf{B} \, dV = 0 \tag{9.6}$$

In Eq. (9.6), V denotes the volume enclosed by the surface A. Therefore, we obtain

$$\nabla \cdot \mathbf{B} = 0 \tag{9.7}$$

Finally, we can summarize the four Maxwell's equations in the following differential form:

$$\nabla \cdot \mathbf{E} = \frac{\rho}{\epsilon_0} \tag{9.8}$$

$$\nabla \times \mathbf{E} = -\frac{\partial \mathbf{B}}{\partial t}$$

$$\nabla \times \mathbf{B} = \mu_0 \left(\mathbf{J} + \epsilon_0 \frac{\partial \mathbf{E}}{\partial t} \right)$$

$$\nabla \cdot \mathbf{B} = 0$$

Next, we will consider Maxwell's equations in a medium, which are written in terms of the displacement vector $\mathbf{D}$ and magnetic field strength $\mathbf{H}$, such as

$$\mathbf{E} = \frac{1}{\epsilon_0}(\mathbf{D} - \mathbf{P}) \tag{9.9}$$

$$\mathbf{B} = \mu_0(\mathbf{H} + \mathbf{M})$$

where $\mathbf{P}$ is electric polarization and $\mathbf{M}$ is magnetization vector. The first Maxwell's equation in medium is derived in Chapter 4 (see Eq. (4.79)):

$$\nabla \cdot \mathbf{D} = \rho \tag{9.10}$$

In Eq. (9.10), ρ is the excess charge density in the medium. The second and fourth Maxwell's equations are given by the second and fourth expressions, respectively, in Eq. (9.8).

To obtain the third Maxwell's equation, we start with differential form of Ampére's law:

$$\nabla \times \mathbf{B} = \mu_0 \nabla \times (\mathbf{H} + \mathbf{M}) \tag{9.11}$$

$$= \mu_0 (\nabla \times \mathbf{H}) + \mu_0 (\nabla \times \mathbf{M})$$

$$= \mu_0 \mathbf{J}'$$

where $\mathbf{J}'$ is the equivalent macroscopic current, which is sum of the microscopic current $\mathbf{J}$ and an effective current density due to magnetization:

$$\mathbf{J}_M = \nabla \times \mathbf{M} \tag{9.12}$$

Combining Eqs. (9.11) and (9.12), we obtain

$$\mu_0 \left(\nabla \times \mathbf{H} \right) = \mu_0 \mathbf{J} \tag{9.13}$$

Using the continuity equation for the charge and current, we write

$$\nabla \cdot \mathbf{J} + \frac{\partial \rho}{\partial t} = 0 \tag{9.14}$$

Replacing Eq. (9.10) into Eq. (9.14), we get

$$\nabla \cdot \left(\mathbf{J} + \frac{\partial \mathbf{D}}{\partial t} \right) = 0 \tag{9.15}$$

Replacing the current $\mathbf{J}$ in Ampére's law (Eq. (9.13)) by its generalization

$$\mathbf{J} \to \mathbf{J} + \frac{\partial \mathbf{D}}{\partial t} \tag{9.16}$$

we obtain the third Maxwell's equation:

$$\mu_0 \left(\nabla \times \mathbf{H} \right) = \mu_0 \left(\mathbf{J} + \frac{\partial \mathbf{D}}{\partial t} \right) \tag{9.17}$$

or

$$\nabla \times \mathbf{H} = \mathbf{J} + \frac{\partial \mathbf{D}}{\partial t} \tag{9.18}$$

9.2 Vector and Scalar Potentials of Electromagnetic Field

Maxwell's equations include a set of first-order partial differential equations related to different electric and magnetic field components. In general, they can be solved for simple problems. However, for convenience potentials that include a smaller number of second-order equations are often introduced, which satisfy some of Maxwell's equations identically. We already introduced those potentials when we discussed the electrostatic and magnetostatic fields, separately, as scalar electric potential ϕ and the vector potential $\mathbf{A}$, which relate to the fields as $\mathbf{E} = -\nabla \phi$ and $\mathbf{B} = \nabla \times \mathbf{A}$, as discussed in the previous chapters.

In the following, for convenience, we restrict ourselves to the free space form of Maxwell's equations given by Eq. (9.8). Since $\nabla \cdot \mathbf{B} = 0$, we can, again for the electromagnetic field, write that

$$\mathbf{B} = \nabla \times \mathbf{A} \tag{9.19}$$

Using the second expression in Eqs. (9.8) and (9.19), we can write

$$\nabla \times \mathbf{E} = -\frac{\partial (\nabla \times \mathbf{A})}{\partial t} \tag{9.20}$$

which can be arranged in this form:

$$\nabla \times \left(\mathbf{E} + \frac{\partial \mathbf{A}}{\partial t} \right) = 0 \tag{9.21}$$

Equation (9.21) indicates that the quantity of a vanishing curl can be written as a gradient of a scalar function, namely, the potential scalar ϕ:

$$\mathbf{E} + \frac{\partial \mathbf{A}}{\partial t} = -\nabla \phi \tag{9.22}$$

or

$$\mathbf{E} = -\nabla \phi - \frac{\partial \mathbf{A}}{\partial t} \tag{9.23}$$

Definitions of the electromagnetic fields $\mathbf{E}$ and $\mathbf{B}$ related to the potentials ϕ and $\mathbf{A}$ given by Eq. (9.19) and Eq. (9.23) satisfy identically the two homogeneous Maxwell's equations in Eq. (9.8). The dynamics of ϕ and $\mathbf{A}$ can be determined using the other two non-homogeneous Maxwell's equation in Eq. (9.8).

For instance, using the first Maxwell's equation in Eq. (9.8), we obtain

$$\nabla^2 \phi + \frac{\partial (\nabla \cdot \mathbf{A})}{\partial t} = -\frac{\rho}{\epsilon_0} \tag{9.24}$$

In addition, using the third Maxwell's equation in Eq. (9.8), we get

$$\nabla \times (\nabla \times \mathbf{A}) = \mu_0 \mathbf{J} + \mu_0 \epsilon_0 \left(-\frac{\partial (\nabla \phi)}{\partial t} - \frac{\partial^2 \mathbf{A}}{\partial t^2} \right) \tag{9.25}$$

Using the following relation

$$\nabla \times (\nabla \times \mathbf{A}) = \nabla (\nabla \cdot \mathbf{A}) - \nabla^2 \mathbf{A} \tag{9.26}$$

and that

$$\sqrt{\mu_0 \epsilon_0} = \frac{1}{c} \tag{9.27}$$

with c denoting the speed of light in vacuum, Eq. (9.25) reduces as follows:

$$\nabla^2 \mathbf{A} - \frac{1}{c^2}\frac{\partial^2 \mathbf{A}}{\partial t^2} - \nabla\left(\nabla \cdot \mathbf{A} + \frac{1}{c^2}\frac{\partial \phi}{\partial t}\right) = -\mu_0 \mathbf{J} \tag{9.28}$$

Those derivations indicate that four first-order differential Maxwell's equations given as in Eq. (9.8) for $(\mathbf{E}, \mathbf{B})$ are equivalent to two second-order differential equations given by Eqs. (9.24) and (9.28) for $(\phi, \mathbf{A})$. Note, however, they are still coupled, that is, solutions, $(\phi, \mathbf{A})$, of Eqs. (9.24) and (9.28) are coupled to each equation. To decouple ϕ and $\mathbf{A}$, we note that potentials ϕ and $\mathbf{A}$ are arbitrarily defined. That is, $\mathbf{B}$, defined by Eq. (9.20), remains unchanged by the following transformation:

$$\mathbf{A}' = \mathbf{A} + \nabla\Psi \tag{9.29}$$

where Ψ is an arbitrary scalar function.

Furthermore, electric field $\mathbf{E}$ given by Eq. (9.24) remains unchanged under the following transformation:

$$\phi' = \phi - \frac{\partial\Psi}{\partial t} \tag{9.30}$$

Equations (9.29) and (9.30) imply that in choosing the pair of potentials $(\phi, \mathbf{A})$, we determine the so-called Lorentz condition to be satisfied:

$$\nabla \cdot \mathbf{A} + \frac{1}{c^2}\frac{\partial\phi}{\partial t} = 0 \tag{9.31}$$

Using the transformation given by Eqs. (9.30) and (9.31), it is straightforward to show that equations for ϕ and $\mathbf{A}$ transform as follows:

$$\nabla^2\phi - \frac{1}{c^2}\frac{\partial^2\phi}{\partial t^2} = -\frac{\rho}{\epsilon_0} \tag{9.32}$$

and

$$\nabla^2\mathbf{A} - \frac{1}{c^2}\frac{\partial^2\mathbf{A}}{\partial t^2} = -\mu_0\mathbf{J} \tag{9.33}$$

The forms of the expressions given in Eqs. (9.32) and (9.33) indicate that equations for ϕ and $\mathbf{A}$ are decoupled, and hence they can be solved separately.

9.3 Electromagnetic Field Energy and Conservation Law

As derived separately in the previous chapters, we can write now the total electromagnetic field energy as the sum of the electric field potential energy U_E and the magnetic field energy U_B:

$$U = U_E + U_B = \frac{1}{2} \int_V \mathbf{E} \cdot \mathbf{D} \, d\mathbf{r} + \frac{1}{2} \int_V \mathbf{B} \cdot \mathbf{H} \, d\mathbf{r} \tag{9.34}$$

$$= \frac{1}{2} \int_V (\mathbf{E} \cdot \mathbf{D} + \mathbf{B} \cdot \mathbf{H}) \, d\mathbf{r}$$

In writing Eq. (9.34), we assume that electromagnetic field $(\mathbf{E}, \mathbf{B})$ exists in a finite volume V of the space. Furthermore, we can introduce the electromagnetic field energy density u as

$$u = \frac{1}{2} (\mathbf{E} \cdot \mathbf{D} + \mathbf{B} \cdot \mathbf{H}) \tag{9.35}$$

Thus,

$$U = \int_V u \, d\mathbf{r} \tag{9.36}$$

and hence u gives the energy per unit of volume.

Consider there is a continuous distribution of charge and current in some finite volume V, where the electromagnetic field is different from zero. We know that only the electric field does work on the moving charges, but not the magnetic field (since the magnetic force is perpendicular with direction of the velocity). Therefore, the total element of work done by the electromagnetic field on the charge during a displacement $\Delta\mathbf{s}$ is given as

$$\Delta W = \int_V \rho(\mathbf{r}) \Delta\mathbf{s} \cdot \mathbf{E} \, d\mathbf{r} \tag{9.37}$$

where $dq = \rho \, d\mathbf{r}$ is the charge in a small element volume $dV = d\mathbf{r}$ and ρ is the charge density at some location $\mathbf{r}$ at the center of that small element volume. Using Eq. (9.37), we can calculate the average power of the fields as follows:

$$\frac{\Delta W}{\Delta t} = \int_V \rho(\mathbf{r}) \frac{\Delta\mathbf{s}}{\Delta t} \cdot \mathbf{E} \, d\mathbf{r} \tag{9.38}$$

Taking the limit when $\Delta t \to 0$ of both sides of Eq. (9.38) and knowing that $\lim_{\Delta \to 0}(\Delta\mathbf{s}/\Delta t) = \mathbf{v}$ is the velocity of the charge dq, we obtain the rate of the work done by the electromagnetic field on the charge:

$$\frac{dW}{dt} = \int_V \mathbf{J} \cdot \mathbf{E} \, d\mathbf{r} \tag{9.39}$$

where $\mathbf{J}$ is the usual current density.

Substituting the current density from Eq. (9.18) into Eq. (9.39), we get

$$\frac{dW}{dt} = \int_V \left(\nabla \times \mathbf{H} - \frac{\partial \mathbf{D}}{\partial t} \right) \cdot \mathbf{E} \, d\mathbf{r} \tag{9.40}$$

$$= - \int_V \left(\nabla \cdot (\mathbf{E} \times \mathbf{H}) + \mathbf{E} \cdot \frac{\partial \mathbf{D}}{\partial t} + \mathbf{H} \cdot \frac{\partial \mathbf{B}}{\partial t} \right) d\mathbf{r}$$

where the following vector relationship was used:

$$\nabla \cdot (\mathbf{E} \times \mathbf{H}) = \mathbf{H} \cdot (\nabla \times \mathbf{E}) - \mathbf{E} \cdot (\nabla \times \mathbf{H}) \tag{9.41}$$

and Maxwell's equation for $\nabla \times \mathbf{E}$. Moreover, we assume that macroscopic medium is linear in its magnetic and electric field properties and there are no dispersion or energy losses. Using Eq. (9.35), we get the following form for Eq. (9.40):

$$- \int_V \mathbf{J} \cdot \mathbf{E} \, d\mathbf{r} = \int_V \left(\nabla \cdot (\mathbf{E} \times \mathbf{H}) + \frac{\partial u}{\partial t} \right) d\mathbf{r} \tag{9.42}$$

We introduce the so-called Poynting vector as

$$\mathbf{S} = \mathbf{E} \times \mathbf{H} \tag{9.43}$$

Then, substituting Eq. (9.43) into Eq. (9.42), we get

$$- \int_V \mathbf{J} \cdot \mathbf{E} \, d\mathbf{r} = \int_V \left(\nabla \cdot \mathbf{S} + \frac{\partial u}{\partial t} \right) d\mathbf{r} \tag{9.44}$$

Comparing both sides of Eq. (9.44) and assuming that the volume V is chosen arbitrary, we get the continuity equation in its differential form:

$$\frac{\partial u}{\partial t} + \nabla \cdot \mathbf{S} = -\mathbf{J} \cdot \mathbf{E} \tag{9.45}$$

Equation (9.45) is also known as the energy conservation law of the electromagnetic field; that is, the rate change with time of the sum of the electromagnetic energy in a volume V and the energy leaving the volume through the boundary surface of that volume per unit of time is equal to the total work done by the fields on the sources (current and charges) within the volume. Moreover, the work done per unit of time and per unit of volume by the electromagnetic field (that is, $\mathbf{J} \cdot \mathbf{E}$) is transformation of electromagnetic energy into internal energy (mechanical or heat energy).

9.4 Conservation Law of Momentum

At a microscopic level, the medium comprises charged particles, such as electrons and atomic nuclei. Therefore, we can think of the rate of energy conversion as a rate of energy increase of the charged particles of the medium per unit of volume. Furthermore, Eq. (9.44) can be treated as a conservation law of energy at the microscopic level of the combined system of particles and electromagnetic fields $(\mathbf{E}, \mathbf{B})$.

For that, we denote by E_{mech} the total mechanical energy of the charged particles inside the volume V and assume that the particles do not leave the volume. The rate change of the mechanical energy is

$$\frac{d E_{mech}}{dt} = \int_V \mathbf{J} \cdot \mathbf{E} \, d\mathbf{r} \qquad (9.46)$$

Combining Eqs. (9.44) and (9.46), we express the conservation law of energy of the combined system as

$$\frac{d}{dt} \left(E_{mech} + E_{field} \right) = \int_V \nabla \cdot \mathbf{S} \, d\mathbf{r} \qquad (9.47)$$

Using Gauss's theorem, we get

$$\frac{d}{dt} \left(E_{mech} + E_{field} \right) = \oint_A \mathbf{S} \cdot \mathbf{n} \, dA \qquad (9.48)$$

where A is the surface enclosing the volume V and $\mathbf{n}$ is an outward normal vector to every point in the surface. In Eq. (9.48), E_{field} is the total energy of electromagnetic field given by

$$E_{field} = \frac{1}{2} \int_V \left(\epsilon_0 \mathbf{E}^2 + \frac{1}{\mu_0} \mathbf{B}^2 \right) d\mathbf{r} = \int_V u \, d\mathbf{r} \qquad (9.49)$$

The net electromagnetic force on a small volume charged particle dq moving with velocity $\mathbf{v}$ is sum vector of the electric field force, $dq\mathbf{E}$, and magnetic field force (Lorentz force), $dq(\mathbf{v} \times \mathbf{B})$:

$$d\mathbf{F} = dq \, (\mathbf{E} + \mathbf{v} \times \mathbf{B}) \qquad (9.50)$$

Integrating Eq. (9.50) over a volume V of the charge distribution, we obtain the total force exerted by the electromagnetic field on the volume charge distribution:

$$\mathbf{F} = \int_V (\mathbf{E} + \mathbf{v} \times \mathbf{B}) \, \rho(\mathbf{r}) \, d\mathbf{r} \qquad (9.51)$$

Using the second law of Newton, and denoting by $\mathbf{P}_{mech}$ the total momentum of the particles inside the volume V, we get

$$\frac{d\mathbf{P}_{mech}}{dt} = \int_V (\rho\mathbf{E} + \mathbf{J} \times \mathbf{B})\, d\mathbf{r} \tag{9.52}$$

To eliminate ρ and $\mathbf{J}$ from the equation, we use Maxwell's equations given in Eq. (9.8), and write

$$\rho\mathbf{E} + \mathbf{J} \times \mathbf{B} = \epsilon_0 \left(\mathbf{E}(\nabla \cdot \mathbf{E}) + \mathbf{B} \times \frac{\partial\mathbf{E}}{\partial t} - c^2\mathbf{B} \times (\nabla \times \mathbf{B}) \right) \tag{9.53}$$

Using the following relation:

$$\mathbf{B} \times \frac{\partial\mathbf{E}}{\partial t} = \mathbf{E} \times \frac{\partial\mathbf{B}}{\partial t} - \frac{\partial}{\partial t}(\mathbf{E} \times \mathbf{B}) \tag{9.54}$$

then, Eq. (9.53) can be written as

$$\rho\mathbf{E} + \mathbf{J} \times \mathbf{B} = \epsilon_0 \left(\mathbf{E}(\nabla \cdot \mathbf{E}) + c^2\mathbf{B}(\nabla \cdot \mathbf{B}) \right. \tag{9.55}$$
$$\left. - \mathbf{E} \times (\nabla \times \mathbf{E}) - \frac{\partial}{\partial t}(\mathbf{E} \times \mathbf{B}) - c^2\mathbf{B} \times (\nabla \times \mathbf{B}) \right)$$

where $c^2\mathbf{B}(\nabla \cdot \mathbf{B}) = 0$ is added, and the second Maxwell's equation relation is used as in Eq. (9.8). Substituting Eq. (9.55) into Eq. (9.52), we get

$$\frac{d\mathbf{P}_{mech}}{dt} + \frac{d}{dt} \int_V \epsilon_0(\mathbf{E} \times \mathbf{B})\, d\mathbf{r} \tag{9.56}$$
$$= \epsilon_0 \int_V \left(\mathbf{E}(\nabla \cdot \mathbf{E}) - \mathbf{E} \times (\nabla \times \mathbf{E}) + c^2\mathbf{B}(\nabla \cdot \mathbf{B}) - c^2\mathbf{B} \times (\nabla \times \mathbf{B}) \right) d\mathbf{r}$$

The second term on the left-hand side of Eq. (9.56) defines the electromagnetic momentum $\mathbf{P}_{field}$ in the volume V as

$$\mathbf{P}_{field} = \epsilon_0 \int_V (\mathbf{E} \times \mathbf{B})\, d\mathbf{r} = \frac{1}{c^2} \int_V (\mathbf{E} \times \mathbf{H})\, d\mathbf{r} \tag{9.57}$$

Therefore, the integrand of Eq. (9.57) represents the density of electromagnetic momentum:

$$\mathbf{p}_{field} = \frac{1}{c^2} (\mathbf{E} \times \mathbf{H}) \tag{9.58}$$

The terms on the integrand of the integral on the right-hand side can be simplified as follows:

$$(\mathbf{B}(\nabla \cdot \mathbf{B}) - \mathbf{B} \times (\nabla \times \mathbf{B}))_1 = (\mathbf{B}(\nabla \cdot \mathbf{B}))_1 - (\mathbf{B} \times (\nabla \times \mathbf{B}))_1 \tag{9.59}$$

$$= B_1 \sum_{i=1}^{3} \frac{\partial B_i}{\partial x_i} - B_2 (\nabla \times \mathbf{B})_3 + B_3 (\nabla \times \mathbf{B})_2$$

$$= B_1 \sum_{i=1}^{3} \frac{\partial B_i}{\partial x_i} - B_2 \left(\frac{\partial B_2}{\partial x_1} - \frac{\partial B_1}{\partial x_2} \right) + B_3 \left(-\frac{\partial B_3}{\partial x_1} + \frac{\partial B_1}{\partial x_3} \right)$$

$$= B_1 \frac{\partial B_1}{\partial x_1} + B_1 \frac{\partial B_2}{\partial x_2} + B_1 \frac{\partial B_3}{\partial x_3} - B_2 \left(\frac{\partial B_2}{\partial x_1} - \frac{\partial B_1}{\partial x_2} \right) + B_3 \left(-\frac{\partial B_3}{\partial x_1} + \frac{\partial B_1}{\partial x_3} \right)$$

$$= \frac{\partial B_1^2}{\partial x_1} + \frac{\partial B_1 B_2}{\partial x_2} + \frac{\partial B_1 B_3}{\partial x_3} - \frac{1}{2} \frac{\partial (B_1^2 + B_2^2 + B_3^2)}{\partial x_1}$$

$$= \sum_{i=1}^{3} \frac{\partial}{\partial x_i} \left(B_1 B_i - \frac{1}{2} \mathbf{B} \cdot \mathbf{B} \delta_{1i} \right)$$

In general, the k-th component is

$$(\mathbf{B}(\nabla \cdot \mathbf{B}) - \mathbf{B} \times (\nabla \times \mathbf{B}))_k = \sum_{i=1}^{3} \frac{\partial}{\partial x_i} \left(B_k B_i - \frac{1}{2} \mathbf{B} \cdot \mathbf{B} \delta_{ki} \right) \tag{9.60}$$

Similarly,

$$(\mathbf{E}(\nabla \cdot \mathbf{E}) - \mathbf{E} \times (\nabla \times \mathbf{E}))_k = \sum_{i=1}^{3} \frac{\partial}{\partial x_i} \left(E_k E_i - \frac{1}{2} \mathbf{E} \cdot \mathbf{E} \delta_{ki} \right) \tag{9.61}$$

We can now determine the so-called Maxwell stress tensor as

$$\sigma_{ki} = \epsilon_0 \left(E_k E_i + c^2 B_k B_i - \frac{1}{2} \left(\mathbf{E} \cdot \mathbf{E} + c^2 \mathbf{B} \cdot \mathbf{B} \right) \delta_{ki} \right) \tag{9.62}$$

Therefore, we can rewrite Eq. (9.56) in the following form:

$$\frac{d}{dt} \left(\mathbf{P}_{mech} + \mathbf{P}_{field} \right)_k = \int_V \sum_{i=1}^{3} \frac{\partial}{\partial x_i} \sigma_{ki} \, d\mathbf{r} \tag{9.63}$$

Using Gauss's formula, Eq. (9.63) can be written as

$$\frac{d}{dt} \left(\mathbf{P}_{mech} + \mathbf{P}_{field} \right)_k = \oint_S \sum_{i=1}^{3} (\sigma_{ki} n_i) \, dA \tag{9.64}$$

where A is the surface enclosing the volume V and n_i is the i-th component of the outward unit vector to the surface at every point.

Equation (9.64) represents the conservation law of momentum, where $\sum_{i=1}^{3} \sigma_{ki} n_i$ is the i-th component of the flow per unit area of the momentum across the surface into the volume V of sources. Thus, $\sum_{i=1}^{3} \sigma_{ki} n_i$ is also the force per unit surface area acting along the surface A and applied to the combined system of particles plus fields inside volume V.

9.5   Dynamics of Charged Particles in Electromagnetic Fields

Maxwell's equations form the basis of all classical electromagnetic phenomena in matter. Those equations combined with the Lorentz force equation and Newton's second law of motion provide a complete description of the classical dynamics of interacting charged particles and electromagnetic fields.

Consider a single particle of mass m and charge q moving with velocity $\mathbf{v}$ in an electromagnetic field $(\mathbf{E}, \mathbf{B})$, as shown in Fig. 9.1. We can write the Lagrangian of the particle as

$$L = T - U = \frac{mv^2}{2} - q\phi + q\mathbf{v} \cdot \mathbf{A} \tag{9.65}$$

where $(\phi, \mathbf{A})$ are the potentials of the electromagnetic field, and

$$v = \sqrt{\dot{x}_1^2 + \dot{x}_2^2 + \dot{x}_3^2} \tag{9.66}$$

is the speed. In Eq. (9.65), the first term gives the kinetic energy, and the last two terms on the right-hand side give, respectively, potential energy of a charged particle in the electromagnetic field. The term $q\phi$ is simply electrostatic potential energy, as discussed in Chap. 3, while the term $-q\mathbf{v} \cdot \mathbf{A}$ gives the potential energy of the moving charge in magnetic field. A physical explanation for these two terms follows from that $\mathbf{J} \cdot \mathbf{E}$ gives the work done per unit of time and per unit of volume by the electromagnetic field, and hence the decrease in potential energy of a moving particle in electromagnetic field is

$$\begin{aligned}
-dU &= \mathbf{J} \cdot \mathbf{E} \, dt \, dV \tag{9.67} \\
&= \mathbf{J} \cdot \mathbf{E} dt \, (d\mathbf{s} \cdot d\mathbf{A}) \\
&= I \, dt \, d\mathbf{s} \cdot \mathbf{E} \\
&= dq d\mathbf{s} \cdot \mathbf{E} \\
&= q\delta(\mathbf{r} - \mathbf{r}') \, d\mathbf{s} \cdot \left(-\nabla\phi - \frac{\partial \mathbf{A}}{\partial t} \right)
\end{aligned}$$

Fig. 9.1 A charged particle moving with velocity **v** at the position **r**. The charge density at any position **r'** of the volume V extended by the electromagnetic field is $\rho(\mathbf{r'}) = q\delta(\mathbf{r} - \mathbf{r'})/dV$, where δ is the function: $\delta(\mathbf{r} - \mathbf{r'}) = 1$ if $\mathbf{r} = \mathbf{r'}$, otherwise it is zero. dV is a small cylindrical volume element along the path of particle $dV = d\mathbf{s} \cdot d\mathbf{A}$, where $d\mathbf{s}$ is displacement of particle during a small time interval dt

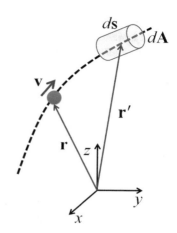

Integrating over the all volume V extended by the electromagnetic field, we get

$$U = - \int_V q\delta(\mathbf{r} - \mathbf{r'}) \, d\mathbf{s} \cdot \left(\nabla\phi + \frac{\partial \mathbf{A}}{\partial t} \right) \tag{9.68}$$

$$= -q \left(\phi(\infty) - \phi(\mathbf{r}) \right) - q\mathbf{v} \cdot \mathbf{A}$$

Taking that the electromagnetic field is zero at the infinity ($\phi(\infty) = 0$), we obtain the potential energy of a charged particle in the electromagnetic field as

$$U(\mathbf{r}) = q\phi(\mathbf{r}) - q\mathbf{v} \cdot \mathbf{A}(\mathbf{r}) \tag{9.69}$$

Lagrangian equations are given as

$$\frac{d}{dt}\left(\frac{\partial L}{\partial \dot{x}_i} \right) - \frac{\partial L}{\partial x_i} = 0, \quad i = 1, 2, 3 \tag{9.70}$$

Substituting Eq. (9.65) into Eq. (9.70), we obtain

$$m\ddot{x}_i + q\frac{\partial A_i}{\partial t} - \left(-q\frac{\partial \phi}{\partial x_i} - q\frac{\partial}{\partial x_i}(\mathbf{v} \cdot \mathbf{A}) \right) = 0, \quad i = 1, 2, 3 \tag{9.71}$$

or

$$m\ddot{x}_i = q\left(-\frac{\partial \phi}{\partial x_i} - \frac{\partial A_i}{\partial t} \right) + q(\mathbf{v} \times (\nabla \times \mathbf{A}))_i, \quad i = 1, 2, 3 \tag{9.72}$$

Using Eqs. (9.19) and (9.23), we can write Eq. (9.71) in the well-known form:

$$m\ddot{x}_i = qE_i + q(\mathbf{v} \times \mathbf{B})_i, \quad i = 1, 2, 3 \tag{9.73}$$

The first term on the right-hand side is the electric field force on q, and the second is the Lorentz force applied on moving charge by a magnetic field. If we consider now a system of N charges, $q_1, q_2, \ldots, q_N$ inside some finite volume V. That could correspond to the system of electrons in the field of the nucleus in an atom. Then, Lagrangian of the system in the presence of the electromagnetic field is given as

$$L = T - U = \sum_{i=1}^{N} \frac{m_i v_i^2}{2} - \sum_{i=1}^{N} (q_i \phi + q_i \mathbf{v_i} \cdot \mathbf{A}) - k_e \sum_{i=1}^{N} \sum_{j=i+1}^{N} \frac{q_i q_j}{r_{ij}} \qquad (9.74)$$

where $(\phi, \mathbf{A})$ are again the potentials of the electromagnetic field, and the last term is the potential interaction energy between charges in the system. In Eq. (9.74), r_{ij} is the distance between two charges i and j.

Using Lagrangian equations, Eq. (9.71), we obtain the following second-order differential equations:

$$m\ddot{x}_i = q_i E_i + q_i (\mathbf{v}_i \times \mathbf{B})_i + k_e \sum_{j=1 \neq i}^{N} \frac{q_i q_j}{r_{ij}^3} (x_i - x_j), \quad i = 1, 2, \cdots, 3N \quad (9.75)$$

where $3N$ also gives the number of degrees of freedom of the system.

9.6 Macroscopic Maxwell Equations

When discussing the electromagnetic fields in a medium made up of the constituents of matter, such as electrons, protons and neutrons, we would have to consider individually those sources of the electric and magnetic moments. However, the number of these sources may go up to 10^{23}, which is a big number. If we denote by small letters $\mathbf{e}, \mathbf{b}, \mathbf{j}$ microscopic fields created by those sources, then microscopic Maxwell's equations will be written as

$$\nabla \cdot \mathbf{e} = \frac{\rho}{\epsilon_0} \qquad (9.76)$$

$$\nabla \times \mathbf{e} = -\frac{\partial \mathbf{b}}{\partial t} \qquad (9.77)$$

$$\nabla \cdot \mathbf{b} = 0 \qquad (9.78)$$

$$\nabla \times \mathbf{b} = \mu_0 \left(\mathbf{j} + \epsilon_0 \frac{\partial \mathbf{e}}{\partial t} \right) \qquad (9.79)$$

Equations (9.76)–(9.79) imply that there will exist up to 10^{23} such equations to be solved numerically for each source. Furthermore, we are not always interested in considering the microscopic fields, but we would like to consider the electromagnetic field behavior on classical length scales, which exceeds the atomic scales.

For that, macroscopic fields are introduced by averaging the microscopic fields as

$$\mathbf{E} = \bar{\mathbf{E}} \tag{9.80}$$

$$\mathbf{B} = \bar{\mathbf{B}} \tag{9.81}$$

where the averaging is defined as

$$\mathbf{A}(\mathbf{r}) = \langle \mathbf{a}(\mathbf{r}) \rangle = \int_V \mathbf{a}(\mathbf{r} - \mathbf{r}') w(\mathbf{r}') \, d\mathbf{r}' \tag{9.82}$$

where w is the weight function at the position of the source, which is normalized to unity. Note that w extends over a sufficiently large region in space.

Therefore,

$$\frac{\partial \mathbf{A}}{\partial t} = \langle \frac{\partial \mathbf{a}}{\partial t} \rangle \tag{9.83}$$

$$\frac{\partial \mathbf{A}}{\partial \mathbf{r}} = \langle \frac{\partial \mathbf{a}}{\partial \mathbf{r}} \rangle \tag{9.84}$$

The averaged Maxwell's equations can be written as follows:

$$\nabla \cdot \mathbf{E} = \frac{\langle \rho \rangle}{\epsilon_0} \tag{9.85}$$

$$\nabla \times \mathbf{E} = -\frac{\partial \mathbf{B}}{\partial t} \tag{9.86}$$

$$\nabla \cdot \mathbf{B} = 0 \tag{9.87}$$

$$\nabla \times \mathbf{B} = \mu_0 \left(\langle \mathbf{j} \rangle + \epsilon_0 \frac{\partial \mathbf{E}}{\partial t} \right) \tag{9.88}$$

To understand the averages of the sources of the fields, such as the charge density and current density, we will have to consider the atomic structure of a typical solid. There are two types of sources of the charge, namely, the nuclei and valence electrons. In the following discussion, we denote by $\mathbf{r}_m$ the position of the center of atom or molecule to a laboratory frame and $\mathbf{a}_{m,i}$ the position of the source i of molecule m to the center of the molecule. Furthermore, in conductors, there exist the free conduction electrons at some position $\mathbf{r}_i(t)$ in the system at time t. We denote by ρ_b charge density contribution from the bound sources of charge (nuclei and valence electrons) and ρ_f the charge density contribution from the free electrons in conduction region. They are written as

$$\rho_b(\mathbf{r}, t) = \sum_{m,i} q_{m,i} \delta(\mathbf{r} - \mathbf{r}_m(t) - \mathbf{a}_{m,i}(t)) \tag{9.89}$$

$$\rho_f(\mathbf{r}, t) = e \sum_{m,i} \delta(\mathbf{r} - \mathbf{r}_i(t)) \tag{9.90}$$

where $q_{m,i}$ is the charge of the source i (nucleus or electron) in molecule m and e is the charge of electron.

The average density of the bound charges over a macroscopic region is given by

$$\langle \rho_b(\mathbf{r}) \rangle = \int_V w(\mathbf{r}') \sum_{m,i} q_{m,i} \delta(\mathbf{r} - \mathbf{r}_m(t) - \mathbf{a}_{m,i}(t)) \, d\mathbf{r}' \tag{9.91}$$

$$= \sum_{m,i} w(\mathbf{r} - \mathbf{r}_m - \mathbf{a}_{m,i}) q_{m,i}$$

$$\approx \sum_m q_m w(\mathbf{r} - \mathbf{r}_m) - \nabla \sum_m w(\mathbf{r} - \mathbf{r}_m) \cdot \sum_i \mathbf{a}_{m,i} q_{m,i}$$

$$= \sum_m q_m \langle \delta(\mathbf{r} - \mathbf{r}_m) \rangle - \nabla \cdot \mathbf{P}(\mathbf{r})$$

The third line gives the Taylor expansion of the weights knowing that the weight function changes exceed the atomic extensions and they decay to zero at distances further than atomic dimensions $\mid \mathbf{a} \mid$. At the zeroth order, Taylor expansion gives the total charge of molecule:

$$q_m = \sum_i q_{m,i} \tag{9.92}$$

The first-order term includes the molecular dipole moments:

$$\mathbf{d}_m = \sum_i \mathbf{a}_{m,i} q_{m,i} \tag{9.93}$$

In Eq. (9.91), $\mathbf{P}$ gives the average polarization of the medium:

$$\mathbf{P}(\mathbf{r}, t) = \langle \sum_m \delta(\mathbf{r} - \mathbf{r}_m(t)) \mathbf{d}_m(t) \rangle \tag{9.94}$$

Equation (9.94) indicates that $\mathbf{P}$ is the density of the dipole moments carried by the molecule in a system.

The average density of the mobile charge carriers in the medium (that is, conduction electrons) is given by

$$\langle \rho_f(\mathbf{r}, t) \rangle = \langle e \sum_i \delta(\mathbf{r} - \mathbf{r}_i(t)) \rangle \tag{9.95}$$

Therefore, the average charge density is

$$\langle \rho(\mathbf{r}, t) \rangle = \sum_m q_m \langle \delta(\mathbf{r} - \mathbf{r}_m) \rangle - \nabla \cdot \mathbf{P}(\mathbf{r}) + \langle e \sum_i \delta(\mathbf{r} - \mathbf{r}_i(t)) \rangle \qquad (9.96)$$

$$= \rho(\mathbf{r}, t) - \nabla \cdot \mathbf{P}(\mathbf{r}, t)$$

where

$$\rho(\mathbf{r}, t) = \sum_m q_m \langle \delta(\mathbf{r} - \mathbf{r}_m) \rangle + \langle e \sum_i \delta(\mathbf{r} - \mathbf{r}_i(t)) \rangle \qquad (9.97)$$

denotes the macroscopic charge density in the system.

Substituting Eq. (9.96) into Eq. (9.85), as we have shown in the previous section, the first Maxwell's equation reduces to

$$\nabla \cdot \mathbf{D}(\mathbf{r}, t) = \rho(\mathbf{r}, t) \qquad (9.98)$$

where $\mathbf{D} = \epsilon_0 \mathbf{E} + \mathbf{P}$.

To compute the average of the microscopic current density, we sum up the bound and free charge current densities: $\mathbf{j}_b + \mathbf{j}_f$. The bound part of the charge current density is given as

$$\mathbf{j}_b(\mathbf{r}) = \sum_{m,i} q_{m,i} \left(\dot{\mathbf{r}} + \dot{\mathbf{a}}_{m,i} \right) \delta \left(\mathbf{r} - \mathbf{r}_m - \mathbf{a}_{m,i} \right) \qquad (9.99)$$

Averaging expression given by Eq. (9.99), we get

$$\langle \mathbf{j}_b(\mathbf{r}) \rangle = \int w(\mathbf{r}') \sum_{m,i} q_{m,i} \left(\dot{\mathbf{r}}_m + \dot{\mathbf{a}}_{m,i} \right) \delta \left(\mathbf{r} - \mathbf{r}_m - \mathbf{a}_{m,i} \right) d\mathbf{r}' \qquad (9.100)$$

$$= \sum_{m,i} q_{m,i} w(\mathbf{r} - \mathbf{r}_m - \mathbf{a}_{m,i})(\dot{\mathbf{r}}_m + \dot{\mathbf{a}}_{m,i})$$

$$\approx \sum_{m,i} q_{m,i} \left[w(\mathbf{r} - \mathbf{r}_m) - \sum_m \nabla w(\mathbf{r} - \mathbf{r}_m) \cdot \mathbf{a}_{m,i} \right] (\dot{\mathbf{r}}_m + \dot{\mathbf{a}}_{m,i})$$

$$= \langle \mathbf{j}_b(\mathbf{r}) \rangle^{(0)} + \langle \mathbf{j}_b(\mathbf{r}) \rangle^{(1)} + \langle \mathbf{j}_b(\mathbf{r}) \rangle^{(2)}$$

where

$$\langle \mathbf{j}_b(\mathbf{r}) \rangle^{(0)} = \sum_m q_m w(\mathbf{r} - \mathbf{r}_m) \dot{\mathbf{r}}_m = \langle \sum_m q_m \dot{\mathbf{r}}_m \delta(\mathbf{r} - \mathbf{r}_m) \rangle \qquad (9.101)$$

gives the current carried by a molecule in the zeroth-order approximation, where the molecule is considered a point-like charge.

The first-order approximation term is given as

$$\langle \mathbf{j}_b(\mathbf{r}) \rangle^{(1)} = \sum_m \left(w(\mathbf{r} - \mathbf{r}_m) \dot{\mathbf{d}}_m - (\nabla w(\mathbf{r} - \mathbf{r}_m) \cdot \mathbf{d}_m) \dot{\mathbf{r}}_m \right) \tag{9.102}$$

Let us compare it with the time derivative of a polarization vector:

$$\dot{\mathbf{P}} = \frac{d}{dt} \left(\sum_m w(\mathbf{r} - \mathbf{r}_m) \mathbf{d}_m \right) \tag{9.103}$$

$$= \sum_m \left(w(\mathbf{r} - \mathbf{r}_m) \dot{\mathbf{d}}_m - (\nabla w(\mathbf{r} - \mathbf{r}_m) \cdot \mathbf{d}_m) \dot{\mathbf{r}}_m \right)$$

Substituting Eq. (9.103) into Eq. (9.102)

$$\langle \mathbf{j}_b(\mathbf{r}) \rangle^{(1)} = \dot{\mathbf{P}} + \sum_m \nabla w(\mathbf{r} - \mathbf{r}_m) \times (\mathbf{d}_m \times \dot{\mathbf{r}}_m) \approx \dot{\mathbf{P}} \tag{9.104}$$

The second-order contribution term is given by

$$\langle \mathbf{j}_b(\mathbf{r}) \rangle^{(2)} = -\sum_{m,i} q_{m,i} \mathbf{a}_{m,i} \left(\dot{\mathbf{a}}_{m,i} \cdot \nabla w(\mathbf{r} - \mathbf{r}_m) \right) \tag{9.105}$$

which relates to the magnetic dipole moments carried by the molecules:

$$\mathbf{M} = \frac{1}{2} \sum_{m,i} q_{m,i} \mathbf{a}_{m,i} \times \dot{\mathbf{a}}_{m,i} w(\mathbf{r} - \mathbf{r}_m) \tag{9.106}$$

Calculating the curl of the expression given by Eq. (9.106), we get

$$\nabla \times \mathbf{M} = \langle \mathbf{j}_b \rangle^{(2)} + \frac{1}{2} \frac{d}{dt} \left(\sum_{m,i} q_{m,i} \mathbf{a}_{m,i} (\mathbf{a}_{m,i} \cdot \nabla w(\mathbf{r} - \mathbf{r}_m)) \right) \tag{9.107}$$

In the right-hand side of Eq. (9.107), the second expression is the electric quadrupole contribution, which can be ignored since it is a small term. Therefore,

$$\langle \mathbf{j}_b \rangle^{(2)} \approx \nabla \times \mathbf{M} \tag{9.108}$$

The contribution from the free charge carrier is

$$\langle \mathbf{j}_f(\mathbf{r}) \rangle = e \langle \sum_i \dot{\mathbf{r}}_i \delta(\mathbf{r} - \mathbf{r}_i) \rangle \tag{9.109}$$

Combining terms in Eqs. (9.101), (9.104), (9.108), and (9.109), we obtain the expression for the average current density as

$$\langle \mathbf{j}(\mathbf{r}) \rangle = \mathbf{J}(\mathbf{r}) + \dot{\mathbf{P}}(\mathbf{r}) + \nabla \times \mathbf{M}(\mathbf{r}) \tag{9.110}$$

where

$$\mathbf{J}(\mathbf{r}) = \langle e \sum_i \dot{\mathbf{r}}_i \delta(\mathbf{r} - \mathbf{r}_i) \rangle + \langle \sum_m q_m \dot{\mathbf{r}}_m \delta(\mathbf{r} - \mathbf{r}_m) \rangle \tag{9.111}$$

gives the current carried by the free charge carrier and the point-like approximated molecule. $\mathbf{M}$ is the density of the magnetic dipole moments in the system:

$$\mathbf{M} = \langle \sum_m \delta(\mathbf{r} - \mathbf{r}_m) \frac{1}{2} \sum_i q_{m,i} \mathbf{a}_{m,i} \times \dot{\mathbf{a}}_{m,i} \rangle \tag{9.112}$$

Substituting Eq. (9.110) into Eq. (9.88), as we have shown in the previous section, the last Maxwell's equation reduces to

$$\nabla \times \left(\frac{1}{\mu_0} \mathbf{B} - \mathbf{M} \right) = \mathbf{J} + \frac{\partial \mathbf{D}}{\partial t} \tag{9.113}$$

Since $\mathbf{H} = \dfrac{1}{\mu_0} \mathbf{B} - \mathbf{M}$, we finally get the inhomogeneous Maxwell's equation:

$$\nabla \times \mathbf{H} = \mathbf{J} + \frac{\partial \mathbf{D}}{\partial t} \tag{9.114}$$

9.7 Exercises

Exercise 9.1 Show that $-\nabla(\mathbf{v} \cdot \mathbf{A}) = \mathbf{v} \times (\nabla \times \mathbf{A})$, where $\mathbf{v}$ is the velocity of a charged particle and $\mathbf{A}$ is potential vector of the electromagnetic field.

Solution 9.1 We start with the following vector identity:

$$\mathbf{a} \times (\nabla \times \mathbf{b}) = \mathbf{a}(\mathbf{b} \cdot \nabla) - \nabla(\mathbf{a} \cdot \mathbf{b}) \tag{9.115}$$

Then,

$$\mathbf{v} \times (\nabla \times \mathbf{A}) = (\mathbf{A} \cdot \nabla)\mathbf{v} - \nabla(\mathbf{v} \cdot \mathbf{A}) = -\nabla(\mathbf{v} \cdot \mathbf{A}) \tag{9.116}$$

because the velocity components do not depend on the coordinates, and thus

$$\frac{\partial v_i}{\partial x_j} = 0, \quad i, j = 1, 2, 3 \tag{9.117}$$

Exercise 9.2 An electric dipole with moment $\mathbf{p}$ and fixed direction is located at the position $\mathbf{r}_0(t)$ with respect to the origin. Its velocity is $\mathbf{v} = d\mathbf{r}_0/dt$. Find the dipole's charge and current densities.

Solution 9.2 We start with definition of the charge density

$$\rho(\mathbf{r}) = \sum_{i=1}^{2} \frac{q_i}{d\mathbf{r}} \delta\,(\mathbf{r} - \mathbf{r}_i) \tag{9.118}$$

We denote by $\Delta\mathbf{r}$ the vector from positive to negative charge, then

$$\rho(\mathbf{r}) = \frac{q}{d\mathbf{r}}\delta\left(\mathbf{r} - \mathbf{r}_0 + \frac{\Delta\mathbf{r}}{2}\right) - \frac{q}{d\mathbf{r}}\delta\left(\mathbf{r} - \mathbf{r}_0 - \frac{\Delta\mathbf{r}}{2}\right) \tag{9.119}$$

$$= \frac{q}{d\mathbf{r}}\Delta\mathbf{r}\frac{1}{\Delta\mathbf{r}}\left(\delta\left(\mathbf{r} - \mathbf{r}_0 + \frac{\Delta\mathbf{r}}{2}\right) - \delta\left(\mathbf{r} - \mathbf{r}_0 - \frac{\Delta\mathbf{r}}{2}\right)\right)$$

$$= -\mathbf{p} \cdot \frac{1}{d\mathbf{r}}\delta'(\mathbf{r} - \mathbf{r}_0)$$

$$= -(\mathbf{p} \cdot \nabla)\delta(\mathbf{r} - \mathbf{r}_0)$$

The current density is

$$\mathbf{J} = \mathbf{v}\rho(\mathbf{r}) = -\mathbf{v}(\mathbf{p} \cdot \nabla)\delta(\mathbf{r} - \mathbf{r}_0) \tag{9.120}$$

Exercise 9.3 Express $\mathbf{A}$ and ϕ in terms of the magnetic field vector $\mathbf{B}$ and electric field vector $\mathbf{E}$ for static potentials and homogeneous fields.

Solution 9.3 We start with Eqs. (9.19) and (9.23), and since the potentials are static, then

$$\frac{\partial \mathbf{A}}{\partial t} = 0 \tag{9.121}$$

and hence

$$\mathbf{E} = -\nabla\phi = -\frac{\partial \phi(\mathbf{r})}{\partial \mathbf{r}} \tag{9.122}$$

Since the field is homogeneous, then

$$\phi(\mathbf{r}) = -\mathbf{E} \cdot \mathbf{r} \tag{9.123}$$

Using Eq. (9.19) and multiplying both sides by $\mathbf{r}$, we get

$$\mathbf{B} \times \mathbf{r} = \mathbf{r} \times (\mathbf{A} \times \nabla) = 2\mathbf{A} \tag{9.124}$$

Or

$$\mathbf{A} = \frac{1}{2}(\mathbf{B} \times \mathbf{r}) \tag{9.125}$$

Exercise 9.4 Show that $(\mathbf{A} \cdot \nabla)\mathbf{r} = \mathbf{A}$, where $\mathbf{A}$ is the vector potential.

Solution 9.4 We write

$$(\mathbf{A} \cdot \nabla)x = A_x \nabla_x x + A_y \nabla_y x + A_z \nabla_z x = A_x \tag{9.126}$$
$$(\mathbf{A} \cdot \nabla)y = A_x \nabla_x y + A_y \nabla_y y + A_z \nabla_z y = A_y \tag{9.127}$$
$$(\mathbf{A} \cdot \nabla)z = A_x \nabla_x z + A_y \nabla_y z + A_z \nabla_z z = A_z \tag{9.128}$$

Therefore, combining those results, we write that

$$(\mathbf{A} \cdot \nabla)\mathbf{r} = \mathbf{A} \tag{9.129}$$

Exercise 9.5 Inside the volume V the potential vector $\mathbf{A}$ satisfies the condition $\nabla \cdot \mathbf{A} = 0$, while at the boundary surface S of that volume $A_n = 0$. Show that

$$\int_V \mathbf{A} \, dV = 0 \tag{9.130}$$

Solution 9.5 Using Gauss's formula, we write

$$\int_V \mathbf{A} \, dV = \oint_S \nabla \cdot \mathbf{A} \, dS \tag{9.131}$$

where S is the surface enclosing the volume V. From the conditions, we obtain

$$\oint_S \nabla \cdot \mathbf{A} \, dS = 0 \tag{9.132}$$

Exercise 9.6 Find the charge density ρ if the electric field is $\mathbf{E} = (\mathbf{b} \cdot \mathbf{r})\mathbf{b}$, where $\mathbf{b}$ is a constant vector.

Solution 9.6 Using Maxwell's equation

$$\nabla \cdot \mathbf{E} = \frac{\rho}{\epsilon_0} \tag{9.133}$$

Using the definition of the divergence, we get

$$\nabla \cdot (\mathbf{b} \cdot \mathbf{r})\mathbf{b} = (\nabla(\mathbf{b} \cdot \mathbf{r})) \cdot \mathbf{b} + (\mathbf{b} \cdot \mathbf{r})(\nabla \cdot \mathbf{b}) = \left(b_x \mathbf{i} + b_y \mathbf{j} + b_z \mathbf{k}\right) \cdot \mathbf{b} = b^2 \tag{9.134}$$

because $\nabla \cdot \mathbf{b} = 0$. Therefore, we get

$$\rho = \epsilon_0 b^2 \tag{9.135}$$

Exercise 9.7 Find the current charge density $\mathbf{J}$ if the magnetic field is $\mathbf{B} = f(r)(\mathbf{a} \times \mathbf{r})$, where $\mathbf{a}$ is a constant vector and $f(r)$ is a scalar function and $\mathbf{a}$ is a constant vector.

Solution 9.7 Using Maxwell's equation

$$\nabla \times \mathbf{B} = \mu_0 \mathbf{J} \tag{9.136}$$

Consider the left-hand side:

$$
\begin{aligned}
\nabla \times \mathbf{B} &= \nabla \times (f(r)(\mathbf{a} \times \mathbf{r})) \tag{9.137}\\
&= \nabla f(r) \times (\mathbf{a} \times \mathbf{r}) + f(r)\nabla \times (\mathbf{a} \times \mathbf{r})\\
&= \frac{f'(r)}{r}\mathbf{r} \times (\mathbf{a} \times \mathbf{r}) + f(r)\nabla \times (\mathbf{a} \times \mathbf{r})\\
&= f(r)\nabla \times (\mathbf{a} \times \mathbf{r})\\
&= f(r)\left[\mathbf{a}(\nabla \cdot \mathbf{r}) - \mathbf{r}(\nabla \cdot \mathbf{a}) + (\mathbf{r} \cdot \nabla)\mathbf{a} - (\mathbf{a} \cdot \nabla)\mathbf{r}\right]\\
&= f(r)[3\mathbf{a} - \mathbf{a}] = 2f(r)\mathbf{a}
\end{aligned}
$$

where

$$\nabla f(r) = f'(r)\frac{\mathbf{r}}{r} \tag{9.138}$$

where $f'(r) = df(r)/dr$. Therefore, we get

$$\mathbf{J} = \frac{2}{\mu_0} f(r)\mathbf{a} \tag{9.139}$$

Exercise 9.8 Prove that if the current density is zero outside some finite volume and continuously differential everywhere, then

$$\int_V \mathbf{J} \, dV = 0 \tag{9.140}$$

Solution 9.8 Since the current density is differential function inside the volume, then using Gauss's formula

$$\int_V \mathbf{J} \, dV = \oint_A (\nabla \cdot \mathbf{J}) \, dA \tag{9.141}$$

where A is the surface enclosing the volume V. Using Maxwell's equation, we write

$$\oint_A (\nabla \cdot \mathbf{J}) \, dA = \frac{1}{\mu_0} \oint_A \nabla \cdot (\nabla \times \mathbf{B}) \, dA = 0 \tag{9.142}$$

because $\nabla \cdot (\nabla \times \mathbf{a}) = 0$ for any arbitrary vector $\mathbf{a}$.

Exercise 9.9 Suppose that $\mathbf{J} = (\mathbf{a} \times \nabla)\delta(\mathbf{r} - \mathbf{r}_0)$. Determine the magnetic moment.

Solution 9.9 Magnetic moment is

$$\mathbf{m} = \frac{1}{2} \int_V (\mathbf{r} \times \mathbf{J}) \, dV \tag{9.143}$$

Substituting the expression for $\mathbf{J}$, we get

$$\mathbf{m} = \frac{1}{2} \int_V (\mathbf{r} \times (\mathbf{a} \times \nabla) \delta(\mathbf{r} - \mathbf{r}_0)) \, dV \tag{9.144}$$

$$= \frac{1}{2} \mathbf{r}_0 \times (\mathbf{a} \times \nabla)$$

$$= \frac{1}{2} ((\mathbf{r}_0 \cdot \nabla)\mathbf{a} - (\mathbf{r}_0 \cdot \mathbf{a})\nabla)$$

$$= -\frac{1}{2} (\mathbf{r}_0 \cdot \mathbf{a})\nabla$$

where $\mathbf{a}$ is assumed to be a constant vector.

Exercise 9.10 Show that magnetic moment does not depend on the choice of the coordinate system origin.

Solution 9.10 Let $\mathbf{r}_0$ be a displacement of the origin of the coordinate system, then magnetic dipole to new origin is written as

$$\mathbf{m}' = \frac{1}{2} \int_V ((\mathbf{r} - \mathbf{r}_0) \times \mathbf{J}) \, dV \tag{9.145}$$

Simplifying that expression, we get

$$\mathbf{m} = \frac{1}{2} \int_V (\mathbf{r} \times \mathbf{J}) \, dV - \frac{1}{2} \int_V (\mathbf{r}_0 \times \mathbf{J}) \, dV \tag{9.146}$$

$$= \mathbf{m} - \frac{1}{2} \mathbf{r}_0 \times \int_V \mathbf{J} \, dV$$

$$= \mathbf{m} - \frac{1}{2} \mathbf{r}_0 \times \oint_A \nabla \cdot \mathbf{J} \, dA$$

$$= \mathbf{m}$$

because $\nabla \cdot \mathbf{J} = 0$ the current density continuity equation.

References

Griffiths DJ (1999) Introduction to electrodynamics, 3rd edn. Prentice Hall
Jackson JD (1999) Classical electrodynamics, 3rd edn. John Wiley and Sons
Landau LD, Lifshitz EM (1971) The classical theory of fields. Pergamon Press
Lorentz LV (1867) Phil Mag Ser 3(34):287
Sykja H (2006) Bazat e Elektrodinamikës. SHBUT

Chapter 10
More About Faraday's Law of Induction

This chapter aims to extend the law of induction and introduce alternating current circuits and the phenomena of resonance, useful in wireless applications of electromagnetic theory.

In this chapter, we introduce some more information about Faraday's law of induction. For further reading, one can also consider other available literature (Holliday et al., 2011).

10.1 Moving Conductor in a Closed Circuit

Consider a conductor moving in a magnetic field $\mathbf{B}$, which points toward the page, along two fixed conductors. The conductor is part of a closed circuit as shown in Fig. 10.1. The length of the stationary resistance (R) part is l. In Fig. 10.1, $\mathbf{F}_{app}$ denotes the applied force, which is perpendicular to the moving conductor part with a constant velocity $\mathbf{v}$. On the moving electrons and ions of the moving conductor is acting Lorenz force: $\mathbf{F}_q = q\,(\mathbf{v} \times \mathbf{B})$, where q is the charge. The directions of these two forces are depicted in Fig. 10.1. As a result, positive charges collect on the upper side of the moving conductor and electrons (negative charges) on the lower side, creating a difference of the electric potential $\Delta\Phi \equiv \epsilon_{ind} = \Phi_+ - \Phi_-$.

Furthermore, as the conductor moves to the right, the magnetic field flux Φ_B through the closed circuit path changes. When the moving conductor is at the position x, then the magnetic flux through the closed surface of the circuit is

$$\Phi_B = B \cdot (l \cdot x) = Blx \tag{10.1}$$

© The Author(s), under exclusive license to Springer Nature Switzerland AG 2022
H. Kamberaj, *Electromagnetism*, Undergraduate Texts in Physics,
https://doi.org/10.1007/978-3-030-96780-2_10

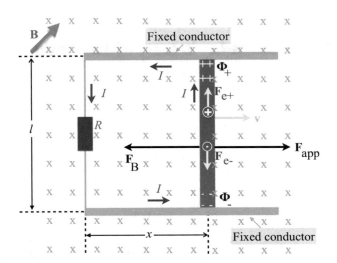

Fig. 10.1 A moving conductor in a magnetic field **B** pointing toward the page part of a closed conducting path. $\mathbf{F}_{app}$ is the applied force to the moving conductor gaining the velocity **v** and $\mathbf{F}_B$ is the magnetic force acting on the current I flowing in the moving conductor

Using Faraday's law, the induced electric potential is

$$\epsilon_{ind} = -\frac{d\Phi_B}{dt} = -Blv \tag{10.2}$$

where the speed is $v = dx/dt$, which is the speed of the moving conductor. The induced electric potential ϵ_{ind} creates a current flowing in the circuit, as shown in Fig. 10.1. Therefore, upon the moving conductor of length l is exerted a magnetic force $\mathbf{F}_B$ pointing to the left (see Fig. 10.1):

$$\mathbf{F}_B = I\,(\mathbf{L} \times \mathbf{B}) \tag{10.3}$$

where **L** is a vector of length l pointing along the direction of the current flow. It has a magnitude of

$$F_B = lIB \tag{10.4}$$

and direction is opposite to the direction of applied force, $\mathbf{F}_{app}$. Besides, since the conductor is moving with constant velocity, the resultant force must be equal to zero, based on Newton's first law; therefore, ignoring the mass of the conductor, we have

$$F_{app} = lIB \tag{10.5}$$

Also, since the resistance of the circuit is R, the voltage drop across the resistance equals the magnitude of the induced voltage, and hence

Fig. 10.2 Equivalent circuit

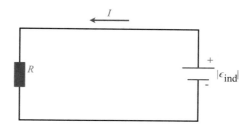

$$I = \frac{|\epsilon_{ind}|}{R} = \frac{Blv}{R} \tag{10.6}$$

The equivalent electric circuit is shown in Fig. 10.2. Furthermore, the work done by applied force for moving the conductor by a displacement **d** to the right is

$$W_{app} = \mathbf{F}_{app} \cdot \mathbf{d} \tag{10.7}$$

which equals the work done on the charges moving through the magnetic field. Note that the magnetic field force is moving the electrons downward with a certain average drift velocity that establishes the current I in the circuit (see also Fig. 10.2).

Applying the conservation law of the energy, the work done by the applied force on the moving conductor during the interval of time Δt is equal to the electric energy supplied by the induced electric potential ϵ_{ind} during the same time interval Δt. The power delivered by the applied field is

$$P = \frac{d W_{app}}{dt} = \mathbf{F}_{app} \cdot \mathbf{v} = (IlB)v = I(Blv) = I \mid \epsilon_{ind} \mid \tag{10.8}$$

where Eqs. (10.6) and (10.7) are used. Therefore, we also write the conservation law of energy as

$$P = \underbrace{\frac{|\epsilon_{ind}|^2}{R}}_{\text{Mechanical energy}} = \underbrace{I \mid \epsilon_{ind} \mid}_{\text{Electrical energy supplied by } \epsilon_{ind}} = \underbrace{I^2 R}_{\text{Internal energy in } R} \tag{10.9}$$

In Eq. (10.9), the first term represents the mechanical energy supplied by the applied force, the second term represents the electrical energy supplied by induced emf, and the last term is the internal energy in resistance.

10.1.1 Induced Electric Potential and Electric Field

As discussed above, the change in the magnetic flux induces electric potential in a conducting loop. Therefore, an electric field is created in the conductor as a result

Fig. 10.3 Induced electric field in a closed loop placed in a uniform magnetic field pointing toward the page

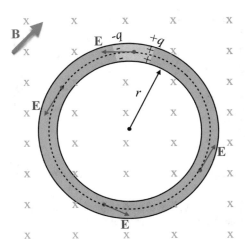

of the changing magnetic flux. The induced electric field is varying in time and it is non-conservative.

Consider a loop in a magnetic field perpendicular to the surface of the loop pointing toward the page. If the magnetic field varies with time, based on Faraday's law, an induced electric voltage creates in the loop (see also Fig. 10.3):

$$\epsilon_{ind} = -\frac{d\Phi_B}{dt} \tag{10.10}$$

where $\Phi_B = \pi r^2 B$ is the magnetic flux through the surface of the loop (r is the loop's radius) and $A = \pi r^2$ is the area of surface enclosed by the loop. Thus,

$$\epsilon_{ind} = -\pi r^2 \frac{dB}{dt} \tag{10.11}$$

Polarity of the induced potential difference is indicated in Fig. 10.3; therefore, the induced electric field is tangent to the loop, and it has the same magnitude since all the points on the loop are equivalent, as shown in Fig. 10.3.

The work done to move a test charge (q) around the loop equals to

$$W_q = q \cdot \epsilon_{ind} \tag{10.12}$$

The force acting on the charge q is

$$\mathbf{F} = q \cdot \mathbf{E} \tag{10.13}$$

The work done by electrical force to move the test charge around the loop is

$$W_q = \int_{\mathcal{L}} \mathbf{F}_q \cdot d\mathbf{l} = qE \int_{\mathcal{L}} dl = 2\pi rqE \tag{10.14}$$

where the integration is over the closed-loop line $\mathcal{L}$.
  Combing Eqs. (10.12) and (10.14), we obtain

$$q\epsilon_{ind} = 2\pi rqE \tag{10.15}$$

From Eq. (10.15), we find the magnitude of the electric field

$$E = \frac{\epsilon_{ind}}{2\pi r} \tag{10.16}$$

Substituting Eq. (10.11) into Eq. (10.16), we obtain

$$E = \frac{1}{2\pi r}\left(-\pi r^2 \frac{dB}{dt}\right) = -\frac{r}{2}\frac{dB}{dt} \tag{10.17}$$

  Equation 10.17 indicates that the induced electric field due to time-varying magnetic field is a function of time. The minus sign indicates that the electric field opposes the change in magnetic field.
  Furthermore, the induced electric voltage in any closed loop can be expressed as

$$\epsilon_{ind} = \oint_{\mathcal{L}} \mathbf{E} \cdot d\mathbf{l} = -\pi r^2 \frac{dB}{dt} \tag{10.18}$$

In general, the path may not be a closed loop and $\mathbf{E}$ may not be constant. However, Eqs. (10.10) and (10.18) indicate that induced electric force is non-conservative because its work along a closed path is not zero, or equivalently $\oint_{\mathcal{L}} \mathbf{E} \cdot d\mathbf{l} \neq 0$.

10.1.2 Generators and Motors

Figure 10.4 shows an alternative current (AC) generator, which can be used to convert mechanical energy to electrical energy. It comprises a loop of wire rotated using some external mechanical rotor in a magnetic field. When the loop rotates in the magnetic field, the magnetic flux passing through the area enclosed by the loop changes with time, which induces emf and a current in the loop, based on Faraday's law. If A is the area enclosed by the loop and θ the angle between the magnetic field vector $\mathbf{B}$ and surface area vector $\mathbf{A} = A\mathbf{n}$ (where $\mathbf{n}$ is a unit vector perpendicular to the loop's

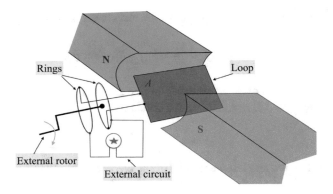

Fig. 10.4 A generator design

surface), and ω the angular speed of rotation of the loop (which is assumed to be constant). Then, the magnetic flux is

$$\Phi_{B_1} = BA \cos \theta \qquad (10.19)$$

where $\theta = \omega t$; therefore,

$$\Phi_{B_1} = BA \cos(\omega t) \qquad (10.20)$$

Assuming N turns, then

$$\Phi_B = N\Phi_{B_1} = NBA \cos(\omega t) \qquad (10.21)$$

Thus, the induced electric voltage is

$$\epsilon_{\text{ind}} = -\frac{d\Phi_B}{dt} \qquad (10.22)$$
$$= NBA\omega \sin(\omega t)$$

Equation 10.22 implies that maximum induced emf is

$$\epsilon_{\text{ind,max}} = NBA\omega \qquad (10.23)$$

Motors are devices that convert electrical energy into mechanical energy, that is, the reverse mode of a generator. In this case, a current supplied to the loop by a battery and the torque acting on the current-carrying loop causes it to rotate.

Fig. 10.5 A simple electric
circuit containing a switch
(S), source emf (ϵ), induced
emf (ϵ_L), and a resistance
(R)

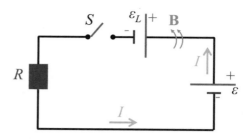

10.2 Inductance

10.2.1 Self-inductance

There is a difference between the emfs and the current produced by a battery or other sources and those induced by changing the magnetic field flux.

In general, the source emf and source current describe parameters associated with a physical source. In contrast, the induced emf and induced current describe parameters associated with changing magnetic field flux.

Consider the simple electric circuit shown in Fig. 10.5. It consists of a source emf, ϵ, a resistance, R, and a switch, S. When the switch is closed the source current does not instantly increase to its maximum value I_{max}:

$$I_{max} = \frac{\epsilon}{R} \tag{10.24}$$

At some instance of time t the current in the circuit is $I(t)$. Besides, the current passing through straight wire line produces a magnetic field $B = \mu_0 I/2\pi r$ (where r is the distance from the wire), and thus $B \sim I$. Since the current increases to reach its maximum value I_{max}, so does the magnetic field. Furthermore, the magnetic flux passing through the surface area enclosed by the circuit is

$$\Phi_B = \int_S \mathbf{B} \cdot d\mathbf{S} \tag{10.25}$$

where S is the surface area enclosed by the circuit. Therefore, since B increases, Φ_B increases with time, that is, $d\Phi_B/dt \neq 0$, which in turn creates an induced emf in the circuit:

$$\epsilon_L = -\frac{d\Phi_B}{dt} \tag{10.26}$$

Based on Lenz's law, the polarization of induced emf ϵ_L is such that it would create an induced current in the circuit, such that this induced current would create a magnetic field that would oppose the change of the magnetic field created by the source current.

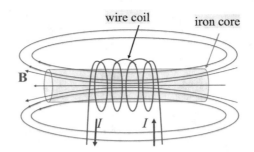

Fig. 10.6 An iron core and wire coil in which is passing the current I

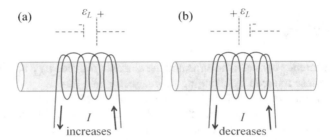

Fig. 10.7 **a** The induced emf polarity when the current I increases; **b** The induced emf polarity when the current I decreases

Therefore, the direction of the induced emf, ϵ_L, is opposite to the direction of source emf, ϵ. As a result, a gradual increase of the current source occurs to its maximum value I_{max} (equilibrium value) rather than an instantaneous. This effect is called self-induction, and ϵ_L is called self-induced emf or back emf.

Consider the solenoid in Fig. 10.6, which consists of the cylindrical iron coil and the wire coil of N turns. A current I flows along the wire coil. Magnetic field direction **B** to the left is created due to the current in the wire coil.

Increasing the current yields an increase of the magnetic flux passing through the coil, and hence $d\Phi_B/dt \neq 0$ and $-d\Phi_B/dt < 0$. Then, the induced emf has a polarity as shown in Fig. 10.7a. In contrast, with decreasing the current I, the magnetic flux through the coil decreases, and $d\Phi_B/dt \neq 0$, but $-d\Phi_B/dt > 0$. Therefore, the induced emf has a polarity as shown in Fig. 10.7b.

Polarization of ϵ_L is defined based on Lenz's law; the polarity of the induced emf ϵ_L opposes the change of the magnetic field from the source current. Using Faraday's law, the induced emf is given by Eq. (10.26), with Φ_B given by Eq. (10.25). Therefore, $\Phi_B \sim |\mathbf{B}|$ and $|\mathbf{B}| \sim I$. Here, I is the source current in the circuit. Therefore, the self-induced emf, ϵ_L, is proportional to the rate change of the source current, $\epsilon_L \sim dI/dt$.

Suppose there is a closed spaced coil of N turns (for example, either a solenoid or a toroid) carrying a source current I. Then, we write

$$\epsilon_L = -N\frac{d\Phi_B}{dt} = -L\frac{dI}{dt} \tag{10.27}$$

where L is a proportionality constant, called the inductance of the coil. In general, L depends on the geometry of the coil and other physical parameters. Equation 10.27 can also be arranged as follows:

$$\epsilon_L = -\frac{d(N\Phi_B)}{dt} = -\frac{d(LI)}{dt} \tag{10.28}$$

Thus, we can find that

$$N\Phi_B = LI \tag{10.29}$$

Or,

$$L = \frac{N\Phi_B}{I} \tag{10.30}$$

Furthermore, using (10.27), we can write that

$$L = -\frac{\epsilon_L}{\left(\dfrac{dI}{dt}\right)} \tag{10.31}$$

Equation 10.31 is in analogy with Ohm's law: $R = \epsilon/I$. Note that L is a measure of the opposition to the change in the circuit source current. The SI units of L are *Henry* (H):

$$1\,\mathrm{H} = 1\,\frac{\mathrm{V}}{\mathrm{A/s}} = 1\,\frac{\mathrm{V \cdot s}}{\mathrm{A}} \tag{10.32}$$

10.2.2 *Mutual Inductance*

The magnetic flux passing through the surface area of a loop is (see also Fig. 10.8):

$$\Phi_B = \int_S \mathbf{B} \cdot d\mathbf{S} \tag{10.33}$$

where $|\mathbf{B}| \sim I$. Therefore, Φ_B varies with time because I varies with time. Thus, an induced emf occurs through the process of mutual inductance. This is related with the fact that it depends on the interaction between two circuits.

Consider two parallel coils of N_1 and N_2 turns, respectively, as shown in Fig. 10.9. Through the coil I is passing the current I_1 and coil II the current I_2. Suppose the current I_1 is creating a magnetic field with magnetic field lines as depicted in Fig. 10.9. Some of these lines pass through the coil II. We denote by Φ_{12} the magnetic flux of the magnetic field created by coil I through the coil II. The mutual inductance, namely, M_{12}, of coil II with respect to coil I is

Fig. 10.8 The magnetic field created by a current flowing in a loop

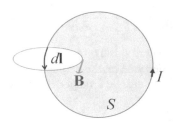

Fig. 10.9 The magnetic field created by two parallel coils of N_1 and N_2 turns, respectively, in which is passing the current I_1 and I_2

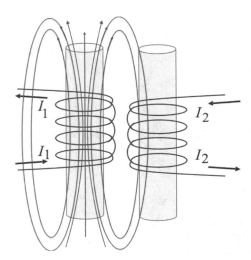

$$M_{12} = N_2 \frac{\Phi_{12}}{I_1} \tag{10.34}$$

Assuming that the current I_1 is varying with time, then an induced emf is created at coil II, given as

$$\epsilon_{ind,2} = -N_2 \frac{d\Phi_{12}}{dt} \tag{10.35}$$

where Φ_{12} is calculated from Eq. (10.34) as

$$\Phi_{12} = I_1 \frac{M_{12}}{N_2} \tag{10.36}$$

Therefore, substituting Eq. (10.36) into Eq. (10.35), we get

$$\epsilon_{ind,2} = -N_2 \frac{d}{dt} \left(I_1 \frac{M_{12}}{N_2} \right) \tag{10.37}$$

$$= -M_{12} \frac{dI_1}{dt}$$

Similarly, if I_2 is the current passing through the coil II, which varies on time, then there exists a mutual inductance M_{21} such that

$$M_{21} = N_1 \frac{\Phi_{21}}{I_2} \tag{10.38}$$

In Eq. (10.38), Φ_{21} is the magnetic flux through coil I due to magnetic field created by coil II. The induced emf at coil I is

$$\epsilon_{ind,1} = -N_1 \frac{d\Phi_{21}}{dt} \tag{10.39}$$

Substituting Φ_{21} from Eq. (10.38) into Eq. (10.39), we obtain

$$\epsilon_{ind,1} = -M_{21} \frac{dI_2}{dt} \tag{10.40}$$

Assuming that coil I and coil II are identical, then $M_{12} = M_{21} \equiv M$ because they depend only on the coil's geometry. Therefore, we obtain the following:

$$\epsilon_{ind,1} = -M \frac{dI_2}{dt}, \tag{10.41}$$

$$\epsilon_{ind,2} = -M \frac{dI_1}{dt}$$

A circuit containing a large self-inductance is called an *inductor*. Based on Lenz's law, the self-inductance prevents the current in the circuit from increasing or decreasing instantaneously.

10.3 Oscillations in an LC Circuit

Consider a capacitor connected to an inductor, as shown in Fig. 10.10. That is also called an *LC circuit*. Assuming initially that the capacitor is fully charged (that it, its maximum charge is Q_{max}), then the switch is closed. The current and the charge will oscillate between its maximum positive and negative values. There is no resistance in the circuit (that is, $R = 0$); therefore, there is no internal energy losses. Besides, we assume that there is no energy radiated away. Q_{max} denotes the initial charge in capacitor. Furthermore, the switch is closed at $t = 0$. Therefore, the energy of capacitor at $t = 0$ is

$$U_C = \frac{Q_{max}^2}{2C} \tag{10.42}$$

Moreover, at $t = 0$, the current in the circuit is zero, $I = 0$, and hence there is no energy stored in the inductor. When the switch closes, the rate of charge leaving

Fig. 10.10 The LC circuit

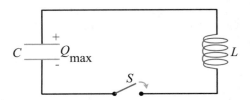

capacitor equals the current in the circuit. As the capacitor gets discharged, its electric field gets decreased. During this process the current in the circuit increases, and thus the magnetic field of the inductor increases (note that $B \sim I$). As a result of that, the energy transfers from electric energy of the capacitor into magnetic field energy of the inductor.

When capacitor is fully discharged, the energy in the capacitor becomes zero, the current in the circuit reaches its maximum value, and hence all the energy is stored in the inductor. Then, the current continues going to capacitor, and thus increasing its charge and the current decreases. At the end, the capacitor gets back fully charged; however, its polarity is opposite. This process repeats by transferring electrical energy of the capacitor into magnetic field energy of inductor, and vice versa.

As analogy, one can consider the mechanical oscillations of a mass m around its equilibrium position $x = 0$. The oscillations of charge in the capacitor of an LC circuit are analogue of the oscillations in the displacement x of the mass m from its equilibrium position, and oscillations in the current value of an LC circuit are analogue of the oscillations in the velocity of the mass m.

Now, we consider an intermediate step at time t, when the charge in the capacitor is Q ($Q < Q_{max}$), and thus the electrical energy of capacitor is

$$U_C = \frac{Q^2}{2C} \tag{10.43}$$

At the same time t, the current in the circuit is I, and the magnetic field energy stored in the inductor is

$$U_L = \frac{LI^2}{2} \tag{10.44}$$

Then, the total energy in the LC circuit at the time t is the sum of the two terms given by Eqs. (10.43) and (10.44), given by the expression:

$$U = U_C + U_L = \frac{Q^2}{2C} + \frac{LI^2}{2} \tag{10.45}$$

Since the resistance is zero in the circuit, there is no energy transformation into internal energy (which means no losses of the initial energy given by Eq. (10.42)). Therefore, U is constant at any time t, and equal to its value at $t = 0$:

$$U = \frac{Q_{max}^2}{2C} \qquad (10.46)$$

Furthermore, $dU/dt = 0$:

$$0 = 2\frac{Q}{2C}\frac{dQ}{dt} + 2\frac{LI}{2}\frac{dI}{dt} \qquad (10.47)$$

Alternatively, using the definition of the current $I = dQ/dt$, we obtain

$$\frac{d^2Q}{dt^2} + \frac{1}{LC}Q = 0 \qquad (10.48)$$

where the relation $dI/dt = d^2Q/dt^2$ is used. Denoting by

$$\omega^2 = \frac{1}{LC} \qquad (10.49)$$

we obtain the following second-order differential equation:

$$\frac{d^2Q}{dt^2} + \omega^2 Q = 0 \qquad (10.50)$$

The solution of Eq. (10.50) is given as

$$Q(t) = Q_{max} \cos(\omega t + \phi) \qquad (10.51)$$

where ω represents the angular frequency that depends on L and C (see Eq. (10.49)):

$$\omega = \sqrt{\frac{1}{LC}} \qquad (10.52)$$

which is also called *natural frequency*. In Eq. (10.51), ϕ is the phase angle, determined by the initial conditions, at $t = 0$. For instance, at $t = 0$, we have $Q(t = 0) = Q_{max}$, and thus from Eq. (10.51):

$$Q_{max} = Q_{max} \cos(\phi) \qquad (10.53)$$

Thus, using Eq. (10.53), we find that $\phi = 0$, and Eq. (10.51) can be written as

$$Q(t) = Q_{max} \cos(\omega t) \qquad (10.54)$$

The current I flowing in the LC circuit is determined as

$$I = \frac{dQ(t)}{dt} = -\omega Q_{max} \sin(\omega t) \equiv -I_{max} \sin(\omega t) \qquad (10.55)$$

where $I_{max} = \omega Q_{max}$.

Fig. 10.11 The oscillations of charge and current in the LC circuit. In the numerical calculations, $\omega = \pi$ and $Q_{max} = 1$

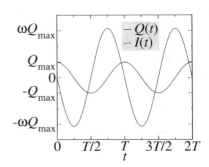

The period of oscillations is

$$T = \frac{2\pi}{\omega} \tag{10.56}$$

Thus, we can also write

$$\omega t = \frac{2\pi}{T} t \tag{10.57}$$

In Fig. 10.11, we show the oscillations of charge and current in the LC circuit. In the numerical calculations, $\omega = \pi$ and $Q_{max} = 1$, used as illustration.

Using Eq. (10.45), the total energy of the system is

$$U = \underbrace{\frac{1}{2C} Q_{max}^2 \cos^2(\omega t)}_{\text{electric energy}} + \underbrace{\frac{L}{2} I_{max}^2 \sin^2(\omega t)}_{\text{magnetic energy}} \tag{10.58}$$

As it can be seen, Eq. (10.58) indicates that the energy in the LC circuit continuously oscillates between the electric energy stored in capacitor and magnetic field energy stored in inductor, as shown in Fig. 10.12. Because there is no loss of energy in resistance, then

$$\frac{1}{2C} Q_{max}^2 = \frac{L}{2} I_{max}^2 \tag{10.59}$$

Therefore, Eq. (10.58) can also be written as

$$U = \frac{Q_{max}^2}{2C} \left(\cos^2(\omega t) + \sin^2(\omega t) \right) = \frac{Q_{max}^2}{2C} \tag{10.60}$$

which is constant.

Fig. 10.12 The oscillations of the electrical energy stored in capacitor and magnetic field energy stored in inductor for the LC circuit. In the numerical calculations: $\omega = \pi$, $C = 1$, and $Q_{max} = 1$

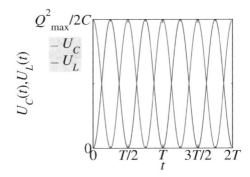

10.4 The RL Circuit

Consider a circuit consisting of the resistance R and the inductance L, as shown in Fig. 10.13. This is also called an *RL circuit*. The switch S is closed as $t = 0$. Because of the inductance of the inductor a back emf is produced. Furthermore, an inductor in a circuit opposes the change in the current passing through the circuit. The source current I starts increasing, and back emf that opposes the increase of the current is induced in the inductor:

$$\epsilon_L = -L\frac{dI}{dt} \tag{10.61}$$

Since the current I is increasing, $dI/dt > 0$, and thus $\epsilon_L < 0$. Therefore, the polarity of ϵ_L is opposite of source emf ϵ (see also Fig. 10.13).

Using Kirchhoff's law, we write

$$\epsilon + V_R + \epsilon_L = 0 \tag{10.62}$$

where $V_R = -IR$ ($\mid V_R \mid = IR$) is the voltage drop in the resistance. Combining Eqs. (10.61) and (10.62), we get

Fig. 10.13 The RL circuit

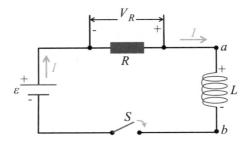

$$\epsilon - IR - L\frac{dI}{dt} = 0 \tag{10.63}$$

Rearranging Eq. (10.63), we can write that

$$L\frac{dI}{dt} + IR = \epsilon \tag{10.64}$$

For finding the general solution of Eq. (10.64) concerning I, we first solve the following homogeneous equation:

$$L\frac{dI}{dt} + IR = 0 \tag{10.65}$$

which can be arranged as

$$\frac{dI}{I} = -\frac{R}{L}dt \tag{10.66}$$

Integrating both sides of Eq. (10.66), we obtain

$$\int_{I_0}^{I} \frac{dI}{I} = -\int_{0}^{t} \frac{R}{L}dt \tag{10.67}$$

Or,

$$\ln\left(\frac{I}{I_0}\right) = -\frac{R}{L}t \tag{10.68}$$

which can be solved for I to give

$$I = I_0 \exp\left(-t/\tau\right) \tag{10.69}$$

where

$$\tau = \frac{L}{R} \tag{10.70}$$

is a time constant. Assuming that the integration constant I_0 is a function of time t, then substituting Eq. (10.69) into Eq. (10.64), we find

$$L\left(\frac{dI_0}{dt}\exp\left(-t/\tau\right) - \frac{I_0}{\tau}\exp\left(-t/\tau\right)\right) + RI_0\exp\left(-t/\tau\right) = \epsilon \tag{10.71}$$

which can be simplified as

$$\frac{dI_0}{dt} = \frac{\epsilon}{L}\exp\left(t/\tau\right) \tag{10.72}$$

Solving Eq. (10.72) for I_0, we find

$$I_0 = \frac{\epsilon}{L} \int \exp(t/\tau)\, dt = \frac{\epsilon}{R} \exp(t/\tau) + C \tag{10.73}$$

where C is an integration constant determined by the initial conditions at $t = 0$ and Eq. (10.70) is used. First, we substitute Eq. (10.73) into Eq. (10.69):

$$I(t) = \frac{\epsilon}{R} + C \exp(-t/\tau) \tag{10.74}$$

Using the initial conditions that $I(t = 0) = 0$, from Eq. (10.74), we get

$$0 = I(0) = \frac{\epsilon}{R} + C \exp(-0/\tau) \tag{10.75}$$

Or,

$$C = -\frac{\epsilon}{R} \tag{10.76}$$

Substituting the expression for C (Eq. (10.76)) into the expression for I (Eq. (10.74)), we obtain the current as

$$I(t) = \frac{\epsilon}{R}(1 - \exp(-t/\tau)) \tag{10.77}$$

In Eq. (10.77), τ is the time that takes for the current in the circuit to become

$$I(\tau) = \frac{\epsilon}{R}(1 - \exp(-1)) \approx 0.63\frac{\epsilon}{R} \tag{10.78}$$

In Fig. 10.14, we show $I(t)$ as a function of time for the RL circuit. The induced emf is given by Eq. (10.61), where the rate change of the current is

$$\frac{dI}{dt} = \frac{\epsilon}{R}\left(\frac{1}{\tau}\exp(-t/\tau)\right) = \frac{\epsilon}{L}\exp(-t/\tau) \tag{10.79}$$

Therefore, we find that

$$\epsilon_L = -\epsilon \exp(-t/\tau) \tag{10.80}$$

Graphically ϵ_L versus time t is shown in Fig. 10.14. It can be seen that as t increases, the induced ϵ_L approaches zero, which corresponds with the time when the current in the RL circuit reaches the equilibrium maximum value, $I_{max} = \epsilon/R$.

Based on Lenz's law, the induced emf in an inductor prevents the source emf (that is, battery) from establishing an instantaneous current. In that case, the source emf has to do work against the inductor to create the current. Therefore, part of the energy supplied by the source emf appears as internal energy in the resistor:

$$U_{int} = I^2 Rt \tag{10.81}$$

Fig. 10.14 The current and
the induced emf as a function
of time for the RL circuit

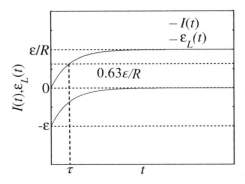

where t is the time during which the current flows in the resistance. The other part
of energy is stored as the magnetic field energy of the inductor:

$$U_L = I \mid \epsilon_L \mid t \tag{10.82}$$

The rate of the total energy is then written as

$$\frac{dU_{\text{tot}}}{dt} = \frac{dU_{\text{int}}}{dt} + \frac{dU_L}{dt} \tag{10.83}$$

Therefore, we find

$$I\epsilon = I^2 R + IL\frac{dI}{dt} \tag{10.84}$$

Here, U_L, which represents the energy stored in the inductor at any instance of time
t, is such that its rate is given as

$$\frac{dU_L}{dt} = IL\frac{dI}{dt} \tag{10.85}$$

Or, we can write that

$$dU_L = ILdI \tag{10.86}$$

Integrating both sides of Eq. (10.86), we obtain

$$U_L = \int_0^{U_L} dU_L = \int_0^I ILdI = \frac{LI^2}{2} \tag{10.87}$$

which represents the energy stored in the inductor in the form of the magnetic field
when the current is I.

Consider a solenoid of N turns, with self-inductance given as $L = N\Phi_B/I$, which can be written as

$$L = \mu_0 n^2 V = \mu_0 n^2 A\ell \tag{10.88}$$

where A is the cross-sectional area of the solenoid and ℓ its length, and V the volume of the solenoid. Here, Φ_B is the magnetic field flux through the solenoid given as

$$\Phi_B = B \cdot A \tag{10.89}$$
$$= \frac{\mu_0 N I}{\ell} A$$
$$= \mu_0 n I A$$

The magnetic field of the solenoid is

$$B = \frac{\mu_0 N I}{\ell} = \mu_0 n I \tag{10.90}$$

and thus

$$I = \frac{B}{\mu_0 n} \tag{10.91}$$

Then, the energy stored in magnetic field is

$$U_B = \frac{L I^2}{2} \tag{10.92}$$
$$= (\mu_0 n^2 A\ell) \frac{I^2}{2}$$
$$= (\mu_0 n^2 A\ell) \frac{B^2}{2\mu_0^2 n^2}$$
$$= \frac{B^2}{2\mu_0} A\ell = \frac{B^2}{2\mu_0} V$$

Furthermore, the magnetic field energy density is

$$u_B = \frac{U_B}{V} = \frac{B^2}{2\mu_0} \tag{10.93}$$

10.5   The RLC Circuit

Now, we consider the RLC circuit, which contains a resistance (R), inductor (L), and a capacitor (C), as shown in Fig. 10.15. Also, we assume that the capacitor is initially fully charged, that is, its charge at $t = 0$ is Q_{max}, which corresponds to the state before the switch is closed. When the switch is closed, a current is established in the circuit. The total energy stored in capacitor and inductor is given as

$$U_{LC} = U_L + U_C = \frac{LI^2}{2} + \frac{Q^2}{2C} \tag{10.94}$$

However, U_{LC}, given by Eq. (10.94), is not constant because of the losses of the energy as internal energy in the resistance R. The rate of the energy transformation into internal energy within the resistor is given as

$$\frac{dU_R}{dt} = -I^2 R \tag{10.95}$$

where minus sign indicates that is an energy lost, and hence the rate decreases with time t (explaining the sign minus in front of I^2R in Eq. (10.95)). Therefore, we can write

$$\frac{dU_{LC}}{dt} = \frac{dU_R}{dt} \tag{10.96}$$

Using Eqs. (10.94) and (10.95), we obtain

$$LI\frac{dI}{dt} + \frac{Q}{C}\frac{dQ}{dt} = -I^2 R \tag{10.97}$$

where Q is the charge in the capacitor at the time t and I is the current in the circuit $I = dQ/dt$. The expression in Eq. (10.97) can further be written as

$$L\frac{d^2Q}{dt^2} + R\frac{dQ}{dt} + \frac{Q}{C} = 0 \tag{10.98}$$

which is a non-homogeneous second-order differential equation for the variable Q.

Fig. 10.15 The RLC circuit

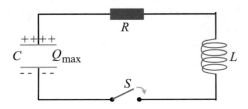

For solving Eq. (10.98), we assume the general form of the solution as follows:

$$Q(t) = \exp(\rho t) \tag{10.99}$$

where ρ is a parameter to be defined. First, we substitute the expression for Q (given by Eq. (10.99)) into Eq. (10.98), and obtain

$$L\rho^2 e^{\rho t} + R\rho e^{\rho t} + \frac{1}{C} e^{\rho t} = 0 \tag{10.100}$$

or

$$\rho^2 + \frac{R}{L}\rho + \frac{1}{LC} = 0 \tag{10.101}$$

which is a quadratic equation for ρ. The roots of Eq. (10.101) are given as

$$\rho_{1,2} = -\frac{R}{2L} \pm \sqrt{\frac{R^2}{4L^2} - \frac{1}{LC}} \tag{10.102}$$

$$= -\frac{R}{2L} \pm \frac{1}{\sqrt{LC}}\sqrt{\frac{R^2 C}{4L} - 1}$$

In Eq. (10.102), we do the following substitutions:

$$\omega_0 = \frac{1}{\sqrt{LC}} \tag{10.103}$$

and

$$\xi = \sqrt{\frac{R^2 C}{4L}} \tag{10.104}$$

The first term in Eq. (10.102) is rewritten as

$$\frac{R}{2L} = \sqrt{\frac{R^2 C}{4L}} \cdot \frac{1}{\sqrt{LC}} \equiv \xi \omega_0 \tag{10.105}$$

Then, Eq. (10.102) takes the form

$$\rho_{1,2} = -\xi \omega_0 \pm \omega_0 \sqrt{\xi^2 - 1} \tag{10.106}$$

where ω_0 is the angular frequency of the free oscillations in the LC circuit. In analogy with classical mechanics, the RLC circuit corresponds to the dumping motion of the oscillator in presence of the friction force.

10.5.1   Case 1

First, we consider the case for $R = 0$, and so $\xi = 0$. That is, the LC circuit, and hence free oscillations establish where the energy is transformed between the electrical energy stored in the capacitor and magnetic field energy stored in the inductor. From Eq. (10.106), we have

$$\rho_{1,2} = \pm i\omega_0 \tag{10.107}$$

where $i = \sqrt{-1}$. Therefore, the charge at time t is

$$Q(t) = A \exp(i\omega_0 t) + B \exp(-i\omega_0 t) \tag{10.108}$$

Using the initial conditions that at $t = 0$, $Q(0) = Q_{max}$, and so $A + B = Q_{max}$. Then, Eq. (10.108) simplifies as

$$\begin{aligned} Q(t) &= A \exp(i\omega_0 t) + Q_{max} \exp(-i\omega_0 t) - A \exp(-i\omega_0 t) \\ &= (2A - Q_{max})i \sin(\omega_0 t) + Q_{max} \cos(\omega_0 t) \end{aligned} \tag{10.109}$$

where the relation $\exp(\pm i\phi) = \cos\phi \pm i \sin\phi$ is used.

Moreover, at some finite time $t = 3T/4$ (where $T = 2\pi/\omega_0$ is the period), the charge $Q(3T/4) = 0$ (see also Fig. 10.11); therefore, the first term must have zero amplitude, $2A - Q_{max} = 0$ (because $\cos(\omega_0 3T/4) = 0$). Then, we obtain

$$Q(t) = Q_{max} \cos(\omega_0 t) \tag{10.110}$$

which is the solution obtained for LC circuit.

10.5.2   Case 2

Consider the case when $| \xi | < 1$, hence $\xi^2 - 1 < 0$. Therefore,

$$\rho_{1,2} = -\xi\omega_0 \pm \omega_0 i \sqrt{1 - \xi^2} \tag{10.111}$$

Substituting expression for ρ from Eq. (10.111) into Eq. (10.99), we obtain

$$\begin{aligned} Q(t) &= A \exp(-\xi\omega_0 t) \exp\left(i\omega_0\sqrt{1 - \xi^2}\,t\right) \\ &\quad + B \exp(-\xi\omega_0 t) \exp\left(-i\omega_0\sqrt{1 - \xi^2}\,t\right) \\ &= \exp(-\xi\omega_0 t)\left[A \exp\left(i\omega_0\sqrt{1 - \xi^2}\,t\right) + B \exp\left(-i\omega_0\sqrt{1 - \xi^2}\,t\right)\right] \\ &= \exp(-\xi\omega_0 t)\left[a \cos\left(\omega_0\sqrt{1 - \xi^2}\,t\right) + b \sin\left(\omega_0\sqrt{1 - \xi^2}\,t\right)\right] \end{aligned} \tag{10.112}$$

Fig. 10.16 The dumping
oscillations in a RLC circuit.
Here, $\omega_0 = \pi$, $\xi = 0.05$,
$A = 1$, and $\alpha \equiv \sqrt{1 - \xi^2}$

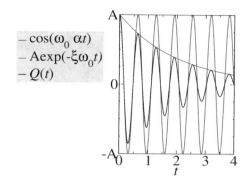

It can be seen that the amplitude is not constant, but it varies with time according to
$\exp(-\xi\omega_0 t)$, which decays to zero for $t \to \infty$. That factor is also called the *dumping exponential factor*. The dumping oscillations in the RLC circuit are illustrated
in Fig. 10.16, where we can distinguish the exponential decay of the amplitude,
oscillations of the cosine factor, and the dumping oscillations of $Q(t)$.

10.5.3 Case 3

The third case is for $| \xi | > 1$. Then, the roots are

$$\rho_{1,2} = -\xi\omega_0 \pm \omega_0\sqrt{\xi^2 - 1} \tag{10.113}$$

and the charge in the capacitor at any time t is

$$Q(t) = A \exp(-\rho_1 t) + B \exp(+\rho_2 t) \tag{10.114}$$

10.6 Alternating Current Circuits

10.6.1 AC Sources and Phases

In general, the AC circuit consists of circuit elements and a generator (see also
Fig. 10.4). It provides an alternative current (AC). As discussed previously, the principle of a generator is based on Faraday's law of induction. From Fig. 10.4, a conducting loop rotates at constant speed ω, then a sinusoidal voltage (or emf) is induced
in the loop with an instantaneous value of

$$\Delta V = V_0 \sin(\omega t) \tag{10.115}$$

In Eq. (10.115), V_0 is the maximum voltage output of the AC generator (or alternating voltage amplitude). Furthermore, the relationship between angular frequency ω and linear frequency f (or period $T = 1/f$) is

$$\omega = 2\pi f = \frac{2\pi}{T} \qquad (10.116)$$

The analysis of the circuit is performed using the so-called *phasor diagram*.

10.6.2   Resistors in an AC Circuit

An AC generator has a symbol in the circuit as shown in Fig. 10.17a.

Now, we consider a circuit consisting of a resistor R connected to a generator with a sinusoidal voltage as given in Eq. (10.115) (see also Fig. 10.17b). The voltage drop in the resistance ΔV_R equals the difference of the electric potential between the points a and b:

$$\Delta V_R = V_{ab} \qquad (10.117)$$

Using Kirchhoff's law, the algebraic sum of the voltages around a closed loop in a circuit is equal to zero, and hence we write

$$\Delta V - \Delta V_R = 0 \qquad (10.118)$$

Combining Eqs. (10.115) and (10.118), we have

$$\Delta V = \Delta V_R = V_0 \sin(\omega t) \qquad (10.119)$$

Fig. 10.17 a AC generator symbol and **b** the circuit consisting of a resistor and an AC generator of voltage $\Delta V = V_0 \sin(\omega t)$

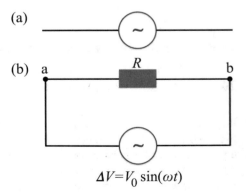

Fig. 10.18 The plots of the instantaneous current i_R and voltage ΔV_R for $R = 2\,\Omega$, $V_0 = 1$ V, and $\omega = \pi$ rad/s

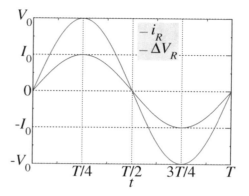

Therefore, instantaneous current in the resistor is

$$i_R = \frac{\Delta V_R}{R} = \frac{V_0}{R} \sin(\omega t) \tag{10.120}$$

$$= I_0 \sin(\omega t)$$

where I_0 is the maximum value of current (or the amplitude):

$$I_0 = \frac{V_0}{R} \tag{10.121}$$

Then, we can write the instantaneous voltage across the resistor as

$$\Delta V_R = R I_0 \sin(\omega t) \tag{10.122}$$

The plots of the instantaneous current i_R and voltage ΔV_R are shown in Fig. 10.18, for $R = 2\,\Omega$, $V_0 = 1$ V, and $\omega = \pi$ rad/s.

The phasor diagram can be used to represent current-voltage phase relationship. For the circuit in Fig. 10.17, the phasor diagram is shown in Fig. 10.19. It can be seen that

$$I_{av}^{(R)} = \frac{1}{T} \int_0^T i_R(t)\,dt \tag{10.123}$$

$$= \frac{1}{T} \int_0^T I_0 \sin(\omega t)\,dt$$

$$= \frac{1}{T} \frac{I_0}{\omega} (\cos(0) - \cos(\omega T))$$

$$= \frac{I_0}{\omega T} \cdot 0 = 0$$

Fig. 10.19 The phasor
diagram

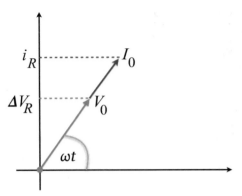

Thus, the average value of the current during one period T is equal to zero. That is, the current for a period of time $T/2$ maintains one sign (e.g., negative) and for the same period of time $T/2$ the opposite sign (e.g., positive). Hence, the sum is zero.

The direction of the current has no effect on the behavior of the resistance. That is, as the current flows through the resistor, electrons will collide with the fixed atoms of the material of the resistor, and if the current changes the direction, the effect remains the same. That is, the temperature of the resistor will increase due to these collisions.

Quantitatively, the rate at which the electrical energy is converted into internal energy is

$$\mathcal{P} = i_R^2 R \tag{10.124}$$

where $\mathcal{P}$ is the power, i_R is the instantaneous current in the resistor, and R is the resistance. Since $\mathcal{P}$ is proportional to i_R^2, then $\mathcal{P}$ does not depend on the sign of i_R. However, the temperature increase of the AC current having a maximum value of I_0 is not the same that produced by a direct current equal to I_0. Let us consider an instantaneous current as in Eq. (10.112). The average power $\mathcal{P}_{\text{av}}^{(AC)}$ over a period T is

$$\mathcal{P}_{\text{av}}^{(AC)} = \frac{1}{T} \int_0^T \mathcal{P}(t)dt \tag{10.125}$$

$$= \frac{1}{T} \int_0^T i_R^2(t) R\, dt$$

$$= \frac{R}{T} \int_0^T I_0^2 \sin^2(\omega t)dt$$

$$= \frac{R}{T} I_0^2 \int_0^T \frac{1 - \cos(2\omega t)}{2} dt$$

$$= \frac{R}{T} I_0^2 \frac{T}{2}$$

$$= \frac{I_0^2 R}{2}$$

$$= I_{rms}^2 R$$

where I_{rms} is the so-called the *root mean square* current:

$$I_{rms} = \sqrt{\frac{1}{T} \int_0^T i_R^2(t) dt} \qquad (10.126)$$

which can be calculated as

$$I_{rms}^2 = \frac{1}{T} \int_0^T i_R^2(t) dt \qquad (10.127)$$

$$= \frac{1}{T} \int_0^T I_0^2 \sin^2(\omega t) \, dt$$

$$= \frac{I_0^2}{T} \int_0^T \frac{1 - \cos(2\omega t)}{2} dt$$

$$= \frac{I_0^2}{2}$$

Or,

$$I_{rms} = \frac{I_0}{\sqrt{2}} \approx 0.707 I_0 \qquad (10.128)$$

On the other hand, for a direct current (DC) with $I = I_0$:

$$\mathcal{P}_{av}^{(DC)} = \mathcal{P} = I_0^2 R = 2 \mathcal{P}_{av}^{(AC)} \qquad (10.129)$$

which implies that the average power of a DC is two times larger of an AC working at the same maximum current, I_0.

Note that the root mean square voltage, V_{rms}, is

$$V_{rms} = \sqrt{\frac{1}{T} \int_0^T \Delta V_R^2(t) dt} \qquad (10.130)$$

which can be evaluated as follows:

$$V_{rms}^2 = \frac{1}{T} \int_0^T \Delta V_R^2(t) dt \qquad (10.131)$$

$$= \frac{1}{T} \int_0^T V_0^2 \sin^2(\omega t) dt$$

$$= \frac{V_0^2}{T} \int_0^T \frac{1 - \cos(2\omega t)}{2} dt$$

$$= \frac{V_0^2}{2}$$

Alternatively,

$$V_{rms} = \frac{V_0}{\sqrt{2}} \approx 0.707 V_0 \qquad (10.132)$$

10.6.3   Inductors in an AC Circuit

In Fig. 10.20, an inductor in the AC circuit is shown. The voltage drop across the inductor equals the electric voltage between a and b endpoints: $V_{ab} = \Delta V_L$. On the other hand, the self-induced voltage across the inductor is given as

$$\Delta V_L \equiv \epsilon_L = -L \frac{di_L}{dt} \qquad (10.133)$$

Using Kirchhoff's law:
$$\Delta V + \Delta V_L = 0 \qquad (10.134)$$

and thus
$$V_0 \sin(\omega t) = \Delta V = -\Delta V_L = L \frac{di_L}{dt} \qquad (10.135)$$

From Eq. (10.135), we find that

Fig. 10.20 The inductor in an AC circuit

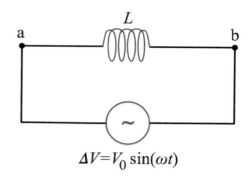

$$i_L(t) = \int \frac{V_0}{L} \sin(\omega t)\, dt \tag{10.136}$$

$$= -\frac{V_0}{L\omega} \cos(\omega t)$$

$$= \frac{V_0}{L\omega} \sin\left(\omega t - \frac{\pi}{2}\right)$$

where the trigonometric relation $\cos\phi = -\sin(\phi - \pi/2)$ is used. Comparing the expression given by Eq. (10.136) with the general form of i_L:

$$i_L(t) = I_0 \sin(\omega t \pm \phi) \tag{10.137}$$

we write that

$$I_0 = \frac{V_0}{L\omega} \equiv \frac{V_0}{X_L} \tag{10.138}$$

where

$$X_L = L\omega \tag{10.139}$$

is the so-called *inductive reactance*.

Using Eq. (10.133), we can determine self-induced voltage drop across the inductor as

$$\epsilon_L(t) = -L\frac{di_L}{dt} \tag{10.140}$$

$$= -L\frac{V_0}{L\omega}\omega \sin(\omega t)$$

$$= -V_0 \sin(\omega t)$$

$$= -I_0 X_L \sin(\omega t)$$

where

$$V_0 = X_L I_0 \tag{10.141}$$

Fig. 10.21 The phasor
diagram of an inductor in an
AC circuit

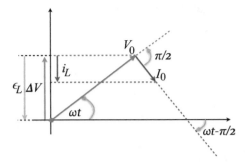

In Fig. 10.21, we show the phasor diagram of an inductor in an AC circuit.
Note that

$$I_{\text{rms}} = \frac{V_{\text{rms}}}{X_L} \tag{10.142}$$

10.6.4 Capacitors in an AC Circuit

Consider a capacitor in an AC circuit, as shown in Fig. 10.22. The output voltage of
the AC generator is

$$\Delta V = V_0 \sin(\omega t) \tag{10.143}$$

The instantaneous voltage across the capacitor is

$$\Delta V_C = V_{ab} \tag{10.144}$$

Applying Kirchhoff's law, we write

$$\Delta V_C = \Delta V = V_0 \sin(\omega t) \tag{10.145}$$

Using the following relationship:

$$C = \frac{q}{\Delta V_C} \tag{10.146}$$

Alternatively, the instantaneous charge in the capacitor is

$$q = C \Delta V_C \tag{10.147}$$

and furthermore we have the instantaneous current as

Fig. 10.22 The capacitor in
an AC circuit

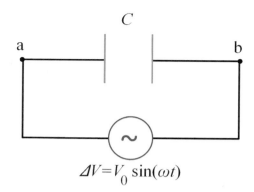

$$\Delta V = V_0 \sin(\omega t)$$

$$i_C = \frac{dq}{dt} = C\frac{d}{dt}(\Delta V_C) = C\frac{d}{dt}(\Delta V) \qquad (10.148)$$
$$= C V_0 \omega \cos(\omega t)$$
$$= C V_0 \omega \sin\left(\omega t + \frac{\pi}{2}\right)$$

Hence, we obtain

$$i_C = I_0 \sin\left(\omega t + \frac{\pi}{2}\right), \qquad (10.149)$$

where

$$I_0 = V_0 (C\omega) \equiv \frac{V_0}{X_C} \qquad (10.150)$$

where

$$X_C = \frac{1}{C\omega} \qquad (10.151)$$

is the so-called *capacitive reactance*.

In Fig. 10.23, we present the phasor diagram of the capacitor in an AC circuit.
Besides, we can write that

$$I_{rms} = \frac{V_{rms}}{X_C} \qquad (10.152)$$

10.6.5   The RLC Series in an AC Circuit

Next, consider the RLC series in an AC circuit, as presented in Fig. 10.24. The series
consists of resistor (R), inductor (L), and the capacitor (C). The generator produces
an alternating voltage given as

Fig. 10.23 The phasor
diagram of an capacitor in an
AC circuit

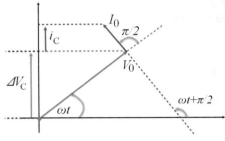

Fig. 10.24 The RLC series
in an AC circuit

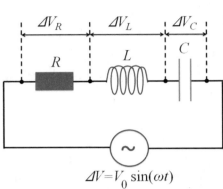

$$\Delta V = V_0 \sin(\omega t) \tag{10.153}$$

The current varies with time as

$$i(t) = I_0 \sin(\omega t - \phi) \tag{10.154}$$

where ϕ is a phase angle shift between the voltage and the current, determined in the
following.

Note that the current passing across each element of the series has the same amplitude. Therefore, the voltage drops in resistor, inductor, and capacitor are, respectively,
given as

$$\Delta V_R = (I_0 R) \sin(\omega t) = V_{0,R} \sin(\omega t) , \tag{10.155}$$
$$\Delta V_L = (I_0 X_L) \sin\left(\omega t - \frac{\pi}{2}\right) = -V_{0,L} \cos(\omega t) ,$$
$$\Delta V_C = (I_0 X_C) \sin\left(\omega t + \frac{\pi}{2}\right) = V_{0,C} \cos(\omega t)$$

where

$$V_{0,R} = I_0 R , \tag{10.156}$$

Fig. 10.25 The phasor diagram of the RLC series in an AC circuit

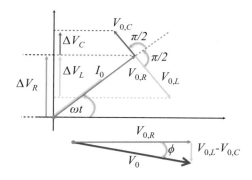

$$V_{0,L} = I_0 X_L \,,$$
$$V_{0,C} = I_0 X_C$$

Using Kirchhoff's law, we write

$$\Delta V = \Delta V_R + \Delta V_L + \Delta V_C \equiv V_0 \sin (\omega t + \phi) \tag{10.157}$$

where ϕ is a phase angle shift between the voltage and the current, which has to be determined. The phasor diagram is used to determine V_0 and ϕ, which is shown in Fig. 10.25. From the phasor diagram, the amplitude of the resultant voltage drop in the circuit is

$$V_0 = \sqrt{V_{0,R}^2 + \left(V_{0,L} - V_{0,C} \right)^2} \tag{10.158}$$
$$= \sqrt{(I_0 R)^2 + (I_0 X_L - I_0 X_C)^2}$$
$$= I_0 \sqrt{R^2 + (X_L - X_C)^2}$$

Denoting by Z the so-called *impedance*:

$$Z = \sqrt{R^2 + (X_L - X_C)^2} \tag{10.159}$$

we obtain Ohm's law for the RLC series in an AC circuit as

$$V_0 = I_0 Z \tag{10.160}$$

Alternatively,

$$I_0 = \frac{V_0}{Z} \tag{10.161}$$

Furthermore, the phase angle shift is found from the tangent as

Fig. 10.26 The variations of
the voltage drops in each
element of the RLC series in
an AC circuit

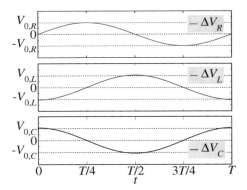

$$\tan \phi = \frac{X_L - X_C}{R} \tag{10.162}$$

as

$$\phi = \tan^{-1}\left(\frac{X_L - X_C}{R}\right) \tag{10.163}$$

In general, the sign of the angle ϕ is defined by the following rules. If $X_L > X_C, \phi < 0$ because it is measured clockwise; if $X_L < X_C, \phi > 0$ because it is an angle measured counterclockwise. For $X_L = X_C$, $\tan \phi = 0$, and thus $\phi = 0$; that is, $Z = R$, and so here is no phase angle shift between the voltage and the current:

$$I_0 = \frac{V_0}{R} \tag{10.164}$$

The frequency for which $X_L = X_C$ is called the *resonance frequency*.

In Fig. 10.26, we show the plots of ΔV_R, ΔV_L, and ΔV_C to indicate the phase shifts between the alternating voltage drops in every element of the circuit.

10.7   Power in the AC Circuit

In the case of the AC circuit, the instantaneous power delivered by the generator is

$$\mathcal{P} = i(t) \cdot \Delta V(t) \tag{10.165}$$
$$= (I_0 \sin(\omega t - \phi)) \cdot (V_0 \sin(\omega t))$$

which corresponds to the general case of the RLC series in an AC circuit. Using the trigonometric relationship:

$$\sin(\alpha \pm \beta) = \sin(\alpha)\cos(\beta) \pm \cos(\alpha)\sin(\beta) \tag{10.166}$$

we can write Eq. (10.165) as follows:

$$
\begin{aligned}
\mathcal{P} &= I_0 V_0 \left(\sin(\omega t) \cos(\phi) - \cos(\omega t) \sin(\phi) \right) \sin(\omega t) \\
&= I_0 V_0 \left(\sin^2(\omega t) \cos(\phi) - \sin(\omega t) \cos(\omega t) \sin(\phi) \right) \\
&= I_0 V_0 \left(\sin^2(\omega t) \cos(\phi) - \frac{1}{2} \sin(2\omega t) \sin(\phi) \right)
\end{aligned}
\tag{10.167}
$$

The average delivered power for a period T is then calculated as

$$
\begin{aligned}
\mathcal{P}_{av} &= \frac{1}{T} \int_0^T \mathcal{P}(t) dt \\
&= \frac{I_0 V_0}{T} \int_0^T \left(\sin^2(\omega t) \cos(\phi) - \frac{1}{2} \sin(2\omega t) \sin(\phi) \right) dt \\
&= \frac{I_0 V_0}{T} \left(\cos(\phi) \int_0^T \sin^2(\omega t) dt - \frac{1}{2} \sin(\phi) \int_0^T \sin(2\omega t) dt \right) \\
&= \frac{I_0 V_0}{T} \left(\cos(\phi) \int_0^T \frac{1 - \cos(2\omega t)}{2} dt \right. \\
&\quad + \left. \frac{1}{4\omega} \sin(\phi) \left(\cos(2\omega T) - \cos(0) \right) \right) \\
&= \frac{I_0 V_0}{T} \left(\cos(\phi) \frac{T}{2} + 0 \right) \\
&= \frac{I_0 V_0}{2} \cos(\phi)
\end{aligned}
\tag{10.168}
$$

Therefore, the average power is

$$
\begin{aligned}
\mathcal{P}_{av} &= \frac{I_0 V_0}{2} \cos(\phi) \\
&= \frac{\left(\sqrt{2} I_{rms} \right) \left(\sqrt{2} V_{rms} \right)}{2} \cos(\phi) \\
&= I_{rms} V_{rms} \cos(\phi) \\
&\equiv F I_{rms} V_{rms}
\end{aligned}
\tag{10.169}
$$

where

$$
F = \cos(\phi)
\tag{10.170}
$$

is the so-called *power factor*.

Moreover, from Fig. 10.25, $V_{0,R} = V_0 \cos(\phi) = I_0 R$; therefore,

$$\cos(\phi) = \frac{I_0 R}{V_0} \tag{10.171}$$

and the average power is

$$\mathcal{P}_{av} = \frac{I_0 V_0}{2} \frac{I_0 R}{V_0} = \frac{I_0^2 R}{2} \tag{10.172}$$

Alternatively,

$$\mathcal{P}_{av} = I_{rms}^2 R \tag{10.173}$$

As a result, the average power delivered by a generator of AC circuit is converted into the internal energy in the resistor (which is similar to the DC circuit). Furthermore, if there is no resistor, then there is no power loss.

Moreover, if the load is just a resistor (that is, $X_C = X_L = 0$), then

$$\mathcal{P}_{av} = I_{rms} V_{rms} \tag{10.174}$$

which indicates that the power factor is one, $F = 1$.

10.8   Resonance in the RLC Series Circuit

The current rms is given by

$$I_{rms} = \frac{V_{rms}}{Z} \tag{10.175}$$

where Z is given by Eq. (10.159). Therefore, we can also write that

$$I_{rms} = \frac{V_{rms}}{\sqrt{R^2 + (X_L - X_C)^2}} \tag{10.176}$$

From Eq. (10.176), if $X_L = X_C$, we say that there is a resonance; that is,

$$I_{rms} = \frac{V_{rms}}{R} . \tag{10.177}$$

That indicates that I_{rms} has maximum value. The frequency for which that occurs can be found using the following relation:

$$\omega_0 L = \frac{1}{\omega_0 C} \tag{10.178}$$

where ω_0 denotes the resonance frequency, which can be obtained as

Fig. 10.27 The current rms
versus frequency ω for
different resistance values
$R_4 < R_3 < R_2 < R_1$. ω_0
represents the resonance
frequency

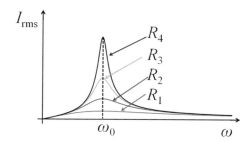

$$\omega_0 = \frac{1}{\sqrt{LC}} \tag{10.179}$$

Besides, using Eq. (10.176), we obtain

$$I_{\text{rms}} = \frac{V_{\text{rms}}}{\sqrt{R^2 + \left(\omega L - \dfrac{1}{\omega C}\right)^2}} \tag{10.180}$$

$$= \frac{V_{\text{rms}}}{\sqrt{R^2 + \left(\dfrac{\omega^2 LC - 1}{\omega C}\right)^2}}$$

In Fig. 10.27, we plot the current rms versus frequency ω for different resistance values $R_4 < R_3 < R_2 < R_1$. ω_0 represents the resonance frequency. That indicates that I_{rms}, given by Eq. (10.180), is maximum at $\omega = \omega_0$, and hence at this frequency, the generator must deliver maximum power.

We can express the average power as a function of the frequency ω as

$$\mathcal{P}_{\text{av}} = I_{\text{rms}}^2 R = \left(\frac{V_{\text{rms}}}{Z}\right)^2 R \tag{10.181}$$

$$= \frac{V_{\text{rms}}^2}{R^2 + (X_L - X_C)^2} R$$

$$= \frac{V_{\text{rms}}^2}{R^2 + \left(\omega L - \dfrac{1}{\omega C}\right)^2} R$$

$$= \frac{V_{\text{rms}}^2}{R^2 + \dfrac{L^2}{\omega^2}\left(\omega^2 - \omega_0^2\right)^2} R$$

$$= \frac{V_{\text{rms}}^2 R \omega^2}{R^2 \omega^2 + L^2\left(\omega^2 - \omega_0^2\right)^2}$$

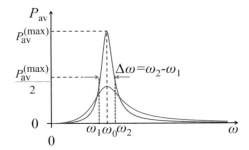

Fig. 10.28 The average power versus frequency ω for two different resistance values. ω_0 represents the resonance frequency. $\Delta\omega = \omega_2 - \omega_1$ is the width measured at half of the maximum power

From Eq. (10.181), if $\omega = \omega_0$ (that is, at resonance condition), then power delivered by the generator is maximum, and given as

$$\mathcal{P}_{av}^{(max)} = \frac{V_{rms}^2}{R} \tag{10.182}$$

In Fig. 10.28, we plot the average power versus frequency ω. The sharpness of the average power curve is characterized by the dimensionless parameter η:

$$\eta = \frac{\omega_0}{\Delta\omega} \tag{10.183}$$

which is also called *quality factor*. In Eq. (10.183), $\Delta\omega$ is the width of the curve calculated as the following difference:

$$\Delta\omega = \omega_2 - \omega_1 \tag{10.184}$$

where ω_1 and ω_2 are the frequencies for which $\mathcal{P}_{av} = \mathcal{P}_{av}^{(max)}/2$ (see also Fig. 10.28). Here,

$$\Delta\omega = \frac{R}{L} \tag{10.185}$$

and thus

$$\eta = \frac{\omega_0 L}{R} \tag{10.186}$$

10.9 Exercises

Exercise 10.1 Consider a motor in which the coils have a total resistance of $10\,\Omega$ and the applied electric voltage is $120\,V$. When the motor is running at its maximum speed, the back emf is $70\,V$. Find the current in the coil when a) the motor is turned on and b) it has reached maximum speed.

Solution 10.1 (a) When the motor is turned on, the back emf is zero, and thus

$$I = \frac{\epsilon_{ind}}{R} = \frac{120\,V}{10\,\Omega} = 12\,A. \qquad (10.187)$$

(b) At maximum speed, the back emf has its maximum value, and hence

$$I = \frac{\epsilon_{ind} - \epsilon_{back}}{R} = \frac{120\,V - 70\,V}{10\,\Omega} = 5.0\,A. \qquad (10.188)$$

Exercise 10.2 Consider an AC generator consisting of 8 turns, each with area $A = 0.09\,m^2$, and the total resistance of the wire is $R = 12\,\Omega$. The loop rotates in a $B = 0.5\,T$ magnetic field at constant frequency of $f = 60\,Hz$. a) Find the maximum induced emf and b) what is the maximum induced current?

Solution 10.2 (a) The angular speed is

$$\omega = 2\pi f = 2\pi(60\,Hz) \approx 377\,rad/s. \qquad (10.189)$$

Then, the induced emf is

$$\epsilon_{ind} = NAB\omega \sin(\omega t) \qquad (10.190)$$

and

$$\epsilon_{ind,max} = NAB\omega = 8(0.09\,m^2)(0.5\,T)(377\,rad/s) \approx 136\,V. \qquad (10.191)$$

(b) The maximum induced current is

$$I_{ind,max} = \frac{\epsilon_{ind,max}}{R} = \frac{136\,V}{12\,\Omega} \approx 11.3\,A. \qquad (10.192)$$

Exercise 10.3 A long solenoid of radius R has n turns of wires per unit of length and carries a time-varying current that as a function of time is

$$I = I_{max} \cos(\omega t) , \tag{10.193}$$

where I_{max} is the maximum current and ω is the angular frequency. (a) Find the magnitude of the induced electric field outside solenoid, at distance $r > R$ from its long central axis. (b) What is magnitude of the induced electric field inside solenoid, at distance r from central axis.

Solution 10.3 (a) Using the following relationship:

$$\oint_{\mathcal{L}} \mathbf{E} \cdot d\mathbf{l} = -\frac{d\Phi_B}{dt} \tag{10.194}$$

where $\Phi_B = BA = B\pi R^2$ and $\mathbf{E}$ is constant. Therefore,

$$\oint_{\mathcal{L}} \mathbf{E} \cdot d\mathbf{l} = \oint_{\mathcal{L}} E dl = E \oint_{\mathcal{L}} dl = E 2\pi r \tag{10.195}$$

Combining Eqs. (10.194) and (10.195), we get

$$E 2\pi r = -\frac{d(B\pi R^2)}{dt} = -\pi R^2 \frac{dB}{dt} \tag{10.196}$$

The magnitude of the electric field is

$$E = -\frac{R^2}{2r}\frac{dB}{dt} \tag{10.197}$$

Using Ampére's law, we write

$$\oint_{\mathcal{L}} \mathbf{B} \cdot d\mathbf{l} = \mu_0 I \tag{10.198}$$

Or,

$$B_1 \cdot l = \mu_0 I \tag{10.199}$$

Then, the magnetic field strength is

$$B_1 = \frac{\mu_0 I}{l} \tag{10.200}$$

For N turns, we have

$$B = N\frac{\mu_0 I}{l} = n\mu_0 I \tag{10.201}$$

Writing the maximum magnetic field as $B_{max} = n\mu_0 I_{max}$, then

$$B = n\mu_0 I_{max} \cos(\omega t) \tag{10.202}$$

The derivative of B for time t is

$$\frac{dB}{dt} = -n\mu_0 I_{max}\omega \sin(\omega t) \tag{10.203}$$

Substituting expression given by Eq. (10.203) into Eq. (10.197), we get

$$E = \frac{R^2}{2r} n\mu_0 I_{max}\omega \sin(\omega t) \tag{10.204}$$

(b) For $r < R$,

$$\Phi_B = B\pi r^2 \tag{10.205}$$

Then

$$E 2\pi r = -\pi r^2 \frac{dB}{dt} \tag{10.206}$$

Therefore,

$$E = \frac{r}{2} n\mu_0 I_{max}\omega \sin(\omega t) \tag{10.207}$$

Exercise 10.4 The conducting bar shown in Fig. 10.1 of mass m and length l moves on two parallel frictionless rails in presence of a uniform magnetic field directed into the page. The bar is given an initial velocity $\mathbf{v}_i$ to the right and it is released at $t = 0$. Find the speed of the bar as a function of time.

Solution 10.4 The induced current is counterclockwise and the magnetic force is

$$\mathbf{F}_B = I\,(\mathbf{L} \times \mathbf{B}) = -Il B\mathbf{i} \tag{10.208}$$

where $\mathbf{i}$ is the unit vector along the directions of motion (to the right). Thus,

$$\mathbf{v}_i = v_i \mathbf{i} \tag{10.209}$$

Using the second law of Newton:

$$\mathbf{F}_B = m\mathbf{a} = m\frac{d\mathbf{v}}{dt} \tag{10.210}$$

Combining Eqs. (10.208) and (10.210), we get

$$- Il B\mathbf{i} = m\frac{d\mathbf{v}}{dt} = m\frac{dv}{dt}\mathbf{i} \tag{10.211}$$

Or,

$$- Il B = m\frac{dv}{dt} \tag{10.212}$$

Using Faraday's law, we write

$$\epsilon_{\text{ind}} = -\frac{d\Phi_B}{dt} = -\frac{d(Blx)}{dt} = -Bl\frac{dx}{dt} = -Blv \tag{10.213}$$

Then, the induced current is

$$I = \frac{\mid \epsilon_{\text{ind}} \mid}{R} = \frac{Blv}{R} \tag{10.214}$$

Or,

$$Bl = \frac{RI}{v} \tag{10.215}$$

Substituting Eq. (10.215) into Eq. (10.212), we obtain

$$- Bl\left(\frac{Blv}{R}\right) = m\frac{dv}{dt} \tag{10.216}$$

Or,

$$-\left(\frac{B^2l^2v}{R}\right) = m\frac{dv}{dt} \tag{10.217}$$

After rearranging it, we write

$$\frac{dv}{v} = -\left(\frac{B^2l^2}{mR}\right)dt \tag{10.218}$$

Integrating both sides of Eq. (10.218), we obtain

$$\int_{v_i}^{v}\frac{dv}{v} = -\left(\frac{B^2l^2}{mR}\right)\int_{0}^{t}dt \tag{10.219}$$

Or,

$$\ln\left(\frac{v}{v_i}\right) = -\left(\frac{B^2l^2}{mR}\right)t \tag{10.220}$$

Denoting by τ a time characteristic as

$$\tau = \frac{mR}{B^2l^2} \tag{10.221}$$

Then, we get

$$v(t) = v_i \exp\left(-\frac{t}{\tau}\right) \tag{10.222}$$

The induced current is (from Eq. (10.214)

$$I = \frac{Bl}{R}v_i \exp\left(-\frac{t}{\tau}\right) \tag{10.223}$$

and using Eq. (10.213), the induced emf is

$$\epsilon_{ind} = -Blv_i \exp\left(-\frac{t}{\tau}\right) \tag{10.224}$$

Exercise 10.5 Consider a conducting bar of length l rotates with constant angular speed ω about a pivot at one end. A uniform magnetic field **B** is directed perpendicular to the plane of rotation (see also Fig. 10.29). Find the motion emf induced between the ends of the bar.

Solution 10.5 We consider a segment of length dr having a linear velocity **v** at a distance r from the pivot, as shown in Fig. 10.29. The linear speed is given as $v = \omega r$. The magnitude of the induced emf in this segment is

$$d\epsilon_{ind} = Bvdr \tag{10.225}$$

The total induced emf is then calculated as

$$\epsilon_{ind} = \int_0^l Bvdr = \int_0^l B(\omega r)dr \tag{10.226}$$

After integration, we find that

$$\epsilon_{ind} = \frac{B\omega l^2}{2} \tag{10.227}$$

Fig. 10.29 A conducting bar
of length l rotates with
constant angular speed about
a pivot at one end. A uniform
magnetic field **B** is directed
perpendicular to the plane of
rotation

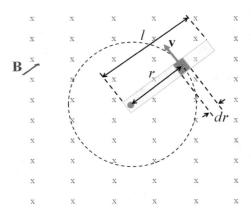

Exercise 10.6 Find the inductance of a uniformly wound solenoid having N
turns and length l. Assume that $l \gg r$, with r being the radius of the winding.
The core of the solenoid is air.

Solution 10.6 We can use Ampére's law to find the magnetic field strength as

$$\oint_{\mathcal{L}} \mathbf{B} \cdot d\mathbf{l} = N\mu_0 I \tag{10.228}$$

where I is the current passing through the coil. Alternatively,

$$Bl = N\mu_0 I \tag{10.229}$$

Therefore, the magnetic field strength is

$$B = \frac{N\mu_0 I}{l} = n\mu_0 I \tag{10.230}$$

where n is the number of turns per unit length of solenoid. The magnetic flux through
each turn is then calculated as follows:

$$\Phi_B = B \cdot A \tag{10.231}$$

where A is the cross-sectional area of the solenoid, given as $A = \pi r^2$. Therefore,

$$\Phi_B = \pi B r^2 = \pi n \mu_0 I r^2 \tag{10.232}$$

Then, the inductance is

$$L = N\frac{\Phi_B}{I} = \frac{N}{I}\frac{N}{l}\pi\mu_0 I r^2 = \mu_0\pi\frac{N^2 r^2}{l} \tag{10.233}$$

Thus, inductance L depends on the N and on the geometry of the solenoid, namely, l and r. Furthermore, we can write Eq. (10.46) as follows:

$$L = \mu_0(l\pi r^2)n^2 = \mu_0 n^2 V \tag{10.234}$$

where $V = l\pi r^2$ is the volume of the solenoid.

Exercise 10.7 Consider an air-core solenoid consisting of $N = 300$ turns of length $l = 25.0$ cm, and its cross-sectional area is $A = 4.00$ cm^2. (a) Calculate the inductance of solenoid. (b) Calculate self-induced emf in the solenoid if the current through it is decreasing at a rate of 50.0 A/s.

Solution 10.7 (a) Using the derived formula for the inductance of a solenoid given by Eq. (10.233), we have

$$L = \mu_0\frac{N^2 A}{l} \tag{10.235}$$

Replacing the numerical values, we get

$$L = \left(4\pi \times 10^{-7}\,\text{T m/A}\right)\frac{(300)^2 \cdot (4.00 \times 10^{-4}\,\text{m}^2)}{25.0 \times 10^{-2}\,\text{m}} \tag{10.236}$$
$$= 1.81 \times 10^{-4}\,\text{Tm}^2/\text{A} = 0.181\,\text{mH}$$

(b) The self-induced emf is

$$\epsilon_L = -L\frac{dI}{dt} \tag{10.237}$$
$$= -\left(1.81 \times 10^{-4}\,\text{Tm}^2/\text{A}\right)\cdot(50.0\,\text{A/s})$$
$$= 9.05 \times 10^{-3}\,\text{V} = 9.05\,\text{mV}$$

Exercise 10.8 Consider the DC circuit consisting of RL series, shown in Fig. 10.30. The switch is closed at $t = 0$. (a) Calculate the time constant τ of the circuit. (b) Calculate the current in the circuit at $t = 2.00$ ms. c) Compare the potential difference across the resistor with that across the inductor.

Fig. 10.30 A DC circuit of
series of RL

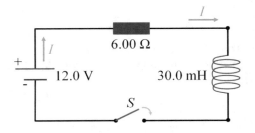

Solution 10.8 (a) The time-varying current is

$$I(t) = \frac{\epsilon}{R} \left(1 - e^{-t/\tau} \right) . \tag{10.238}$$

The time constant τ is given as

$$\tau = \frac{L}{R} = \frac{30.0 \times 10^{-3} \, \text{H}}{6.00 \, \Omega} = 0.005 \, \text{s} = 5.00 \, \text{ms} . \tag{10.239}$$

(b) The current I at $t = 2.00$ ms is

$$I(2.00 \, \text{ms}) = \frac{12.0 \, \text{V}}{6.00 \, \Omega} \left(1 - e^{-(2.00 \, \text{ms})/(5.00 \, \text{ms})} \right) = 0.659 \, \text{A} . \tag{10.240}$$

(c) For the potential difference across the inductor, we use the expression given by
Eq. (10.80):

$$\epsilon_L(t) = -\epsilon e^{-t/\tau} . \tag{10.241}$$

The voltage drop across the resistor is

$$\Delta V_R = I(t)R = \epsilon \left(1 - e^{-t/\tau} \right) . \tag{10.242}$$

Then, the difference between them is

$$\Delta V_R - \epsilon_L = \epsilon , \tag{10.243}$$

which is a constant equal to the source emf.

Fig. 10.31 A DC circuit of
the RL series

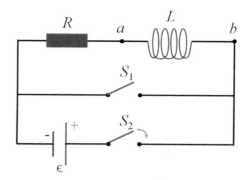

Exercise 10.9 Consider a DC circuit of RL series, as shown in Fig. 10.31. When the switch S_2 is closed at the instance S_1 is open ($t = 0$). The current in the upper loop varies as

$$I(t) = I_0 e^{-t/\tau} \tag{10.244}$$

where $I_0 = \epsilon/R$ is the initial current and $\tau = L/R$ is the time constant. Show that the energy initially stored in the magnetic field of the inductor appears as internal energy in the resistor as the current decays to zero.

Solution 10.9 The rate of energy change dU/dt at which the energy is delivered to the resistor is $E_R = I^2 R$, where I is the instantaneous current:

$$\frac{dU}{dt} = I^2 R = I_0^2 R e^{-2t/\tau} \tag{10.245}$$

Thus, we can write

$$dU = I_0^2 R e^{-2t/\tau} dt \tag{10.246}$$

Integrating both sides of Eq. (10.246), we obtain

$$\int dU = I_0^2 R \int_0^\infty e^{-2t/\tau} dt \tag{10.247}$$

Therefore, we find that

$$U = -I_0^2 R \frac{\tau}{2} \int_0^\infty e^{-2t/\tau} d\left(-\frac{2t}{\tau}\right) \tag{10.248}$$

$$= -\frac{I_0^2 R \tau}{2} \left(e^{-2\cdot\infty/\tau} - e^0 \right)$$

$$= \frac{I_0^2 L}{2}$$

Exercise 10.10 Consider the CD circuit in Fig. 10.32. The capacitor is initially charged when the switch S_1 is open and S_2 is closed. The switch S_1 is then closed and at the same instant S_2 is opened, such that the capacitor is connected to the inductor. (a) Calculate the frequency of oscillation in the circuit. (b) What is the maximum Q_{max} and I_{max}? (c) Determine the current and charge as the function of time t.

Solution 10.10 (a) The linear frequency is given as

$$f = \frac{\omega}{2\pi} = \frac{1}{2\pi\sqrt{LC}} \tag{10.249}$$

$$= \frac{1}{2\pi\sqrt{(2.80 \times 10^{-3} \text{ H})(9.00 \times 10^{-12} \text{ F})}}$$

$$= 1.00 \times 10^6 \text{ Hz}.$$

(b) The initial charge in capacitor equals the maximum charge on the capacitor:

$$C = \frac{Q_{max}}{\epsilon}. \tag{10.250}$$

Alternatively,

$$Q_{max} = \epsilon C = (12.0 \text{ V})(9.00 \times 10^{-12} \text{ F}) = 1.08 \times 10^{-10} \text{ C} \tag{10.251}$$

The maximum current is then

$$I_{max} = Q_{max}\omega \tag{10.252}$$

$$= 2\pi(1.08 \times 10^{-10} \text{ C})(1.00 \times 10^6 \text{ Hz})$$

$$= 6.79 \times 10^{-4} \text{ A}.$$

(c) The variation of charge Q is

$$Q = Q_{max} \cos(\omega t) \tag{10.253}$$

$$= (1.08 \times 10^{-10} \text{ C}) \cos\left(2\pi \times 10^6 \text{ (rad/s)} \cdot t\right).$$

Fig. 10.32 A DC circuit of the LC

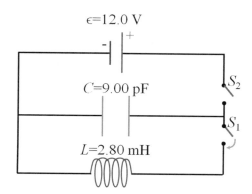

The current is varying according to

$$I = -I_{\text{max}} \sin(\omega t) \qquad (10.254)$$
$$= -(6.79 \times 10^{-4} \text{ A}) \sin(2\pi \times 10^6 \text{ (rad/s)} \cdot t) .$$

Reference

Holliday D, Resnick R, Walker J (2011) Fundamentals of physics. John Wiley and Sons

Chapter 11
Some Applications of Electromagnetic Theory

This chapter aims to introduce two applications of electromagnetic theory to electrostatic properties of macromolecular solutions and wireless charging, important in bio-nanotechnology and wireless technology development.

In this chapter, we discuss the applications of electromagnetic theory to electrostatic properties of macromolecular solutions and wireless charging. For further reading, one can also consider other available literature (Kamberaj 2020; Deuflhard et al. 1997) [Philips et al. (2013)].

11.1 Electrostatic Properties of Macromolecular Solutions

11.1.1 The pH and Equilibrium Constant

The dissociation constant of water molecules manifests the competition between the energy of binding and the entropy of charge liberation. It requires considering the interchange of energetic (E) and entropic (S) effects in controlling the separation to examine pH from a quantitative viewpoint. The competition between energy minimization (bound water molecule state) and entropy maximum (ionic dissociation of water) sets the stage for many biological reactions.

The reaction for dissociation of a water molecule is

$$H_2O \rightleftharpoons H^+ + OH^- . \tag{11.1}$$

© The Author(s), under exclusive license to Springer Nature Switzerland AG 2022
H. Kamberaj, *Electromagnetism*, Undergraduate Texts in Physics,
https://doi.org/10.1007/978-3-030-96780-2_11

The problem is to find the fraction of water molecules that are in dissociated state in a sample of water. Equilibrium constant is gives as

$$K_{eq} = \left(\prod_i^N c_{i0}^{\nu_i} \right) \exp \left(-\beta \sum_i^N \mu_{i0} \nu_i \right) . \tag{11.2}$$

In Eq. (11.2), $\beta = 1/k_B T$ (where k_B is Boltzmann's constant and T is the temperature), μ_{i0} is the standard chemical potential, and

$$\mu_i = \mu_{i0} + k_B T \ln \left(\frac{c_i}{c_{i0}} \right) , \tag{11.3}$$

where c_{i0} is the standard state concentration of the i component. ν_i is the stoichiometric coefficient that equals the change in the number of particles of the i-th component during reaction:

$$\text{A+B} \rightleftharpoons \text{AB} . \tag{11.4}$$

In Eq. (11.4), $\nu_A = \nu_B = -1$ and $\nu_{AB} = 1$. On the other hand, for reaction given by Eq. (11.1), $\nu_{H^+} = \nu_{OH^-} = +1$ and $\nu_{H_2O} = -1$.
   The dissociation constant is defined as

$$K_d = \frac{1}{K_{eq}} . \tag{11.5}$$

The law of mass action implies that

$$K_{eq} = \prod_i^N c_i^{\nu_i} , \tag{11.6}$$

which is known as the *law of mass action*.
   Therefore, for the reaction given by Eq. (11.1), we write

$$K_d = \frac{c_{H^+} \cdot c_{OH^-}}{c_{H_2O}} \tag{11.7}$$

$$= \frac{c_{H^+,0} \cdot c_{OH^-,0}}{c_{H_2O,0}} \exp \left(-\beta \left(\mu_{H^+,0} + \mu_{OH^-,0} - \mu_{H_2O,0} \right) \right) .$$

In Eq. (11.7), the left-hand side is the ratio of the concentrations of the products to reactions at a given temperature and the right-hand side is related with the dissociation constant. Furthermore, $\mu_{X,0}$ denotes the standard state chemical potential. Moreover, note that pH of solution is given as

$$\text{pH} = -\log_{10} \left(c_{H^+} \right) . \tag{11.8}$$

$c_{i,0}$ represents the concentration of the species of the type i in some standard state. Assuming that H^+ is due to only dissociation, then $c_{H^+} = c_{OH^-}$. Furthermore, $c_{H_2O} = c_{H_2O,0}$ and $c_{H^+,0} = c_{OH^-,0} = 1\,M$. Therefore, we obtain

$$\frac{c_{H^+}^2}{c_{H_2O,0}} = \frac{1\,M \cdot 1\,M}{c_{H_2O,0}} \exp\left(-\beta\left(\mu_{H^+,0} + \mu_{OH^-,0} - \mu_{H_2O,0}\right)\right) . \tag{11.9}$$

Using the fact, the change in energy of reaction is

$$\mu_{H^+,0} + \mu_{OH^-,0} - \mu_{H_2O,0} = 79.9\,\frac{kJ}{mol} . \tag{11.10}$$

We obtain that

$$c_{H^+}^2 = \exp\left(-\beta 79.9\,\frac{kJ}{mol}\right) = 1.0 \times 10^{-14}\,M^2 . \tag{11.11}$$

Or,

$$c_{H^+} = 1.0 \times 10^{-7}\,M . \tag{11.12}$$

That is, using Eq. (11.8), we obtain

$$pH = 7 . \tag{11.13}$$

This is a result for water under standard conditions (that is, $T = 300\,K$).

11.1.2 Charge on DNA and Proteins

The charge state of a macromolecule depends on pH of the solution. On the other hand, the charge state of the macromolecule is important in determining both their structure and function in solution. Here, we discuss how the charge state is tuned, by considering the following reaction:

$$HM \rightleftharpoons H^+ + M^- . \tag{11.14}$$

Here, M is the macromolecule of interest. From the equation of the law of mass action, the dissociation constant becomes:

$$K_d = \frac{c_{H^+} \cdot c_{M^-}}{c_{HM}} . \tag{11.15}$$

We introduce pK to measure the tendency of the macromolecule to undergo the dissociation reaction as

$$pK = -\log_{10} K_d . \tag{11.16}$$

Or,

$$+ \log_{10} K_d = \log_{10} c_{H^+} + \log_{10} c_{M^-} - \log_{10} c_{HM} \tag{11.17}$$
$$= -pH + \log_{10} c_{M^-} - \log_{10} c_{HM} .$$

Combining Eqs. (11.16) and (11.17), we obtain

$$pH = pK + \log_{10} \left(\frac{c_{M^-}}{c_{HM}} \right) , \tag{11.18}$$

which is known as *Hendersen-Hasselbalch equation*.

If $c_{M^-} = c_{HM}$, then $pH = pK$. That corresponds to the pH at which half of HM molecules are dissociated. Therefore, pK equals pH at which half of macromolecules have been dissociated.

Consider a DNA, which has a pK such that $pK \approx 1$. Then,

$$pH = 1 + \log_{10} \left(\frac{c_{DNA^-}}{c_{HDNA}} \right) . \tag{11.19}$$

Using Eq. (11.19), at normal pH (that is, $pH = 7$), phosphates on the DNA backbones are fully dissociated. That is, two electronic charges for every base pair, or a linear charge density:

$$\lambda = \frac{2e}{0.34 \, \text{nm}} . \tag{11.20}$$

11.1.3 Charge States of Amino Acids

An important aspect of charge state of proteins is that different side chains of amino acids have different dissociation tendency. That results in that at different pH values, different side chains will be in different states of dissociation. Usually, the titration curve is used to characterize the charge of a macromolecule at different pH values. In the following, we introduce a theoretical framework for prediction of the titration curve.[1] If we assume that the concentrations are proportional to probabilities, then Eq. (11.18) takes the following form:

$$pH = pK + \log_{10} \left(\exp \left(-\frac{\Delta G}{k_B T} \right) \right) = pK + \gamma \frac{\Delta G}{2.303 k_B T} , \tag{11.21}$$

where ΔG equals the free energy of the ionized state of a protein relative to neutral state and $\ln 10 \approx 2.303$. In Eq. (11.21), $\gamma = +1$ for bases and $\gamma = -1$ for acids. For a protein with n ionizable groups:

---

[1] Deuflhard et al. (1997).

$$\Delta G = \sum_{i=1}^{n} \Delta G_i \,, \tag{11.22}$$

where ΔG_i is expressed from Eq. (11.21), and it gives the contribution of each ionizable group:

$$\Delta G_i = 2.303\gamma_i k_B T \left(\text{pH} - \text{pK}_i\right) . \tag{11.23}$$

Here, an ionizable group can be in an isolated amino acid in solvent, and we have a model compound environment with $\text{pK}_{i,\text{model}}$, or in one of the protein's amino acids in solvent, and we have a protein environment with $\text{pK}_{i,\text{protein}}$. We write

$$\text{pK}_{i,\text{protein}} = \frac{\Delta G^0_{i,\text{protein}}}{2.303 k_B T} \,, \tag{11.24}$$

$$\text{pK}_{i,\text{model}} = \frac{\Delta G^0_{i,\text{model}}}{2.303 k_B T} \,,$$

where $\Delta G^0_{i,X}$ is the standard free energy change for a state with dissociated molecule and proton in solution relative to the state with proton bound to a molecule (such as the protein).

For a protein with n ionizable groups, Eq. (11.22), we have

$$\Delta G\left(x_1, x_2, \ldots, x_n\right) = 2.303 k_B T \sum_{i=1}^{n} x_i \gamma_i \left(\text{pH} - \text{pK}_{i,\text{protein}}\right), \tag{11.25}$$

where $x_i = 0$, if the group i is neutral; otherwise, $x_i = 1$, if the group is ionized. Introducing the standard free energy difference as

$$\Delta G^0_{i,\text{m}\rightarrow\text{p}} = \Delta G^0_{i,\text{protein}} - \Delta G^0_{i,\text{model}} \tag{11.26}$$

$$= 2.303 k_B T \left(\text{pK}_{i,\text{protein}} - \text{pK}_{i,\text{model}}\right) .$$

Alternatively,

$$\text{pK}_{i,\text{protein}} = \frac{\Delta G^0_{i,\text{m}\rightarrow\text{p}}}{2.303 k_B T} + \text{pK}_{i,\text{model}} . \tag{11.27}$$

Then, substituting the expression given by Eq. (11.27) into (11.25), we have:

$$\Delta G\left(x_1, x_2, \ldots, x_n\right) = 2.303 k_B T \sum_{i=1}^{n} x_i \gamma_i \left(\text{pH} - \text{pK}_{i,\text{model}} \right. \tag{11.28}$$

$$\left. - \frac{\Delta G^0_{i,\text{m}\rightarrow\text{p}}}{2.303 k_B T} \right) .$$

In general, $\Delta G^0_{i,\mathrm{m}\to\mathrm{p}}$ has an electrostatic nature, and hence only the electrostatic contribution to $\Delta G^0_{i,\mathrm{m}\to\mathrm{p}}$ is practically calculated. Therefore, Eq. (11.28) can also be written as

$$\Delta G (x_1, x_2, \ldots, x_n) = 2.303 k_B T \sum_{i=1}^{n} x_i \gamma_i \left(\mathrm{pH} - \mathrm{pK}_{i,\mathrm{intrinsic}}\right) \tag{11.29}$$

$$+ \sum_{i=1}^{n-1} \sum_{j=i+1}^{n} x_i x_j \Psi_{ij} \,,$$

where

$$\mathrm{pK}_{i,\mathrm{intrinsic}} = \mathrm{pK}_{i,\mathrm{model}} + \frac{\Delta G^0_{i,\mathrm{m}\to\mathrm{p}}}{2.303 k_B T} \tag{11.30}$$

$$= \mathrm{pK}_{i,\mathrm{model}} - \frac{\Delta\Delta G^{(\mathrm{elec})}_i}{2.303 k_B T}$$

$$= \mathrm{pK}_{i,\mathrm{model}} + \mathrm{pK}_{i,\mathrm{shift}} \,,$$

and Ψ_{ij} is the interaction energy between the ionized groups. Furthermore, according to this approach,

$$\mathrm{pK}_{i,\mathrm{shift}} = -\frac{\Delta\Delta G^{(\mathrm{elec})}_i}{2.303 k_B T} \,, \tag{11.31}$$

and

$$\Delta\Delta G^{(\mathrm{elec})}_i = \Delta G^{(\mathrm{elec})}_{i,\mathrm{protein}} - \Delta G^{(\mathrm{elec})}_{i,\mathrm{model}} \,, \tag{11.32}$$

In Eq. (11.32), the first and second terms in the right-hand side are given as

$$\Delta G^{(\mathrm{elec})}_{i,\mathrm{protein}} = \Psi^{\mathrm{ionized}}_{i,\mathrm{protein}} - \Psi^{\mathrm{neutral}}_{i,\mathrm{protein}} \,, \tag{11.33}$$

$$\Delta G^{(\mathrm{elec})}_{i,\mathrm{model}} = \Psi^{\mathrm{ionized}}_{i,\mathrm{model}} - \Psi^{\mathrm{neutral}}_{i,\mathrm{model}} \,,$$

where $\Psi^{\mathrm{ionized}}_{i,X}$ and $\Psi^{\mathrm{neutral}}_{i,X}$ give the electrostatic solvation free energies of the ionized and neutral states, respectively, for protein and model compound environments of the group i. Those terms are calculated using Poisson-Boltzmann approach (which is introduced in the following):

$$\Psi = \frac{1}{2} \sum_{i=1}^{N} q_i \phi_i \,, \tag{11.34}$$

where ϕ_i is the electrostatic potential of the protein assembly of partial charges at the position of charge q_i.

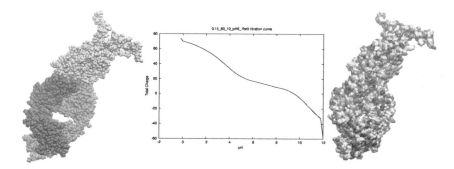

Fig. 11.1 (Middle) Titration Curve for the Entire Spike Protein COVID-19 bound to Antibodies. (Left) Three-dimensional structure colored according to segment name; (Right) Three-dimensional structure colored according to partial charge associate to each atom. Structure was prepared using CHARMM36 force field and titration curve was produced using H+ web server

Figure 11.1 shows the Titration Curve for the Entire Spike Protein COVID-19-Antibodies complex (middle). Three-dimensional structure is colored according to segment name (left) and colored according to partial charge associate to each atom (right). Structure was prepared using CHARMM36 force field[2] and titration curve was produced using H+ web server.[3]

11.1.4 Salt Binding

In cells, many aspects of biological molecules involve recognition of one molecule by another. For instance, binding of RNA polymerase to DNA promotor sites, receptors binding to their ligands, protein-membrane binding, and protein-protein binding.

The surface charge distribution at these macromolecules plays an important role in determining specificity of these interactions and their strength. The importance of electrostatics in macromolecular binding is observed in *vitro* experiments because such interactions are almost strongly dependent on the concentration of ions in the solution in which they are measured. For example, protein-DNA binding interactions is affected by salt. In particular, regulatory proteins transcription factors binding to DNA is modulated by salt concentration.

One can explain that the ions in salty solutions assemble in a screening cloud around macromolecules that have the potential to bind. When the reaction binding occurs, then ions of the screening clouds release, and hence that results in an increase of entropy. Secondly, increasing ion concentration, the ions can interact with receptor and so they compete with ligand.

---

[2] MacKerell et al. (1998).

[3] Anandakrishman et al. (2012).

Fig. 11.2 Interacting
charges q_1 and q_2 in a
medium with dielectric
constant ε. The size of
charges d_1 and d_2 are such
that $d_1, d_2 \ll r$, where r is
their separation; therefore,
the charges are point-like
charges

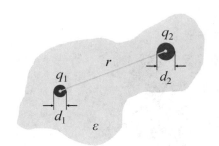

The equilibrium or dissociation constants can lead toward understanding of charge
state of macromolecules in solutions. We need to discuss the electrostatics to understand the macroscopic level. The charges in solution move around. The extent
to which charges leave their macromolecular hosts is dictated by the competition
between energy and entropy. The energy of a positive charge and negative charge
favors the reduction of separation, that is, the binding of the charges with opposite
sign of the charges. On the other hand, the entropic contribution favors the charges
being separated and moving freely in a solution, and hence it drives the system to
the charge liberation.

In Fig. 11.2 are shown two interacting charges q_1 and q_2 in a medium with dielectric constant ε. The size of charges d_1 and d_2 are such that $d_1, d_2 \ll r$, where r is
their separation; therefore, the charges are point-like charges. The magnitude of force
between the charges is given as

$$F = \frac{1}{4\pi\varepsilon_0\varepsilon} \frac{q_1 q_2}{r^2} . \tag{11.35}$$

In Eq. (11.35), the dielectric constant of the medium for air and nonpolar medium
is $\varepsilon \approx 1$, and for water (a polar medium) is $\varepsilon \approx 80$. Therefore, the dielectric constant
of the medium is larger for polar medium. From the microscopic point of view, it
represents the screening of the electrostatic interactions between the charges due
to permanent dipoles of polar molecules. Figure 11.3 illustrates the formation of
permanent dipole in the water molecule due to the distribution of the electronic
charges in the water molecule (which is a polar molecule).

Fig. 11.3 Illustration of the
permanent dipole of a water
molecule

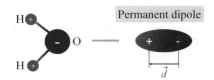

Fig. 11.4 Charge-charge interaction screening by permanent dipoles of the polar molecules of the medium

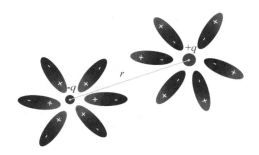

Figure 11.4 represents the charge-charge screening because of the reorientation of permanent dipoles of polar molecules of the medium. This is equivalent to saying that the electric field $\mathbf{E}$ of the charge q at a point P at the distance r from the charge is reduced due to the screening by the permanent dipoles of the medium polar molecules by ε:

$$\mathbf{E}(r) = \frac{1}{4\pi\varepsilon_0\varepsilon} \frac{q\hat{\mathbf{r}}}{r^2}, \tag{11.36}$$

where $\hat{\mathbf{r}}$ is a unit vector pointing to the point P.

Using the first Maxwell's equation given by Eq. (4.79), where $\mathbf{D}$ is given by Eq. (4.63), we can write

$$\varepsilon_0 \nabla \left(\varepsilon(\mathbf{r})\mathbf{E}(\mathbf{r})\right) = \rho(\mathbf{r}), \tag{11.37}$$

where, in general, ε can also be a function of $\mathbf{r}$, which is the general case of the non-homogeneous medium, and the polarized charges created at the dielectric boundaries must be taken into account as well. In Eq. (11.37), ρ is the external charge density; for example, a macromolecule is immersed in a solvent medium. Using the relationship $\mathbf{E} = -\nabla\phi(\mathbf{r})$, where $\phi(\mathbf{r})$ is the scalar potential function, then Eq. (11.37) becomes

$$\nabla \cdot \left(\varepsilon(\mathbf{r})\nabla\phi(\mathbf{r})\right) = -\frac{\rho(\mathbf{r})}{\varepsilon_0}. \tag{11.38}$$

In Eq. (11.38), ρ is the sum of the distribution of the macromolecule fixed charge density $\rho_m(\mathbf{r})$ and ionic charge density $\rho_I(\mathbf{r})$:

$$\rho(\mathbf{r}) = \rho_m(\mathbf{r}) + \rho_I(\mathbf{r}). \tag{11.39}$$

Equation (11.38) is known as Poisson equation. A rigorous solution of that equation is provided using the Poisson-Boltzmann's approach as discussed in the following.

11.1.5   Energy Cost of Assembling a Collection of Charges

Cells spend a great amount of energy in moving charges around. Understanding the energetics associated with charge management, we discuss energy associated with a charge distribution. For that, we calculate the work done to bring isolated charges from infinity, where they do not interact, to a form, such as their distribution in a macromolecular configuration, where they interact with each other. If the electrical energy is positive, then an external agent is doing work to bring charges together; otherwise, it is the system of charges that is doing work.

Consider the potential electrical energy of the spherical ball of radius R and charge Q uniformly distributed, as shown in Fig. 11.5, immersed in a medium with dielectric constant ε. For that, we calculate the work done to assemble the ball. First, we divide the sphere into spherical shells of thickness dr and charge dq, and then calculate the work done to bring together these spherical shells.

The potential electrical energy to bring a shell of thickness dr and charge dq from infinity to a sphere of radius r and charge q uniformly distributed (see also Fig. 11.5) is given as

$$dU_{\text{elec}} = \phi(r)dq ,\tag{11.40}$$

where the scalar electric potential is given as

$$\phi(r) = \int_{r}^{\infty} \frac{q}{4\pi\varepsilon_0\varepsilon r^2} dr .\tag{11.41}$$

After integration in Eq. (11.41), we obtain

$$\phi(r) = \frac{q}{4\pi\varepsilon_0\varepsilon r} .\tag{11.42}$$

Fig. 11.5 A sphere of radius R containing a charge Q. The inner part of the sphere of radius r $(r < R)$ containing the charge q. The spherical ring of thickness dr with charge dq

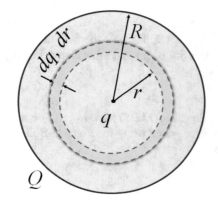

Assuming a uniform distribution of the charge in the sphere of radius R, we write for the charge density of the spherical volume as

$$\rho = \frac{3Q}{4\pi R^3} . \tag{11.43}$$

Then, the total charge q in a sphere of radius r is

$$q = \frac{3Q}{4\pi R^3} \frac{4\pi}{3} r^3 = Q \left(\frac{r}{R}\right)^3 . \tag{11.44}$$

Substituting Eq. (11.44) into (11.42), we obtain an expression for the scalar electric potential at the surface of a sphere of radius r and charge q:

$$\phi(r) = \frac{Qr^2}{4\pi\varepsilon_0\varepsilon R^3} . \tag{11.45}$$

Differentiating Eq. (11.44), we get

$$dq = 3Q \frac{r^2}{R^3} dr . \tag{11.46}$$

Substituting Eqs. (11.46) and (11.45) into (11.40), we find the work done to bring a spherical shell of charge dq from infinity at the surface of a sphere of radius r and charge q:

$$dU_{\text{elec}} = \frac{Qr^2}{4\pi\varepsilon_0\varepsilon R^3} 3Q \frac{r^2}{R^3} dr = \frac{3Q^2}{4\pi\varepsilon_0\varepsilon} \frac{r^4}{R^6} dr . \tag{11.47}$$

Integrating Eq. (11.47) for r from zero to R, we calculate the work done to assemble a charge Q uniformly distributed in a sphere of radius R as

$$U_{\text{elec}} = \frac{3Q^2}{20\pi\varepsilon_0\varepsilon} \frac{1}{R} . \tag{11.48}$$

Equation (11.48) indicates that U_{elec} is positive, which is expected since the assembling charges with the same sign requires external work.

Example 11.1 As an example, we consider the DNA condensation in Bacteriophage $\phi29$. DNA inside a virus is contained in the volume provided by the viral capsid. Since DNA is a strong acid, it is highly charged in solution. Therefore, we need to add energy to bring all charges carried by the DNA into close proximity of the DNA configuration inside capsid.

Fig. 11.6 Force (in pN) versus DNA's length (in nm)

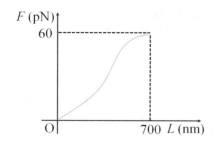

For bacteriophage $\phi 29$, this energy is provided in the form of mechanical work done by portal motor (which is a protein machine that translates DNA), which in turn is fueled by ATP hydrolysis.

The measurements indicate that there is an internal force build-up as DNA is packing. The work against this internal is schematically shown in Fig. 11.6. From here, the internal work done is approximated as

$$W_{\text{int}} \approx \frac{1}{2}(700 \, \text{nm}) \cdot (60 \, \text{pN}) = 210000 \, (\text{pN} \cdot \text{nm}) \,. \tag{11.49}$$

The total charge carried by DNA of $\phi 29$ genome of about 20000 base pairs is calculated using Eq. (11.20) as

$$Q = \frac{2e}{\text{bp}} \cdot 20000 \, \text{bp} = 40000e \,, \tag{11.50}$$

where $e = 1.6 \times 10^{-19} \, \text{C}$ is the magnitude of the charge of one electron. Assuming that this charge is distributed uniformly inside a capsid, considered as spherical with a radius $R \approx 20 \, \text{nm}$. The work done to condense this DNA charge inside the capsid sphere is

$$U_{\text{elec}} = \frac{3Q^2}{20\pi\varepsilon_0\varepsilon} \frac{1}{R} \approx 10^8 \, (\text{pN} \cdot \text{nm}) \,. \tag{11.51}$$

Comparing Eqs. (11.49) and (11.51), we find that $U_{\text{elec}} \gg W_{\text{int}}$. This result indicates that the potential energy value calculated in Eq. (11.51) is not correct when compared with experimental results because it does not include the screening of the forces between charges of DNA by the presence of the counter ions in the medium. Therefore, for a better treatment of the forces between the charges of the macromolecule, we introduce the Poisson-Boltzmann's model in the following.

11.1.6 The Poisson-Boltzmann Equation

Here, we introduce the Poisson-Boltzmann equation as described elsewhere.[4] In the Poisson-Boltzmann approach, all the macromolecular atoms are considered explicitly as particles with partial point charges at the atomic positions, and the dielectric constant of the macromolecule is ε_p (it is often considered to be low, typically, ε_p is in the range of 2–4). The solvent environment surrounding the macromolecule is taken implicitly into account as a dielectric medium with the dielectric constant of ε_w (typically, about 80). The macromolecular dielectric value does not take into account the rearrangement of polar and charged amino acids with external electric fields, which could result into a larger dielectric constants. For example, it is suggested that the increase of the dielectric can compensate for the need for group re-orientations.

In non-homogeneous interacting particles system, density of a particle at any point $\mathbf{r}$ can be written as

$$\sigma_{I,i}(\mathbf{r}) = g_i(\mathbf{r})\sigma_{I,i}^0(\mathbf{r}), \tag{11.52}$$

where $\sigma_{I,i}^0(\mathbf{r})$ is the particle density of the same system considered as ideal gas (i.e., non-interacting particles system), and $g_i(\mathbf{r})$ is the i-th particle distribution, which is taken to follow the Boltzmann distribution

$$g_i(\mathbf{r}) = \exp\left(-\beta W_i(\mathbf{r})\right). \tag{11.53}$$

In Eq. (11.53), $W_i(\mathbf{r})$ is the potential of mean force for the particle i, which is equal to electric potential energy; that is, the average electrostatic potential at the charge's position, $\phi(\mathbf{r})$, multiplied by the charge of particle, q_i:

$$W_i(\mathbf{r}) = q_i\phi(\mathbf{r})$$

where $q_i = z_i e$ with z_i being its valency and e being the charge of proton.

Thus, Eq. (11.52) can be written as

$$\sigma_{I,i}(\mathbf{r}) = \sigma_{I,i}^0(\mathbf{r})\exp\left(-\beta q_i\phi(\mathbf{r})\right). \tag{11.54}$$

Then, the charge density is given as

$$\rho_I(\mathbf{r}) = \sum_i q_i\sigma_{I,i}(\mathbf{r}) = \sum_i q_i\sigma_{I,i}^0(\mathbf{r})\exp\left(-\beta q_i\phi(\mathbf{r})\right), \tag{11.55}$$

where

$$\sigma_{I,i}^0(\mathbf{r}) = c_i^\infty\lambda(\mathbf{r}),$$

---

[4] Kamberaj (2020).

where c_i^∞ is the bulk constant concentration of the i-th ionic species, satisfying the condition of the electrostatic neutrality:

$$\sum_i q_i c_i^\infty = 0$$

$\lambda(\mathbf{r})$ is the accessibility of ions at point $\mathbf{r}$ (i.e., $\lambda(\mathbf{r}) = 0$ in the region inside the macromolecule and $\lambda(\mathbf{r}) = 1$ in the solvent region). Therefore, we can write

$$\rho_I(\mathbf{r}) = \lambda(\mathbf{r}) \sum_i q_i c_i^\infty \exp\left(-\beta q_i \phi(\mathbf{r})\right). \tag{11.56}$$

Using Eq. (11.56), the Poisson equation (see Eq. (11.38)) takes the form of the so-called nonlinear Poisson-Boltzmann equation

$$\nabla \cdot (\varepsilon(\mathbf{r})\nabla\phi(\mathbf{r})) + \frac{\lambda(\mathbf{r})}{\varepsilon_0} \sum_i q_i c_i^\infty \exp\left(-\beta q_i \phi(\mathbf{r})\right) = -\frac{\rho_m(\mathbf{r})}{\varepsilon_0}. \tag{11.57}$$

For an electrostatic neutral solvent, we can write

$$\sum_{i=1}^{N_+} q_i^{(+)} c_i^{+,\infty} = \sum_{i=1}^{N_-} q_i^{(-)} c_i^{-,\infty}$$

where two kind of ionic species are assumed to exist in the solution, positive and negative with N_+ and N_- being the number of positive and negative ions, respectively. Assuming that $N_+ = N_- = N_I$, and since $q_i^{(+)} = -q_i^{(-)} \equiv q_i$ and $c_i^{+,\infty} = c_i^{-,\infty} = c_i^\infty/2$, we get from Eq. (11.57) that

$$\nabla \cdot (\varepsilon(\mathbf{r})\nabla\phi(\mathbf{r})) - \frac{\lambda(\mathbf{r})}{\varepsilon_0} \sum_{i=1}^{N_I} q_i c_i^\infty \sinh\left(\beta q_i \phi(\mathbf{r})\right) = -\frac{\rho_m(\mathbf{r})}{\varepsilon_0}, \tag{11.58}$$

which is a form often found in the literature and it represents a nonlinear partial differential equation. In Eq. (11.58), sinh represents the function: $\sinh(x) = \left(e^x - e^{-x}\right)/2$.

Assuming that the potential is small, the linear form of the equation can be obtained as

$$\nabla \cdot (\varepsilon(\mathbf{r})\nabla\phi(\mathbf{r})) = -\frac{\rho_m(\mathbf{r})}{\varepsilon_0} + \varepsilon \left[\frac{\beta}{\varepsilon\varepsilon_0} \sum_i q_i^2 c_i^\infty\right] \lambda(\mathbf{r})\phi(\mathbf{r}). \tag{11.59}$$

We can determine the so-called Debye screening constant κ as

$$\kappa^2 = \frac{\beta}{\varepsilon\varepsilon_0} \sum_i q_i^2 c_i^\infty = \frac{\beta}{\varepsilon\varepsilon_0} I \equiv \frac{1}{l_D^2}, \tag{11.60}$$

which also describes the exponential decay of the potential in the solvent, with l_D being the Debye length, and I

$$I = \sum_i q_i^2 c_i^\infty$$

being the ionic strength. Note that $\kappa = 0$ in the macromolecule region because the mobile ions are present only in the solvent region.

Equation (11.59) can then be written as

$$\nabla \cdot (\varepsilon(\mathbf{r})\nabla\phi(\mathbf{r})) - \varepsilon(\mathbf{r})\kappa^2\lambda(\mathbf{r})\phi(\mathbf{r}) = -\frac{\rho_m(\mathbf{r})}{\varepsilon_0}. \tag{11.61}$$

Although for biological systems ϕ is not small, and therefore the linearization condition does not hold, comparisons between the linear and nonlinear forms of the Poisson-Boltzmann equation show that both forms are in good agreement with each other. Moreover, these comparisons have shown that small differences are related to the charge density, and hence to the electric field magnitude, at the interface solvent-solute.

From solving either the linear Poisson-Boltzmann equation (see Eq. (11.61)) or the nonlinear Poisson-Boltzmann equation (see Eq. (11.57)), the electrostatic potential, $\phi(\mathbf{r})$, will be obtained at any point $\mathbf{r}$ in space. It can be seen, that knowing ϕ, we may calculate the local concentration of ions through the formula

$$c_i(\mathbf{r}) = c_i^\infty \exp\left(-\beta q_i \phi(\mathbf{r})\right), \tag{11.62}$$

which involves the Boltzmann distribution. Moreover, the gradient of the electrostatic potential can give the electric field, $\mathbf{E}(\mathbf{r}) = -\nabla\phi(\mathbf{r})$.

Another quantity of interest calculated using the electrostatic potential is the electrostatic component of the solvation free energy. The electrostatic term of solvation free energy gives the work done for a possible process of charging the macromolecule and ions in an ionic discharged atmosphere. Using these processes in thermodynamic cycles, we can compute the electrostatic component of free energies for real processes such as solvation. The free energy for charging the solute (e.g., a macromolecule) in an ionic environment can be calculated using different approaches, for example, by direct integration of the charge, by considering a variation principle, or using thermodynamic arguments.

Based on Marcus theory, the electrostatic energy $G_{\text{elec}}^{\text{solv}}$ contains three different terms. The first term is the classical electrostatic energy, $G_{\text{elec}}^{\text{cl}}$

$$G_{\text{elec}}^{\text{cl}} = \frac{1}{2}\int d^3\mathbf{r}\rho_m(\mathbf{r})\phi(\mathbf{r}). \tag{11.63}$$

The second term is arising from mixing the mobile species $G_{\text{elec}}^{\text{mob}}$:

$$G_{\text{elec}}^{\text{mob}} = k_B T \int d^3\mathbf{r} \sum_i c_i(\mathbf{r}) \ln \frac{c_i(\mathbf{r})}{c_i^\infty}. \tag{11.64}$$

Combining Eqs. (11.62) and (11.64), we obtain

$$G_{\text{elec}}^{\text{mob}} = - \int d^3\mathbf{r} \left(\sum_i c_i(\mathbf{r}) q_i \right) \phi(\mathbf{r}). \tag{11.65}$$

The third term is the so-called *osmotic* term, which due to nonuniform ionic concentration, and it is calculated as a volume integral:

$$G_{\text{elec}}^{\text{solvent}} = k_B T \int d^3\mathbf{r} \sum_i \left(c_i^\infty - c_i(\mathbf{r}) \right) \tag{11.66}$$

$$= k_B T \int d^3\mathbf{r} \sum_i c_i(\mathbf{r}) \left[\exp\left(\beta q_i \phi(\mathbf{r}) \right) - 1 \right]$$

$$= \int d^3\mathbf{r} \left(\sum_i c_i(\mathbf{r}) q_i \right) \phi(\mathbf{r}),$$

where the linearity of the exponential term is applied for ϕ small.

Therefore, the total electrostatic energy is

$$G_{\text{elec}} = G_{\text{elec}}^{\text{cl}} + G_{\text{elec}}^{\text{mob}} + G_{\text{elec}}^{\text{solvent}}$$

or

$$G_{\text{elec}} = G_{\text{elec}}^{\text{cl}} = \frac{1}{2} \int d^3\mathbf{r} \rho_m(\mathbf{r}) \phi(\mathbf{r}). \tag{11.67}$$

where ρ_m is the charge density of fixed charges (i.e., nonionic charges such as partial atomic charges of macromolecule). It is possible to also calculate the free energy by knowing the electrostatic potential, for example, for the possible charging process of a macromolecule in an ionic environment. That is done by combining different possible processes in a thermodynamic cycle that can lead to the computation of the theoretical free energy of some real process.

For instance, the thermodynamic cycle shown in Fig. 11.7 can be used to calculate the electrostatic component of the solvation free energy. This thermodynamic cycle indicates that the electrostatic component of the solvation free energy is the difference in the free energies related to two purely hypothetical charging processes; one in some reference surrounding environment phase (e.g., with dielectric constant equal to that of the macromolecule) and the other the solvent surrounding environment with dielectric constant ε_w. Thus, using Eq. (11.66), we can write

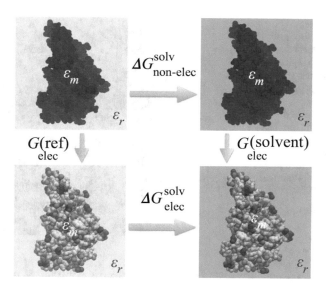

Fig. 11.7 Thermodynamic cycle for calculation of electrostatic solvation free energy. The difference in the charging energy, ΔG_{elec} in reference surrounding phase and in solvent is the electrostatic solvation free energy. Colors in bottom plots indicate the partial atom charges. In the top plots, the blue color indicates no partial charges on atoms

$$\Delta G_{\text{elec}}^{\text{solv}} = G_{\text{elec}}(\text{solvent}) - G_{\text{elec}}(\text{ref}) \tag{11.68}$$
$$= \frac{1}{2} \int d^3\mathbf{r} \rho_m(\mathbf{r}) \left[\phi_w(\mathbf{r}) - \phi_r(\mathbf{r}) \right].$$

Often, $\phi_w(\mathbf{r}) - \phi_r(\mathbf{r})$ is called *reaction* potential, $\phi_{\text{reac}}(\mathbf{r})$, and thus Eq. (11.68) can be written as

$$\Delta G_{\text{elec}}^{\text{solv}} = \frac{1}{2} \int d^3\mathbf{r} \rho_m(\mathbf{r}) \phi_{\text{reac}}(\mathbf{r}). \tag{11.69}$$

In Eq. (11.69), $\Delta G_{\text{elec}}^{\text{solv}}$ represents the work done by electrostatic forces for transferring a set of partial atomic charges of macromolecule from a fixed point in some reference surrounding environment with dielectric constant ε_r to a fixed point in solvent surrounding environment with dielectric constant ε_w.

Depending on the shape and charge distribution of macromolecule, numerical solutions of the Poisson-Boltzmann equation could be difficult. Solvated macromolecular systems are in general modeled by regions with different dielectric constants. Figure 11.8 illustrates a solvated macromolecule occupying the region Ω, where the macromolecule region is represented by Ω_m as a solid surface; the solvent region is represented by Ω_w. The dielectric interface σ is defined by the molecular surface and represents the region not penetrated by mobile ions, and $\mathbf{n}$ will represent an unit vector normal to σ pointing from Ω_m to Ω_w. The transition from solute (with

Fig. 11.8 Different computational regions of interest: The solute (macromolecule) region, Ω_m, with dielectric constant ε_m; solvent region, Ω_w, with dielectric constant ε_w where different mobile ions are denoted; dielectric interface, which characterize the solvent accessible surface are constructed with a probe radius of $r = 1.4\,\text{Å}$ (surface colored in blue)

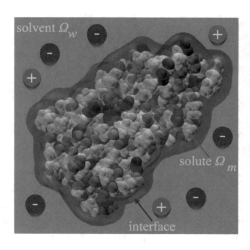

low-dielectric constant) to solvent (with high-dielectric constant) is modeled to be abrupt, giving rise to the dielectric interface σ. There are two conditions on σ that are usually satisfied:

$$(\phi(\mathbf{r}))_{\Omega_m} = (\phi(\mathbf{r}))_{\Omega_w} \qquad (11.70)$$

$$\left(\varepsilon \frac{\partial \phi(\mathbf{r})}{\partial \mathbf{n}}\right)_{\Omega_m} = \left(\varepsilon \frac{\partial \phi(\mathbf{r})}{\partial \mathbf{n}}\right)_{\Omega_w}.$$

These conditions are used in the methods based on *boundary integral equations*, but may not apply to *finite difference methods*. Usually, the boundary of the entire computational domain is also defined, Γ. In addition, approximated Dirichlet boundary condition is imposed in the boundary Γ.

The widely used numerical methods include finite difference method (FDM), the boundary element method (BEM), and finite element method (FEM).

Figure 11.9 shows solvation free energies, $\Delta G_{\text{elec}}^{\text{solv}}$, decomposed for each atom (in kcal/mol) for Bovine Pancreatic Trypsin Inhibitor (BPTI) protein of 58 amino acids.[5] The protein was prepared using CHARMM36 force field. Inset is the three-dimensional structure colored according to solvation free energy values. In the numerical calculations, the dielectric constant of protein was $\varepsilon_p = 2$ and for the water $\varepsilon_w = 80$. The ionic concentration in the solvent was 0.154 M/L.

---

[5] Czapinska et al. (1999).

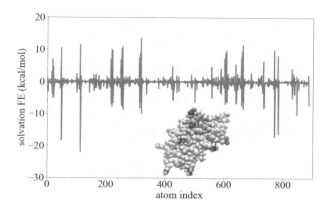

Fig. 11.9 Solvation free energy (in kcal/mol), $\Delta G_{\text{elec}}^{\text{solv}}$, for Bovine Pancreatic Trypsin Inhibitor (BPTI) protein shown for every atom. Inset is the three-dimensional structure colored according to solvation free energy values. The dielectric constant of protein was $\varepsilon_p = 2$ and for the water $\varepsilon_w = 80$. The ionic concentration in the solvent was 0.154 M/L

11.1.7 Calculation of pKa of Amino Acids in Macromolecules

Using Eq. (11.18), and again assuming that the probabilities are proportional to concentrations, we write

$$\frac{c_{M^-}}{c_{HM}} = \exp\left(\frac{-\Delta G}{k_B T}\right) . \tag{11.71}$$

In Eq. (11.71), ΔG is the free energy of the ionized state relative to neutral state of a macromolecule. Therefore, Eq. (11.18) can also be written as

$$pKa = pH - \log_{10}\left(\exp\left(\frac{-\Delta G}{k_B T}\right)\right) . \tag{11.72}$$

Or,

$$\Delta G = 2.303 k_B T \gamma \left(pH - pKa\right) . \tag{11.73}$$

In general, the group can be part of an isolated amino acid in solution (called model compound environment, pKa_{model}), or part of an amino acid in a macromolecule that is solvated (called protein environment model, pKa_{protein}). Solving Eq. (11.73) for pKa, we obtain

$$pKa = pH - \gamma \frac{\Delta G}{2.303 k_B T} , \tag{11.74}$$

because $1/\gamma = \gamma$.

We can consider an amino acid of a protein without N-terminal and C-terminal patches in an isolated state in solution to calculate pKa_{model} in a neutral state, then

$$pKa_{model} = \frac{\Delta G^0_{model}}{2.303 k_B T}, \tag{11.75}$$

where ΔG^0_{model} is the free energy change for the state with dissociated amino acid and proton in solution relative to the state with proton bound to amino acid. The experimental pKa for the compound model can be derived from molecules that are close to the compound model (such as organic molecules).

For the amino acid in a protein environment, we have

$$pKa_{protein} = \frac{\Delta G^0_{protein}}{2.303 k_B T}, \tag{11.76}$$

where $\Delta G^0_{protein}$ is the free energy change for the state with dissociated protein and proton in solution relative to the state with proton bound to protein.

Then, ΔG of a protein assuming only one ionizable group is

$$\Delta G = 2.303 \gamma k_B T \left(pH - pKa_{protein} \right) . \tag{11.77}$$

Furthermore, the standard free energy difference is

$$\Delta \Delta G^0_{m \to p} = \Delta G^0_{protein} - \Delta G^0_{model} . \tag{11.78}$$

Therefore, Eq. (11.77) can also be written as

$$\Delta G = 2.303 \gamma k_B T \left(pH - pKa_{model} - \frac{\Delta G^0_{m \to p}}{2.303 k_B T} \right) . \tag{11.79}$$

Therefore,

$$pKa_{calc} = pKa_{model} + \frac{\Delta G^0_{m \to p}}{2.303 k_B T} . \tag{11.80}$$

It can be seen (see also Eq. (11.80)) that by knowing pKa_{model} and $\Delta G^0_{m \to p}$, we can calculate pKa_{calc}. The second term in Eq. (11.80) represents the so-called pKa_{shift}:

$$pKa_{shift} = \frac{\Delta G^0_{m \to p}}{2.303 k_B T} . \tag{11.81}$$

Therefore,

$$pKa_{calc} = pKa_{model} + pKa_{shift} . \tag{11.82}$$

Note that $\Delta G^0_{m \to p}$ has an electrostatic origin, and hence only the electrostatic contribution is considered, using Poisson-Boltzmann approach. Based on this approach,

$$pKa_{shift} = -\frac{\Delta \Delta G^{(elec)}}{2.303 k_B T}, \tag{11.83}$$

Table 11.1 The values of calculated pKa_{model} for some amino acids using CHARMM36 force field. In addition, the experimental pKa are shown, along with their standard deviations

Amino acid	pKa_{model}	pKa_{shift}	pKa_{calc}	pKa_{exp}
GLU	4.07	−2.63	1.44	2.1 ± 0.1
ASP	3.86	−0.45	3.41	3.1 ± 0.1
HIS	6.10	2.10	8.20	7.75 ± 0.02
CYS	8.23	−1.85	6.38	10.2 ± 0.2
LYS	10.53	−0.011	10.52	10.6 ± 0.1

where

$$\Delta\Delta G^{(elec)} = \Delta G_{protein}^{(elec)} - \Delta G_{model}^{(elec)} . \tag{11.84}$$

In Eq. (11.84), the first and second terms in the right-hand side are given as

$$\Delta G_{protein}^{(elec)} = \Psi_{protein}^{ionized} - \Psi_{protein}^{neutral} , \tag{11.85}$$
$$\Delta G_{model}^{(elec)} = \Psi_{model}^{ionized} - \Psi_{model}^{neutral} .$$

In Eq. (11.85), $\Psi_X^{ionized}$ and $\Psi_X^{neutral}$ give the electrostatic solvation free energies of the ionized and neutral states, respectively, for protein and model compound environments. Those terms are calculated using Poisson-Boltzmann approach, given by Eq. (11.34).

In Table 11.1, we show the values of calculated pKa_{model} for some amino acids using CHARMM36 force field. Besides, we present the predicted values pKa_{calc} for different amino acids in Bernase protein[6] environment, calculated using CHARMM program.[7] The experimental values of pKa are also shown, including the statistical error. In the numerical calculations using Poisson-Boltzmann approach, $\varepsilon_p = 2$, $\varepsilon_w = 80$, and the solvent probe radius is $r_p = 1.4$ Å. The ionic concentration of solvent is 0.154 M/L, and the temperature is 300 K.

11.2 Wireless Charging

There exists an increasing interest on inductive technologies. Furthermore, there are products provided in the markets from companies, such as Powermal and Wireless Power Consortium. These technologies are also called *tightly coupled technologies*. Some of those technologies offer more specific capabilities in the network, such as supervising and measuring the power to each device. In fact, these features may provide some attractive services to providers of charging services (such as airports, Star-

---

[6] Martin et al. (1999).
[7] Brooks et al. (2009).

bucks, and others). Some other efforts include development of dual mode approaches; for instance, wireless charging combining the resonance and inductive technologies together under a single technology. Moreover, some companies are also focusing on newer technologies referred as *highly resonant loosely coupled technology*.

11.2.1 Tightly Coupled Wireless Power Systems

First, we can introduce the transformer model system, as shown in Fig. 11.10, to describe a wireless charging system. The transformer model system comprises the *primary coil* (PC), a *secondary coil* (SC), and the *iron core* that couples the magnetic field lines of the PC into SC (see also Fig. 11.10).

The current flowing through the primary coil creates a magnetic field given by Eq. (7.1), based on Biot-Savart's law. The change in the current passing through the PC produces a magnetic field that changes with time, and hence based on Faraday's law, a change of magnetic flux through the SC, which results in an induced current in the secondary coil. Here, the iron core enables collection of magnetic field lines around the primary coil and pass them to the secondary coil. The induced current in the secondary coil creates a wireless power transfer system. Based on Faraday's law, the voltage V_p across the primary circuit is

$$V_p = -N_1 \frac{d\Phi_B}{dt} , \qquad (11.86)$$

where N_1 is the number of turns at the primary coil. Φ_B is the magnetic flux through every turn. If all magnetic flux remains within the iron, the flux through each turn in

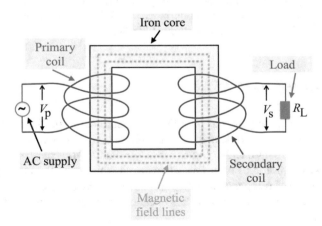

Fig. 11.10 A transformer model system

the primary equals the flux through each turn of the secondary coil, then the voltage across the secondary coil is

$$V_s = -N_2 \frac{d\Phi_B}{dt} , \qquad (11.87)$$

where N_2 is the number of turns in the secondary coil. Then, we obtain

$$V_s = -N_2 \left(-\frac{V_p}{N_1} \right) = \frac{N_2}{N_1} V_p . \qquad (11.88)$$

If $N_2 > N_1$, then $V_s > V_p$, corresponding to the so-called *step up transformer*. If $N_2 < N_1$, then $V_s < V_p$, and we have the so-called *step down transformer*.

If the secondary circuit is open (for example, using some switch), then the primary coil acts as a simple circuit having an alternative source emf and an inductor (assuming no resistance). That is, power factor $F = \cos\phi = 0$ ($\phi = \pi/2$), and the average power delivered from AC generator to primary circuit is $\mathcal{P}_{av} = I_{rms} V_{rms} \cos\phi = 0$.

If the secondary circuit is closed, I_2 current is induced in the secondary circuit. If the load resistance R_L in the secondary circuit is purely resistance, the I_2 is in phase with V_s. If there are no power losses, then

$$I_1 V_p = I_2 V_s , \qquad (11.89)$$

where the first term gives the input power and the second term is the output power. Therefore,

$$I_1 = \frac{V_p}{R_{eq}} , \qquad (11.90)$$

$$I_2 = \frac{V_s}{R_L} .$$

In Eq. (11.90), R_{eq} is the equivalent resistance of R_L when viewed from primary circuit side. Thus,

$$R_{eq} = \frac{V_p}{I_1} = \frac{V_p}{\dfrac{I_2 V_s}{V_p}} \qquad (11.91)$$

$$= \frac{V_p^2}{I_2 V_s} = \frac{V_p^2}{\dfrac{V_s^2}{R_L}}$$

$$= \left(\frac{V_p}{V_s} \right)^2 R_L = \left(\frac{N_1}{N_2} \right)^2 R_L .$$

Now, the tightly coupled wireless charger works in the same way, except that the iron core is removed. Furthermore, the planar coil is used instead of windings,

Fig. 11.11 Tightly coupled
wireless inductive system

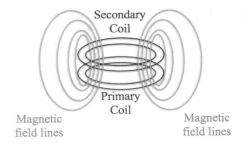

as shown in Fig. 11.11. Since there is no iron core, the magnetic field lines passes through the air, which has a lower permeability compared with iron, and hence there is lower ability to allow magnetic field to pass through. In fact, the factor is around 7000 lower.[8] As a result, the amount of magnetic flux and hence power coupled in the second coil through air is much lower than iron core. Therefore, to obtain a high-power transfer (as efficient as possible, typically 70 % DC into DC in a wireless power system), the primary coil and secondary coil are kept very close proximity to each other, and in a concentric alignment. That is, the secondary coil couples to the strongest and largest part of the primary coil magnetic field. Figure 11.11 illustrates a tightly coupled wireless inductive system. Specifically, any magnetic flux from primary coil that does not couple to secondary coil is a leakage inductance, which causes energy to be wasted because it is equivalent to adding an impedance into the primary coil source that does not induce voltage in the secondary coil. Increasing impedance in the primary coil source causes the energy looses of the amount $I^2 R$, which implies that as the current increases to maintain charging at the target rates in the receiver the wasted energy increases.

For normal use, the leakage inductance should be minimal; that is, the primary coil and secondary coil should be kept far from each other. In contrast, the coil to coil efficiency requires that the primary coil and secondary coil to be close to each other. At lower efficiency charger has to send more power, and hence higher charger losses for the same power delivered. Thus, primary and secondary coils must be arranged tightly coupled. In general, the tightly coupled wireless power systems are concentrically aligned, approximately the same size, and kept in very close proximity (as illustrated in Fig. 11.11).

The laws of physics indicate some limitations on the use of tightly coupled systems. Specifically, they can not be used in many types of consumer devices under the same standards, and they can charge only one device at a time. These limitations on the use of the tightly coupled systems can be avoided by using the phenomena of resonance leading to the development of resonant systems.

---

[8] McGraw-Hill Science and Technology Encyclopedia: Ampére Law.

11.2.2 Loosely Coupled Highly Resonant Systems

For these systems, the secondary coil may be coupled to a fewer magnetic field lines. Besides, they have a larger distance from the primary coil in comparison with tightly coupled inductive systems. As a result, there exists a higher degree of spatial freedom, which is sometimes desired, between the primary and secondary coils. However, we could expect much lower efficiency compared with tightly coupled inductive systems. Therefore, the primary and secondary coils are at high-resonance coupling of the magnetic fields to avoid that. This helps to achieve the desired efficiency power transfer between the coils.

Nikolla Tesla was the first to demonstrate that the resonance principle could be used to transfer power through the wireless systems. That principle is used to develop loosely coupled systems (such as phone devices).

In electrical systems, resonance is achieved by an RLC (resistance R-inductance L-capacitance C) circuit system, which occurs at a specific frequency, determined by the R, L, and C values. The quality factor, η, is defined by the resistance in the circuit (see also Eq. (10.186), Chap. 10). High η indicates a high efficiency of energy transfer. The inductive reactance (ωL) and capacitance reactance ($1/\omega C$) are equal in magnitude at resonance for which η is determined by the resistance in the circuit. Here, the resonance frequency is given by

$$\omega_0 = \frac{1}{\sqrt{LC}} = 2\pi f, \tag{11.92}$$

where f is the frequency at the resonance. High value of η is obtained for high-frequency values. Many resonant systems work at $f = 6.78\,\text{MHz}$. Some resonant wireless power systems work at much lower frequency of $f \approx 100\,\text{kHz}$ with simpler circuit design.

Resonant wireless systems are similar to inductive systems, as illustrated in Fig. 11.12. Both use a primary and secondary coil. However, the difference is that in the loosely coupled resonant systems, the secondary coil is not necessary at close proximity to a large percentage of magnetic field coming from primary coil. For that high η value coils, however, are necessary. Therefore, efficient power transfer is not strictly dependent on the alignment, size, shape, or positioning of primary and secondary coils relative to each other.

The most import thing, multiple secondary coils can also be used to capture power since each coil can share the overall coupling with the primary coil and still have efficient power transfer. The advantages of using the loosely resonant coupled systems include their applicability to different portable electronic devices charged and their use at all times.

Fig. 11.12 Loosely coupled resonant systems

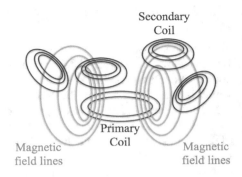

Secondary
Coil

Primary
Coil

Magnetic
field lines

Magnetic
field lines

11.3 Exercises

Exercise 11.1 The voltage output of a generator is given by

$$\Delta V = (200\ \text{V}) \sin(\omega t) . \tag{11.93}$$

Calculate the rms current in the circuit when this generator is connected to a $100\ \Omega$ resistor.

Solution 11.1 First, the maximum voltage is

$$V_0 = 200\ \text{V} . \tag{11.94}$$

Then, the rms is given as

$$V_{\text{rms}} = \frac{V_0}{\sqrt{2}} = \frac{200\ \text{V}}{\sqrt{2}} = 141\ \text{V} . \tag{11.95}$$

The rms current is then

$$I_{\text{rms}} = \frac{V_{\text{rms}}}{R} = \frac{141\ \text{V}}{100\ \Omega} = 1.41\ \text{A} . \tag{11.96}$$

The current passing through the resistor is

$$i_R = \frac{\Delta V_R}{R} = \frac{200\ \text{V}}{100\ \Omega} \sin(\omega t) = (2.00\ \text{A}) \sin(\omega t) . \tag{11.97}$$

From Eq. (11.97), maximum current is $I_0 = 2.00\ \text{A}$.

Exercise 11.2 In a purely inductive AC circuit, $L = 25.0\,\text{mH}$ and the rms voltage is $150\,\text{V}$. Calculate the inductive resistance and rms current in the circuit if the frequency is $60.0\,\text{Hz}$.

Solution 11.2 The inductive reactance is

$$X_L = L\omega = L(2\pi f) \tag{11.98}$$
$$= 2\pi(25.0 \times 10^{-3}\,\text{H})(60.0\,\text{Hz})$$
$$= 9.42\,\Omega\,.$$

The rms current is

$$I_{\text{rms}} = \frac{I_0}{\sqrt{2}} = \frac{\frac{V_0}{X_L}}{\sqrt{2}} = \frac{V_0}{\sqrt{2}X_L} = \frac{V_{\text{rms}}}{X_L} \tag{11.99}$$
$$= \frac{150\,\text{V}}{9.42\,\Omega} = 15.9\,\text{A}\,.$$

Exercise 11.3 An $8.00\,\mu\text{F}$ capacitor is connected to the terminal of a $60.0\,\text{Hz}$ AC generator whose rms voltage is $150\,\text{V}$. Calculate the capacitive reactance and the rms current in the circuit.

Solution 11.3 The angular frequency is

$$\omega = 2\pi f = 2\pi(60.0\,\text{Hz}) = 377\,\text{rad/s}\,. \tag{11.100}$$

The capacitive reactance is

$$X_C = \frac{1}{C\omega} \tag{11.101}$$
$$= \frac{1}{(8.00 \times 10^{-6}\,\text{F})(377\,\text{rad/s})}$$
$$= 332\,\Omega\,.$$

The rms current is

$$I_{rms} = \frac{V_{rms}}{X_C} \qquad (11.102)$$
$$= \frac{150\,V}{332\,\Omega}$$
$$= 0.452\,A = 452\,mA\,.$$

Exercise 11.4 In a series RLC circuit, the applied voltage has a maximum value of 120 V and oscillates with frequency of 60.0 Hz. The circuit consists of an inductor with a varying inductance, a 200 Ω resistor, and a 4.00 μF capacitor. Calculate the value of L such that the voltage across the capacitor lags the applied voltage by 30°.

Solution 11.4 In Fig. 11.13, we present the phasor diagram. From Fig. 11.13, $\phi = -60°$, where the minus sign indicates that the angle is measured clockwise. Furthermore,

$$\tan \phi = \frac{X_L - X_C}{R}\,. \qquad (11.103)$$

Or,

$$2\pi f L = X_L = X_C + R \tan \phi\,. \qquad (11.104)$$

Thus,

Fig. 11.13 The phasor diagram of the problem

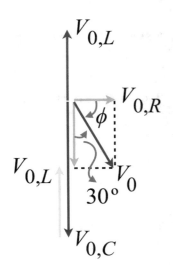

$$L = \frac{1}{(2\pi f)^2 C} + \frac{R}{2\pi f} \tan \phi \tag{11.105}$$

$$= \frac{1}{(2\pi 60.0\,\text{Hz})^2 (4.00 \times 10^{-6}\,\text{F})} + \frac{200\,\Omega}{2\pi 60.0\,\text{Hz}} \tan(-60.0°)$$

$$= 0.840\,\text{H} = 840\,\text{mH}\,.$$

Exercise 11.5 A series of RLC AC circuit has

$$R = 425\,\Omega; \quad L = 1.25\,\text{H}; \quad C = 3.50\,\mu\text{F}; \tag{11.106}$$
$$w = 377\,\text{rad/s}; \quad V_0 = 150\,\text{V}\,.$$

(a) Calculate the inductive reactance, capacitive reactance, and the impedance.
(b) Calculate maximum current.
(c) Calculate the phase angle shift between the current and voltage.
(d) Determine the maximum voltage and instantaneous voltage across any element.

Solution 11.5 (a) The inductive reactance is

$$X_L = L\omega \tag{11.107}$$
$$= (1.25\,\text{H})(377\,\text{rad/s})$$
$$= 471\,\Omega\,.$$

The capacitive reactance is

$$X_C = \frac{1}{C\omega} \tag{11.108}$$
$$= \frac{1}{(3.50 \times 10^{-4},\text{F})(377\,\text{rad/s})}$$
$$= 758\,\Omega\,.$$

The impedance is given by

$$Z = \sqrt{R^2 + (X_L - X_C)^2} \tag{11.109}$$
$$= \sqrt{(425\,\Omega)^2 + (471\,\Omega - 758\,\Omega)^2}$$
$$= 513\,\Omega\,.$$

(b) The maximum current is

$$I_0 = \frac{V_0}{Z} \tag{11.110}$$
$$= \frac{150\,\text{V}}{513\,\Omega}$$
$$= 0.292\,\text{A} = 292\,\text{mA}\,.$$

(c) The phase angle shift is

$$\phi = \tan^{-1}\left(\frac{X_L - X_C}{R}\right) \tag{11.111}$$
$$= \tan^{-1}\left(\frac{471\,\Omega - 758\,\Omega}{425\,\Omega}\right)$$
$$= \tan^{-1}\left(-\frac{287}{425}\right)$$
$$= -34.0°\,.$$

(d) The voltage drop across the resistor is

$$\Delta V_R = V_{0,R}\sin(\omega t)\,. \tag{11.112}$$

Thus,

$$V_{0,R} = I_0 R \tag{11.113}$$
$$= (0.292\,\text{A})(425\,\Omega) = 124\,\text{V}\,.$$

Substituting Eq. (11.113) into (11.112), we have

$$\Delta V_R = (120\,\text{V})\sin(377t)\,. \tag{11.114}$$

The voltage drop across the inductor is

$$\Delta V_L = V_{0,L}\sin\left(\omega t - \frac{\pi}{2}\right) = -V_{0,L}\cos(377t)\,, \tag{11.115}$$

where the amplitude is

$$V_{0,L} = I_0 X_L \tag{11.116}$$
$$= (0.292\,\text{A})(471\,\Omega) = 138\,\text{V}\,.$$

Therefore,

$$\Delta V_L = -(138\,\text{V})\cos(377t)\,, \tag{11.117}$$

Next, we determine the voltage drop across the capacitor as

$$\Delta V_C = V_{0,C} \sin \left(\omega t + \frac{\pi}{2} \right) = V_{0,C} \cos (377t) , \qquad (11.118)$$

with amplitude determined as

$$V_{0,C} = I_0 X_C \qquad (11.119)$$
$$= (0.292 \, \text{A})(758 \, \Omega) = 221 \, \text{V} .$$

The voltage across the capacitor is then

$$\Delta V_C = (221 \, \text{V}) \cos (377t) , \qquad (11.120)$$

Exercise 11.6 Calculate the average power delivered to the series RLC circuit with $R = 425 \, \Omega$, $L = 1.25 \, \text{H}$, $C = 3.50 \, \mu\text{F}$, $\omega = 377 \, \text{rad/s}$, and $V_0 = 150 \, \text{V}$.

Solution 11.6 The rms current is

$$I_{\text{rms}} = \frac{I_0}{\sqrt{2}} = \frac{V_0}{\sqrt{2}Z} , \qquad (11.121)$$

where

$$Z = \sqrt{R^2 + (X_L - X_C)^2} \qquad (11.122)$$
$$= \sqrt{R^2 + \left(2\pi f L - \frac{1}{2\pi f C} \right)^2}$$
$$= 513 \, \Omega .$$

Therefore, the rms current is

$$I_{\text{rms}} = \frac{150 \, \text{V}}{\sqrt{2}(513 \, \Omega)} = 0.206 \, \text{A} = 206 \, \text{mA} . \qquad (11.123)$$

The rms voltage is

$$V_{\text{rms}} = \frac{V_0}{\sqrt{2}} = \frac{150 \, \text{V}}{\sqrt{2}} = 106 \, \text{V} . \qquad (11.124)$$

The phase angle shift is

$$\phi = \tan^{-1}\left(\frac{X_L - X_C}{R}\right) \tag{11.125}$$

$$= \tan^{-1}\left(\frac{471\,\Omega - 758\,\Omega}{425\,\Omega}\right)$$

$$= -34.0°\,.$$

Therefore, the power factor is

$$F = \cos\phi = \cos(-34.0°) = 0.829\,. \tag{11.126}$$

The average power delivered by the AC generator is

$$\mathcal{P}_{av} = I_{rms}V_{rms}\cos\phi \tag{11.127}$$
$$= (0.206\,\text{A})(106\,\text{V})(0.829)$$
$$= 18.1\,\text{W}\,.$$

Exercise 11.7 Let us assume a series RLC circuit in which $R = 150\,\Omega$, $L = 20.0\,\text{mH}$, $V_{rms} = 20.0\,\text{V}$, and $\omega = 5000\,\text{rad/s}$. Calculate the capacitance C for which the current is maximum.

Solution 11.7 The current is maximum at resonance. The resonance frequency is

$$\omega_0 = \frac{1}{\sqrt{LC}}\,. \tag{11.128}$$

Solving it for C, we get

$$C = \frac{1}{L\omega_0^2} \tag{11.129}$$

$$= \frac{1}{(20.0 \times 10^{-3}\,\text{H})(5.00 \times 10^3\,\text{rad/s})^2}$$
$$= 2.00 \times 10^{-6}\,\text{F} = 2.00\,\mu\text{F}\,.$$

Exercise 11.8 An electricity generator needs to deliver 20 MW of power to a city 1.0 km away.

(a) If the resistance of the wire is $2.0\,\Omega$ and the electricity costs about $0.10\,\$/\text{kWh}$ calculate the cost of the utilities to the company to send the

power to the city for one day. A common voltage for commercial power
generators is 22 kV, but a step up transformer is used to boost the voltage
to 230 kV before transmission.

(b) Do the same calculations when the power plant delivers the electricity at
its original voltage of 22 kV.

Solution 11.8 (a) The power losses in the transmission line are due to the line
resistance:

$$P = I^2 R = IV .$$
(11.130)

The current is

$$I = \frac{P}{V} = \frac{20 \times 10^6 \text{ W}}{230 \times 10^3 \text{ V}} = 87 \text{ A} .$$
(11.131)

Therefore,

$$P = (87 \text{ A})^2 (2.0 \, \Omega) = 15 \text{ kW} .$$
(11.132)

The energy loss in 1 day (that is 24 h) is

$$E_{\text{loss}} = P(24 \text{ h}) = (15 \text{ kW})(24 \text{ h}) = 360 \text{ kWh} .$$
(11.133)

The cost in dollars is

$$\text{Cost} = E_{\text{loss}}(0.10 \, \$/\text{kWh}) = (360 \text{ kWh})(0.10 \, \$/\text{kWh}) = 36 \, \$.$$
(11.134)

(b) The current is

$$I = \frac{P}{V} = \frac{20 \times 10^6 \text{ W}}{22 \times 10^3 \text{ V}} = 910 \text{ A} .$$
(11.135)

The power loss is

$$P = (910 \text{ A})^2 (2.0 \, \Omega) = 1700 \text{ kW} .$$
(11.136)

The energy loss in 24 h is

$$E_{\text{loss}} = P(24 \text{ h}) = (1700 \text{ kW})(24 \text{ h}) = 40800 \text{ kWh} .$$
(11.137)

The cost in dollars is

$$\begin{aligned}
\text{Cost} &= E_{\text{loss}}(0.10 \, \$/\text{kWh}) \\
&= (40800 \text{ kWh})(0.10 \, \$/\text{kWh}) \\
&= 4080 \, \$.
\end{aligned}$$
(11.138)

Exercise 11.9 The rms voltage in output of an AC generator is 200 V and operating frequency is 100 Hz. Determine the expression of the output voltage as a function of time t.

Solution 11.9 First, the angular frequency is

$$\omega = 2\pi f = 2\pi(100\,\text{Hz}) = 628\,\text{rad/s} .\qquad(11.139)$$

The amplitude of the output voltage is

$$V_0 = V_{\text{rms}}\sqrt{2} = (200\,\text{V})\sqrt{2} = 288\,\text{V} .\qquad(11.140)$$

Therefore, the instantaneous output voltage is

$$\Delta V(t) = V_0 \sin(\omega t) = (288\,\text{V})\sin(628t) .\qquad(11.141)$$

Exercise 11.10 (a) Calculate the resistance of a light bulb that has an average power of $\mathcal{P}_{\text{av}} = 75$ W, connected to a 60 Hz power source having a maximum voltage of 170 V. (b) Calculate the resistance of a 100 W bulb.

Solution 11.10 (a) For an AC generator, the output voltage is

$$\Delta V(t) = V_0 \sin(\omega t) ,\qquad(11.142)$$

where $V_0 = 170$ V and ω is

$$\omega = 2\pi f = 2\pi(60\,\text{Hz}) = 377\,\text{rad/s} .\qquad(11.143)$$

Thus,
$$\Delta V(t) = (170\,\text{V})\sin(377t) .\qquad(11.144)$$

The average power is
$$\mathcal{P}_{\text{av}} = \frac{V_0^2}{2R} .\qquad(11.145)$$

Therefore, the resistance is

$$R = \frac{V_0^2}{2\mathcal{P}_{\text{av}}} = \frac{(170\,\text{V})^2}{2(75\,\text{W})} = 192.7\,\Omega .\qquad(11.146)$$

(b) If $\mathcal{P}_{\mathrm{av}} = 100\,\mathrm{W}$, then

$$R = \frac{V_0^2}{2\mathcal{P}_{\mathrm{av}}} = \frac{(170\,\mathrm{V})^2}{2(100\,\mathrm{W})} = 144.5\,\Omega \,. \tag{11.147}$$

References

Anandakrishman R, Aguilar B, Onufriev AV (2012) Nucleic acids research 40:W537–W541

Brooks BR, Brooks CL, MacKerell AD, Nilsson L, Petrella RJ, Roux B, Won Y, Archontis G, Bartels C, Boresch S, Caflisch A, Caves L, Cui Q, Dinner AR, Feig M, Fischer S, Gao J, Hodoscek M, Im W, Kuczera K, Lazaridis T, Ma J, Ovchinnikov V, Paci E, Pastor RW, Post CB, Pu JZ, Schaefer M, Tidor B, Venable RM, Woodcock HL, Wu X, Yang W, York DM, Karplus M (2009) CHARMM: The biomolecular simulation program. J Comput Chem 30:1545

Czapinska H, Krzywda S, Sheldrick GM, Otlewski J, Jaskolski M (1999) High-resolution structure of bovine pancreatic trypsin inhibitor with altered binding loop sequence. J Mol Biol 295:1237

Deuflhard P, Hermans J, Leimkuhler B, Mark AE, Reich S, Skeel RD (eds) (1997) Computational molecular dynamics: challenges. Methods, ideas. Springer

Kamberaj H (2020) Molecular dynamics simulations in statistical physics: theory and applications. Springer Nature

MacKerell AD Jr, Brooks B, Brooks CL III, Nilsson L, Roux B, Won Y, Karplus M (1998) CHARMM: the energy function and its parametrization with an overview of the program. Wiley, Chichester

Martin C, Richard V, Salem M, Hartley RW, Mauguen Y (1999) Acta Crystallogr Sect D 55:386

R. Phillips and J. Kondev and J. Theriot and H. G. Garcia (2013) Physical Biology of the Cell, 2nd edn. Garland Science, Taylor and Francis Group

Chapter 12
Electromagnetic Waves in Vacuum and Linear Medium

This chapter aims to derive electromagnetic wave equations in vacuum and a linear medium. Also, this chapter aims to introduce reflection's law and Snell's law of optics, and the Fresnel's equations.

In this chapter, we discuss the electromagnetic wave equations in vacuum and linear media. Furthermore, the reflection's law and the Snell's law are derived, along with the polarization of electromagnetic planar waves and Fresnel's equations. For further reading, one can also consider other available literature (Holliday et al. 2011; Jackson 1999; Landau and Lifshitz 1971; Sykja 2006; Griffiths 1999; Altland and Simons 2010; Protheroe 2013).

12.1 Electromagnetic Wave Equations in Vacuum

Considering the Maxwell's equations (see Eq. (9.8) in Chap. 9), the electromagnetic wave equations in vacuum are derived. Furthermore, we consider charge-free and current-free medium. These are waves traveling with speed c in vacuum.

Taking the curl of the second equation in Eq. (9.8) (Chap. 9), known as Faraday law, we obtain

$$\nabla \times (\nabla \times \mathbf{E}) = -\nabla \times \left(\frac{\partial \mathbf{B}}{\partial t} \right). \qquad (12.1)$$

Using the expression

$$\nabla \times (\nabla \times \mathbf{a}) = \nabla (\nabla \cdot \mathbf{a}) - \nabla^2 \mathbf{a}, \qquad (12.2)$$

© The Author(s), under exclusive license to Springer Nature Switzerland AG 2022
H. Kamberaj, *Electromagnetism*, Undergraduate Texts in Physics,
https://doi.org/10.1007/978-3-030-96780-2_12

we can write that

$$\nabla(\nabla \cdot \mathbf{E}) - \nabla^2 \mathbf{E} = -\frac{\partial}{\partial t}(\nabla \times \mathbf{B}). \tag{12.3}$$

Substituting the third expression of Eq. (9.8) (Chap. 9) into Eq. (12.3), we find

$$\nabla(\nabla \cdot \mathbf{E}) - \nabla^2 \mathbf{E} = -\mu_0 \frac{\partial \mathbf{J}}{\partial t} - \mu_0 \epsilon_0 \frac{\partial^2 \mathbf{E}}{\partial t^2}. \tag{12.4}$$

Since there is vacuum, the current density is zero, $\mathbf{J} = 0$, and the charge density $\rho = 0$; therefore, $\nabla \cdot \mathbf{E} = \rho/\epsilon_0 = 0$ (see also Eq. (9.8), Chap. 9). Thus, we finally obtain

$$\nabla^2 \mathbf{E} - \mu_0 \epsilon_0 \frac{\partial^2 \mathbf{E}}{\partial t^2} = 0. \tag{12.5}$$

Using the relation $c = 1/\sqrt{\mu_0 \epsilon_0}$ for the speed of light in vacuum, we write

$$\nabla^2 \mathbf{E} - \frac{1}{c^2} \frac{\partial^2 \mathbf{E}}{\partial t^2} = 0. \tag{12.6}$$

Now, we can take the curl of the third expression in Eq. (9.8) (Chap. 9), known as Ampére's law:

$$\nabla \times (\nabla \times \mathbf{B}) = \mu_0 \nabla \times \mathbf{J} + \mu_0 \epsilon_0 \frac{\partial}{\partial t}(\nabla \times \mathbf{E}). \tag{12.7}$$

Using Eq. (12.2), the fourth expression in Eq. (9.8) (Chap. 9), and $\mathbf{J} = 0$, we find

$$\nabla^2 \mathbf{B} - \frac{1}{c^2} \frac{\partial^2 \mathbf{B}}{\partial t^2} = 0. \tag{12.8}$$

Equations (12.6) and (12.8) give the wave equations for the electric field vector $\mathbf{E}$ and magnetic field vector $\mathbf{B}$, respectively. Note that these equations represent planar wave equations in three dimensions and they have the same form. Furthermore, the speed of the wave propagation in vacuum is the speed of light in vacuum c. The monochromatic plane wave solutions of Eqs. (12.6) and (12.8) are given as

$$\mathbf{E}(\mathbf{r}, t) = \mathbf{E}_0 \exp(i(\mathbf{k} \cdot \mathbf{r} - \omega t)), \tag{12.9}$$
$$\mathbf{B}(\mathbf{r}, t) = \mathbf{B}_0 \exp(i(\mathbf{k} \cdot \mathbf{r} - \omega t)),$$

where $i = \sqrt{-1}$. In Eq. (12.9), $\mathbf{k}$ is the wave vector, with magnitude given as

$$k = \frac{2\pi}{\lambda}, \tag{12.10}$$

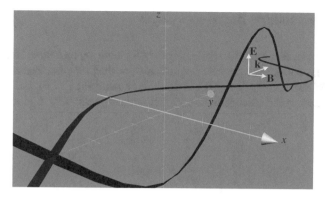

Fig. 12.1 Illustration of the electromagnetic wave propagation along y-axis

where λ is the wavelength, which denotes the distance between adjacency maximum points:

$$\lambda = cT, \tag{12.11}$$

where T is the period. Direction of the vector $\mathbf{k}$ gives the direction of the wave's propagation. The magnetic and electric field vectors are identical at all points at with the phase angle is

$$\mathbf{k} \cdot \mathbf{r} - \omega t = 2m\pi + \text{constant}, \tag{12.12}$$

where $m = 0, 1, 2, \ldots$ That is, the vectors $(\mathbf{E}, \mathbf{B})$ are identical in planes perpendicular to vector $\mathbf{k}$ that are separated one wavelength λ.

Figure 12.1 illustrates propagation of electromagnetic wave along positive y-axis, and relative orientation of vectors $(\mathbf{E}, \mathbf{B})$ and $\mathbf{k}$.

ω is the angular frequency, which is related

$$c = \frac{\omega}{k}, \tag{12.13}$$

where c defines the phase velocity v_p (which is equal to the propagation of the electromagnetic waves velocity in vacuum c). In the dispersion medium, $k = k(\omega)$.

Therefore, combining Eqs. (12.10), (12.11), and (12.13), we write

$$\lambda = \frac{2\pi}{\omega}. \tag{12.14}$$

Note that the expressions in Eq. (12.9) represent the solutions for the wave equations; however, they will also represent electromagnetic waves only they satisfy the Maxwell's equations (see Eq. (9.8), Chap. 9).

12.2   Relationships Between k, E, B

The solutions $(\mathbf{E}, \mathbf{B})$ monochromatic plane waves have the constants $(\mathbf{E}_0, \mathbf{B}_0)$, ω and $\mathbf{k}$ such that $(\mathbf{E}, \mathbf{B})$ satisfy Maxwell's equations. For that, the time derivative, curl and divergence of the vectors $\mathbf{E}$ and $\mathbf{B}$ are calculated using Eq. (12.9). For instance, the time derivatives of the electric field vector $\mathbf{E}$ and magnetic field $\mathbf{B}$ give

$$\frac{\partial \mathbf{E}}{\partial t} = -i\omega \mathbf{E}, \tag{12.15}$$

$$\frac{\partial \mathbf{B}}{\partial t} = -i\omega \mathbf{B}.$$

Taking the curl of $\mathbf{E}$ and $\mathbf{B}$, we write

$$\nabla \times \mathbf{E} = i\,(\mathbf{k} \times \mathbf{E})\,, \tag{12.16}$$

$$\nabla \times \mathbf{B} = i\,(\mathbf{k} \times \mathbf{B})\,.$$

Using the second Maxwell's equation $(\nabla \times \mathbf{E} = -\partial \mathbf{B}/\partial t)$, and combining Eqs. (12.15) and (12.16), we find that

$$\mathbf{k} \times \mathbf{E} = \omega \mathbf{B}. \tag{12.17}$$

Using the fourth Maxwell's equation $(\nabla \times \mathbf{B} = \mu_0 \epsilon_0 \partial \mathbf{E}/\partial t$ for $\mathbf{J} = 0)$, we obtain

$$\mathbf{k} \times \mathbf{B} = -\omega \mu_0 \epsilon_0 \mathbf{E}. \tag{12.18}$$

Taking the divergence of both sides in expressions given by Eq. (12.9), it can be written that

$$\nabla \cdot \mathbf{E} = i\,(\mathbf{k} \cdot \mathbf{E})\,, \tag{12.19}$$

$$\nabla \cdot \mathbf{B} = i\,(\mathbf{k} \cdot \mathbf{B})\,.$$

Using the first Maxwell's equation $(\nabla \cdot \mathbf{E} = 0$ for $\rho = 0)$, we get

$$\mathbf{k} \cdot \mathbf{E} = 0, \tag{12.20}$$

and from the third Maxwell's equation $(\nabla \cdot \mathbf{B} = 0)$, we obtain

$$\mathbf{k} \cdot \mathbf{B} = 0. \tag{12.21}$$

From Eqs. (12.20) and (12.21), we conclude that both electric field vector $\mathbf{E}$ and magnetic field vector $\mathbf{k}$ are perpendicular to vector $\mathbf{k}$. Note that $\mathbf{k}$ represents the direction of propagation of the electromagnetic wave; therefore, $\mathbf{E}$ and $\mathbf{B}$ are perpendicular to the direction of propagation of electromagnetic wave, and hence, the electromagnetic waves are transverse waves.

Furthermore,

$$\mathbf{E} \cdot \mathbf{B} = \mathbf{E} \cdot (\mathbf{k} \times \mathbf{E})/\omega = 0, \tag{12.22}$$

which indicates that **E** and **B** are perpendicular. Therefore, we can write that

$$\mathbf{E}(\mathbf{r}, t) = \mathbf{E}_0 \exp\left(i\left(\mathbf{k} \cdot \mathbf{r} - \omega t\right)\right), \tag{12.23}$$

$$\mathbf{B}(\mathbf{r}, t) = \frac{\mathbf{k} \times \mathbf{E}(\mathbf{r}, t)}{\omega},$$

where Eq. (12.17) is used, and $\mathbf{k} \cdot \mathbf{E}_0 = 0$. Using Eq. (12.13), we write

$$\mathbf{B}(\mathbf{r}, t) = \frac{k E(\mathbf{r}, t)}{\omega} = \frac{1}{c} E(\mathbf{r}, t). \tag{12.24}$$

12.3 Electromagnetic Waves Equations in Linear Medium

Next, we consider the electromagnetic waves in a linear medium of charge-free and current-free medium. For a linear medium, the electric polarization vector and magnetization are considered to depend linearly on the fields:

$$\mathbf{P} = \epsilon_0 \chi_e \mathbf{E} \tag{12.25}$$

$$\mathbf{M} = \chi_m \mathbf{H}$$

and hence

$$\mathbf{D} = \epsilon \epsilon_0 \mathbf{E} \tag{12.26}$$

$$\mathbf{B} = \mu \mu_0 \mathbf{H}$$

where $\epsilon = 1 + \chi_e$ and $\mu = 1 + \chi_m$. For the linear medium, μ_0 and ϵ_0 of the vacuum are considered the limit of the magnetic permeability $\mu_m \equiv \mu \mu_0$ and the electric permittivity of medium $\epsilon_m \equiv \epsilon_0 \epsilon$ of the linear medium, respectively, when $\chi_m \to 0$ and $\chi_e \to 0$. That is, the propagation speed of the electromagnetic wave in a linear medium is

$$v_p = \frac{1}{\sqrt{\mu_m \epsilon_m}} = \frac{c}{\sqrt{\mu \epsilon}}. \tag{12.27}$$

The curl of the second Maxwell's equation for a linear medium becomes

$$\nabla \times (\nabla \times \mathbf{E}) = -\frac{\partial}{\partial t} \left(\mu_0 \mu \nabla \times \mathbf{H}\right). \tag{12.28}$$

Using Eq. (12.2) and the fourth Maxwell's equation for a linear medium with no currents ($\nabla \times \mathbf{H} = \partial \mathbf{D}/\partial t$), we obtain

$$\nabla^2 \mathbf{E} = \epsilon_m \mu_m \frac{\partial^2 \mathbf{E}}{\partial t^2}, \tag{12.29}$$

where $\mathbf{D} = \epsilon_m \mathbf{E}$. Thus, we obtain the wave equation for electric field in a linear medium as

$$\nabla^2 \mathbf{E} - \frac{1}{v_p^2} \frac{\partial^2 \mathbf{E}}{\partial t^2} = 0. \tag{12.30}$$

Similarly, taking the curl of the fourth Maxwell's equation in the medium ($\nabla \times \mathbf{H} = \partial \mathbf{D} / \partial t$ for $\mathbf{J} = 0$), we write

$$\nabla \times (\nabla \times \mathbf{H}) = -\frac{\partial}{\partial t} (\nabla \times \mathbf{D}). \tag{12.31}$$

Or,

$$\frac{1}{\mu_m} \nabla \times (\nabla \times \mathbf{B}) = -\epsilon_m \frac{\partial}{\partial t} (\nabla \times \mathbf{E}). \tag{12.32}$$

Finally, Eq. (12.32) can be written in the form of the wave equation for $\mathbf{B}$ as

$$\nabla^2 \mathbf{B} - \frac{1}{v_p^2} \frac{\partial^2 \mathbf{B}}{\partial t^2} = 0. \tag{12.33}$$

Here, Eqs. (12.30) and (12.33) represent the electromagnetic wave equations in a linear medium propagating with speed v_p.

Note that all the relations derived in the previous section for vacuum between $\mathbf{k}$, $\mathbf{E}$ and $\mathbf{B}$ also hold for the linear medium by replacing c with v_p. In addition, from Eq. (12.24), can write

$$H(\mathbf{r}, t) = \frac{B(\mathbf{r}, t)}{\mu_m} \tag{12.34}$$

$$= \frac{1}{\mu_m} \frac{E(\mathbf{r}, t)}{v_p}$$

$$= \sqrt{\frac{\epsilon_m}{\mu_m}} E(\mathbf{r}, t) \equiv \frac{1}{Z} E(\mathbf{r}, t),$$

where Z is the wave impedance, a characteristic of the medium. For the vacuum,

$$Z = \sqrt{\frac{\mu_m}{\epsilon_m}} = \sqrt{\frac{\mu_0}{\epsilon_0}} \approx 377 \, \Omega. \tag{12.35}$$

12.4   Energy and Momentum of Electromagnetic Waves

Using the relation in Eq. (12.24), the contributions of electric and magnetic field to the total energy density are equal; that is,

$$\epsilon_m \frac{E^2}{2} = \frac{B^2}{2\mu_m}. \tag{12.36}$$

Therefore, the total energy density in the electromagnetic wave is

$$u = \left(\epsilon_m \frac{E^2}{2} + \frac{B^2}{2\mu_m} \right) = \epsilon_m E^2 = \frac{B^2}{\mu_m}. \tag{12.37}$$

Furthermore, we can write that

$$u(\mathbf{r}, t) = \epsilon_m E_0^2 \cos^2 (\mathbf{k} \cdot \mathbf{r} - \omega t + \delta), \tag{12.38}$$

where δ denotes a phase angle shift. The flux of energy is given by Poynting vector $\mathbf{S} = \mathbf{E} \times \mathbf{H}$. For the electromagnetic waves in a linear non-dispersion medium, we have

$$\mathbf{S} = \mathbf{E} \times \left(\frac{\mathbf{k} \times \mathbf{E}}{\mu_m \omega} \right) \tag{12.39}$$

$$= \frac{1}{\mu_m \omega} (\mathbf{k}(\mathbf{E} \cdot \mathbf{E}) - (\mathbf{E} \cdot \mathbf{k})\mathbf{E})$$

$$= \frac{1}{\mu_m \omega} E^2 \mathbf{k}$$

$$= \frac{1}{\mu_m \omega} \frac{\omega}{v_p} E^2 \hat{\mathbf{k}}$$

$$= \frac{\sqrt{\mu_m \epsilon_m}}{\mu_m} E^2 \hat{\mathbf{k}}$$

$$= \sqrt{\frac{\epsilon_m}{\mu_m}} E^2 \hat{\mathbf{k}} = v_p \epsilon_m E^2 \hat{\mathbf{k}},$$

where $\mathbf{E} \cdot \mathbf{k} = 0$ is used, and $\hat{\mathbf{k}}$ is a unit vector along $\mathbf{k}$. Therefore, using Eqs. (12.38) and (12.39), we have

$$\mathbf{S}(\mathbf{r}, t) = v_p u(\mathbf{r}, t)\hat{\mathbf{k}}. \tag{12.40}$$

The intensity of the electromagnetic wave is defined as the time average of the magnitude of the Poynting vector $\mathbf{S}$ as follows:

$$I = v_p \epsilon_m \mid E_0 \mid^2 \langle \cos^2(\mathbf{k} \cdot \mathbf{r} - \omega t + \delta) \rangle = v_p \frac{\epsilon_m \mid E_0 \mid^2}{2}. \tag{12.41}$$

The units of the intensity are W/m^2.

The momentum density is defined as

$$\mathbf{g} = \frac{\mathbf{S}}{c^2}.$$ (12.42)

Then, the momentum flux of an electromagnetic wave in a linear medium is given as

$$\mathbf{p} = v_p \mathbf{g} = v_p \frac{\mathbf{S}}{c^2}.$$ (12.43)

The magnitude of the momentum flux of the electromagnetic wave gives the amount of momentum crossing unit area of a surface perpendicular to the vector $\hat{\mathbf{k}}$ per unit of time. Thus, it can be defined as the radiation pressure. If a parallel beam of radiation is perfectly absorbed by a surface perpendicular to the direction of propagation of the electromagnetic wave, then the radiation pressure is

$$P_{rad,abs} = \frac{v_p}{c^2} \langle S \rangle = \frac{v_p^2}{c^2} \langle u(\mathbf{r}, t) \rangle,$$ (12.44)

where $\langle \cdots \rangle$ denotes a time average. Note that if the radiation is perfectly reflected, then the radiation pressure is twice as large:

$$P_{rad,ref} = 2\frac{v_p^2}{c^2} \langle u(\mathbf{r}, t) \rangle.$$ (12.45)

In the case of the isotropic electromagnetic wave, the radiation pressure is obtained by integrating the momentum flux incident on the area A of one side of a plane surface, as shown in Fig. 12.2. Considering only the component of the momentum density perpendicular to the plane ($\mathbf{g} \cos \theta$) and the projected area $A \cos \theta$ (as depicted in Fig. 12.2), we obtain

$$P_{rad,iso} = \frac{1}{A} \int_0^1 \frac{v_p^2}{c^2} \langle u(\mathbf{r}, t) \rangle \cos \theta A \cos \theta d(\cos \theta) = \frac{1}{3} \frac{v_p^2}{c^2} \langle u(\mathbf{r}, t) \rangle.$$ (12.46)

Fig. 12.2 Illustration of the radiation pressure of an isotropic electromagnetic wave

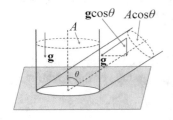

In vacuum, $v_p = c$, and hence from Eq. (12.46), we find that

$$p_{\text{rad,iso}} = \frac{1}{3} \langle u(\mathbf{r}, t) \rangle, \tag{12.47}$$

which represents the state's equation for the gas of photons.

12.5 Coherence of Electromagnetic Waves

The monochromatic plane electromagnetic waves are an idealization picture because they have infinite extent in all directions. Furthermore, they are just sinusoidal waves, and hence their angular frequency spread is zero, $\Delta\omega = 0$. The same would have been true for the monochromatic spherical waves. These waves (that is, having $\Delta\omega = 0$) are called completely coherent. If two identical of such waves superimpose, then the interference can be seen. For instance, consider two electromagnetic waves with magnitudes of electric field vectors as

$$E_1 = E_0 \cos(\mathbf{k} \cdot \mathbf{r}_1 - \omega t), \tag{12.48}$$
$$E_2 = E_0 \cos(\mathbf{k} \cdot \mathbf{r}_2 - \omega t + \delta),$$

where δ is a phase angle shift; moreover, we have assumed they have the same amplitude E_0. The resultant wave is given as follows:

$$
\begin{aligned}
E_R &= E_1 + E_2 \\
&= E_0 \cos(\mathbf{k} \cdot \mathbf{r}_1 - \omega t) + E_0 \cos(\mathbf{k} \cdot \mathbf{r}_2 - \omega t + \delta) \\
&= 2E_0 \cos\left(\frac{\mathbf{k} \cdot \mathbf{r}_2 - \omega t + \delta - \mathbf{k} \cdot \mathbf{r}_1 + \omega t}{2}\right) \\
&\quad \times \sin\left(\frac{\mathbf{k} \cdot \mathbf{r}_2 - \omega t + \delta + \mathbf{k} \cdot \mathbf{r}_1 - \omega t}{2}\right) \\
&= 2E_0 \cos\left(\frac{\mathbf{k} \cdot (\mathbf{r}_2 - \mathbf{r}_1) + \delta}{2}\right) \sin\left(\frac{\mathbf{k} \cdot (\mathbf{r}_2 + \mathbf{r}_1)}{2} - \omega t + \frac{\delta}{2}\right),
\end{aligned}
\tag{12.49}
$$

which is the equation of a traveling wave, and its amplitude is

$$E_{R0} = 2E_0 \cos\left(\frac{\mathbf{k} \cdot (\mathbf{r}_2 - \mathbf{r}_1) + \delta}{2}\right). \tag{12.50}$$

From Eq. (12.50), when

$$\mathbf{k} \cdot (\mathbf{r}_2 - \mathbf{r}_1) + \delta = 2n\pi, \quad n = 0, 1, 2, \ldots, \tag{12.51}$$

the amplitude of the resultant wave is maximum $E_{R0} = \pm 2E_0$, and we have constructive interference, whereas when

$$\mathbf{k} \cdot (\mathbf{r}_2 - \mathbf{r}_1) + \delta = (2n + 1)\pi, \quad n = 0, 1, 2, \ldots, \tag{12.52}$$

the amplitude of the resultant wave is minimum $E_{R0} = 0$, and we have destructive interference. Therefore, the interference pattern will be constructed of alternating bright and dark fringes between 100 % constructive interference with $I_{max} = v_p(\epsilon_m/2)E_{R0}^2 = 4I_0$, and 100 % destructive interference with $I_{min} = 0$. Thus, the so-called visibility of the fringes is

$$V = \frac{I_{max} - I_{min}}{I_{max} + I_{min}} = 1. \tag{12.53}$$

In reality, the electromagnetic waves are only partially coherent; that is, they may exhibit coherence at some particular location over a limited time. Let t_c be the coherence time and the coherence length $L_c = v_p t_c$. Then, electromagnetic waves may display coherence due to the partially coherent wave train of length L_c passing through that location over the time t_c. The width of the wave front across which the coherence is maintained is the so-called coherence width w_c. The time or distance over which the phase is significantly different from the pure sinusoidal wave is used to determine t_c, L_c, and w_c, which can be used to determine the conditions under which a monochromatic electromagnetic planar wave may be a good approximation to the electromagnetic field present.

The real sources of electromagnetic waves will produce waves having a finite extent and have a spread of frequencies. For instance, the atomic transition have a natural line width of about $\Delta\omega \sim \Gamma$, where Γ is the transition rate. Therefore, the spectral line emission can be considered quasi-monochromatic.

It can be shown that

$$\Delta x \, \Delta k_x \sim 1. \tag{12.54}$$

Since $v_p = \omega/k$ and that the wave travels in the direction x so that $k_x = k$, and $\hat{\mathbf{k}} = \hat{\mathbf{x}}$, then

$$\Delta k_x = \Delta\omega/v_p. \tag{12.55}$$

Then, the wave trains associated with emitted photons have a coherence length of

$$L_c = \Delta x \sim \frac{v_p}{\Gamma}. \tag{12.56}$$

In general, the coherence time will be shorter, and so the observed width of a spectral emission line broader. For example, if during the transition, an atom collides with another atom, then a phase angle shift occurs. Therefore, the coherence length will be reduced. That effect becomes more significant when the density and temperature increase because the probability of a collision to occur increases. This effect

Fig. 12.3 Illustration of the coherence width

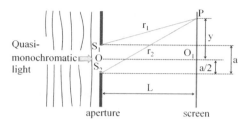

is called pressure broadening. Besides, the frequency of the emitted photon by an individual atom will experience a Doppler shift because of the thermal motion. The Doppler shift amount depend on the velocity component of the atom in the direction toward the observer. The Doppler broadening produces a Gaussian line shape with a line width proportional to $T^{1/2}$, where T is the temperature.

In Fig. 12.3, it is illustrated the experiment on determining the coherence width. For that, consider a quasi-monochromatic plane electromagnetic wave reaching an aperture with diameter a, which is at distance L from a screen. If we consider the electromagnetic field at the point O_1 on the screen, the contributions from the entire source add up (that is, all points between S_1 and S_2). All these contributions will have a finite phase at O_1 given that the difference between the paths $S_1 O_1$ and $S_2 O_1$ is much smaller than $\lambda/2$. That is equivalent to the conditions that $ka^2 \ll L$. If we consider the point P on the screen, which is at distance y from the center of the screen O_1, the coherence will be lost as the difference in the paths r_1 and r_2 is comparable to $\lambda/2$ (or equivalently, a phase difference of $\Delta\phi = \pi$). The distance y_c for which this occurs denotes the radius of the patch of the wave's front called coherence area:

$$A_c = \pi y_c^2. \tag{12.57}$$

This area defines space over which the front of the wave may be considered spatially coherent, with a coherence width $d_c = 2y_c$. If we consider that there is no phase angle difference between the incident electromagnetic waves reaching the point P, then the phase difference is due to the path difference given as (for $a < y \ll L$)

$$\Delta\phi = k(r_2 - r_1) \approx \frac{kya}{L}. \tag{12.58}$$

Therefore,

$$d_c = \frac{2\pi L}{ka} = \frac{\lambda L}{a}. \tag{12.59}$$

There are artificial and natural sources of the coherent electromagnetic waves. For example, a natural source of the coherent electromagnetic wave is the astrophysical MASER (Microwave Amplification by Stimulated Emission of Radiation) in which a molecular line is observed at the microwave region of the spectrum. Astrophysical MASER occur in the shocked regions in molecular clouds in the Galaxy. Long-lived upper level of a molecular species is created by excitation through the collisions or infrared photons giving rise to a population inversion; that is, more molecules exist in the upper level than in lower level. That population inversion give rise to stimulated emission to dominate over the absorption, and hence giving a negative absorption coefficient.

A single photon can initiate MASER action from the spontaneous decay. The photon propagates across the region of the molecular cloud having population inversion and the photon repeated stimulated emission of essentially identical photons in phase with itself, similarly to the photons produced by stimulated emission. Therefore, there exists a cascade where the number of photons build up exponentially with distance. Because the path lengths across the cloud are huge the resulting brightness of coherent microwave radiation is sufficiently high to be observed with radio telescopes.

The artificial sources of coherent electromagnetic waves include MASER and LASER. For instance, in a laboratory setup artificial MASERs using hyperfine splitting of the 21 cm line spin-flip transition of the ground state of neutral atom hydrogen in a weak magnetic field, or Zeeman effect, are used as a frequency standard. The LASER (Light Amplifier by Stimulated Emission of Radiation) has many applications. The working principle of a LASER is the same as the astrophysical MASER, introduced above. However, the main difference is that transitions in the optical or infrared are used in LASERs, and depending on the medium various methods are used to create the upper metastable level. Furthermore, in the LASER, the enormous path lengths needed are achieved by creating a LASER cavity containing LASER medium between two parallel mirrors in which the light is reflected back and forth, where only a small fraction of radiation is escaping away from one of the mirrors (which is allowed to partially transmit light) to form a LASER beam.

It is worthy to note that coherence, including the quantum optics, is essential in physical optics. Furthermore, it is a field with a broad area of research, and the electromagnetic field for studying the LASER beam is a good approximation to the monochromatic electromagnetic plane wave.

12.6 Polarization of Electromagnetic Waves

Consider a monochromatic plane electromagnetic wave propagating along the positive z-axis; that is, $\mathbf{k} = k\hat{\mathbf{z}}$, where $\hat{\mathbf{z}}$ is a unit vector along $+z$-direction. Thus, we can write the electric field vector of the electromagnetic field as

$$\mathbf{E}(\mathbf{r}, t) = \left(E_{0x} e^{i\delta_x} \hat{\mathbf{x}} + E_{0y} e^{i\delta_y} \hat{\mathbf{y}} \right) e^{i(kz - \omega t)}. \tag{12.60}$$

The real component of the electric field is

$$\text{Re} \left(\mathbf{E}(\mathbf{r}, t) \right) = E_{0x} \cos \left(kz - \omega t + \delta_x \right) \hat{\mathbf{x}} + E_{0y} \cos \left(kz - \omega t + \delta_y \right) \hat{\mathbf{y}}. \tag{12.61}$$

Fig. 12.4 Linear
polarization

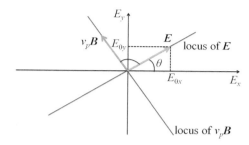

There exist linearly, circularly, and elliptically polarization of electromagnetic plane waves depending on the ration E_{0y}/E_{0x} and the relative phase shift $(\delta_y - \delta_x)$.

12.6.1 Linear Polarization

When $\delta_x = \delta_y \equiv \delta$, the linear polarization is obtained. From Eq. (12.61), we have

$$\mathrm{Re}\,(\mathbf{E}(\mathbf{r}, t)) = \left(E_{0x}\hat{\mathbf{x}} + E_{0y}\hat{\mathbf{y}}\right)\cos\,(kz - \omega t + \delta)\,. \qquad (12.62)$$

The locus of the tip of the vector $\mathbf{E}$ undergoes a simple periodic motion along the line that forms the angle θ (such that $\tan\theta = E_{0y}/E_{0x}$) related to the E_x-axis in a plane of E_y and E_x axes, as depicted in Fig. 12.4. In other words, the electromagnetic wave is linearly polarized where the electric field direction is an angle θ to the E_x-axis, or equivalently a plane polarized electromagnetic wave.

The quasi-monochromatic electromagnetic plane waves can also be linearly polarized, which can be produced by passing the unpolarized light waves across a monochromatic filter and then across a polarized (for example, a Polaroid Film), or by passing a laser beam across a polarizer.

12.6.2 Circular and Elliptical Polarization

In the circular and elliptical polarization, the electric and magnetic field vectors $(\mathbf{E}, \mathbf{B})$ rotate with angular frequency ω. In the case of a circular polarization, $E_{0y} = E_{0x} \equiv E_0$ and $\delta_y - \delta_x = \pm\pi/2$. Therefore, the locus of the tip of vector $\mathbf{E}$ undergoes a circular motion with a radius E_0 in the $x - y$-plane with an angular frequency ω, as shown in Fig. 12.5. Depending on the sign of $\delta_y - \delta_x$ there exist two possibilities, namely the left polarization (or positive helicity) and right polarization (or negative helicity). Practically, one can define $\delta_x = 0$, and discuss the possible polarization in terms of the sign of δ_y. For circularly polarized wave traveling along the positive z-axis (that is, $\mathbf{k} = k\hat{\mathbf{z}}$), we define the basis unit vectors as

Fig. 12.5 Circular
polarization

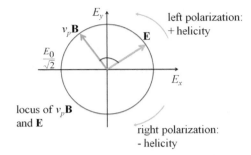

$$\mathbf{e}_R = \frac{1}{\sqrt{2}} \left(\hat{\mathbf{x}} - i\hat{\mathbf{y}} \right), \tag{12.63}$$

$$\mathbf{e}_L = \frac{1}{\sqrt{2}} \left(\hat{\mathbf{x}} + i\hat{\mathbf{y}} \right).$$

The right circular polarized electromagnetic wave has the following electric field vectors:

$$\mathbf{E}(\mathbf{r}, t) = E_0 e^{i(kz-\omega t)} \mathbf{e}_R = \frac{1}{\sqrt{2}} E_0 e^{i(kz-\omega t)} \left(\hat{\mathbf{x}} + e^{-i\pi/2}\hat{\mathbf{y}} \right), \tag{12.64}$$

where $\pm i = \mp e^{-i\pi/2}$ is used. Therefore,

$$\text{Re}\left(\mathbf{E}(\mathbf{r}, t)\right) = \frac{1}{\sqrt{2}} E_0 \left(\cos(kz - \omega t + \delta_x)\hat{\mathbf{x}} + \sin(kz - \omega t + \delta_x)\hat{\mathbf{y}} \right). \tag{12.65}$$

For the left circular polarized electromagnetic wave, we write

$$\mathbf{E}(\mathbf{r}, t) = E_0 e^{i(kz-\omega t)} \mathbf{e}_L = \frac{1}{\sqrt{2}} E_0 e^{i(kz-\omega t)} \left(\hat{\mathbf{x}} + e^{+i\pi/2}\hat{\mathbf{y}} \right). \tag{12.66}$$

Therefore,

$$\text{Re}\left(\mathbf{E}(\mathbf{r}, t)\right) = \frac{1}{\sqrt{2}} E_0 \left(\cos(kz - \omega t + \delta_x)\hat{\mathbf{x}} - \sin(kz - \omega t + \delta_x)\hat{\mathbf{y}} \right). \tag{12.67}$$

Figure 12.5 presents a situation where the observer is at the z-axis and is looking the wave approaching towards the observer. The right-hand rule can be used to determine the direction of rotation of electric field vector $\mathbf{E}$; that is, with the right hand, we let the thumb, which gives direction of the source of the electromagnetic waves, to point in the direction of the negative z-axis (along $-\hat{\mathbf{z}}$), then if the fingers of the right-hand curl in the direction of rotation of the vector $\mathbf{E}$, the polarization is right-handed.

For the elliptical polarization the tip of the electric field vector $\mathbf{E}$ rotates with an angular speed ω along an ellipse in the $x - y$-plane. For simplicity, we are assuming

Fig. 12.6 Elliptical polarization

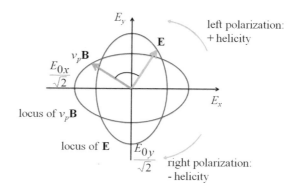

that the major axis of the ellipse is along the y-axis; however, in general, it can point in any arbitrary direction. Besides, we assume that $\delta_y - \delta_x = \pm\pi/2$ as in the case of circular polarization. Then, we have

$$\text{Re}\,(\mathbf{E}(\mathbf{r}, t)) = \frac{E_{0x}}{\sqrt{2}} \cos(kz - \omega t + \delta_x)\hat{\mathbf{x}} + \frac{E_{0y}}{\sqrt{2}} \sin(kz - \omega t + \delta_y)\hat{\mathbf{y}} \quad \text{(right)},$$

(12.68)

$$\text{Re}\,(\mathbf{E}(\mathbf{r}, t)) = \frac{E_{0x}}{\sqrt{2}} \cos(kz - \omega t + \delta_x)\hat{\mathbf{x}} - \frac{E_{0y}}{\sqrt{2}} \sin(kz - \omega t + \delta_y)\hat{\mathbf{y}} \quad \text{(left)}.$$

Figure 12.6 shows the elliptical polarization for $E_{0x} < E_{0y}$.

12.7 Reflection and Refraction of Electromagnetic Waves

Consider the reflection and transmission of electromagnetic waves at the boundaries between linear media. That will derive the so-called laws of reflection and refraction. For a linear medium, the so-called refraction index is defined by

$$n = \frac{c}{v_p}.$$

(12.69)

Using Eq. (12.27), we obtain that

$$n = \sqrt{\mu\epsilon}.$$

(12.70)

Furthermore, we write

$$\lambda_0 = n\lambda,$$

(12.71)

$$k_0 = \frac{k}{n},$$

where λ_0 and k_0 are the wavelength and magnitude of wave number in vacuum, and λ and k the corresponding quantities in the linear medium. Using the relation $k = \omega/v_p$, the electromagnetic planar wave vectors are also related as

$$\mathbf{B}(\mathbf{r}, t) = \frac{1}{\omega} (\mathbf{k} \times \mathbf{E}(\mathbf{r}, t)) \tag{12.72}$$

$$= \frac{1}{v_p} \left(\hat{\mathbf{k}} \times \mathbf{E}(\mathbf{r}, t) \right)$$

$$= \frac{n}{c} \left(\hat{\mathbf{k}} \times \mathbf{E}(\mathbf{r}, t) \right).$$

12.7.1   Laws of Reflection and Refraction

Figure 12.7 presents a monochromatic electromagnetic plane wave incident at the boundary separating two linear media (dielectrics) with (n_1, μ_1, ϵ_1) and (n_2, μ_2, ϵ_2), respectively. The incident wave is partly reflected in the first medium and partly transmitted in the second medium (called refracted wave). The reflected and refracted rays are in the same plane with the incident ray and the normal to the interface boundary, which is also called plane of incidence.

The electric and magnetic fields of the incident, reflected, and refracted waves are

$$\mathbf{E}_1(\mathbf{r}, t) = \mathbf{E}_{10} \exp\left(i\,(\mathbf{k}_1 \cdot \mathbf{r} - \omega_1 t)\right) \quad \text{(incident wave),} \tag{12.73}$$

$$\mathbf{E}_1'(\mathbf{r}, t) = \mathbf{E}_{10}' \exp\left(i\,(\mathbf{k}_1' \cdot \mathbf{r} - \omega_1' t)\right) \quad \text{(reflected wave),}$$

$$\mathbf{E}_2(\mathbf{r}, t) = \mathbf{E}_{20} \exp\left(i\,(\mathbf{k}_2 \cdot \mathbf{r} - \omega_2 t)\right) \quad \text{(refracted wave),}$$

$$\mathbf{B}(\mathbf{r}, t) = \frac{n}{c} (\mathbf{k} \times \mathbf{E}(\mathbf{r}, t)) \quad \text{(for every wave).}$$

Fig. 12.7 Reflection and refraction of the incident electromagnetic planar waves at the boundary between two linear media (dielectrics) with (n_1, μ_1, ϵ_1) and (n_2, μ_2, ϵ_2), respectively. θ_1 is the incident angle, which is the angle between the incident planar wave direction and normal line to the boundary, θ_1' is the reflection angle, and θ_2 is the refraction angle

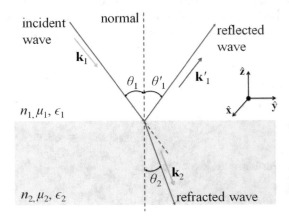

The electromagnetic field vectors $(\mathbf{E}, \mathbf{B})$ satisfy the boundary conditions; that is, at all the times ω is the same for all three waves (namely, incident, reflected, and refracted), and the vectors $\mathbf{E}$, $\mathbf{D}$, $\mathbf{B}$, and $\mathbf{H}$ have the same phase angle at any point at the boundary. Therefore, for any two points $\mathbf{r}_1$ and $\mathbf{r}_2$ at the boundary plane between the two media, we obtain

$$\mathbf{k}_1 \cdot \Delta\mathbf{r} = \mathbf{k}_1' \cdot \Delta\mathbf{r} = \mathbf{k}_2 \cdot \Delta\mathbf{r}, \tag{12.74}$$
$$k_1 \sin\theta_1 = k_1' \sin\theta_1' = k_2 \sin\theta_2,$$

where $\Delta\mathbf{r} = \mathbf{r}_2 - \mathbf{r}_1$ is the difference between any two arbitrary points along the boundary of the two dielectrics. Since the incident and reflected wave are traveling in the same medium, the magnitudes $k_1 = k_1'$; therefore,

$$\theta_1 = \theta_1'. \tag{12.75}$$

This is known as the *reflection's law*.
   Furthermore, from Eq. (12.74), we have

$$k_1 \sin\theta_1 = k_2 \sin\theta_2. \tag{12.76}$$

Using relations in Eq. (12.71), we obtain

$$k_0 n_1 \sin\theta_1 = k_0 n_2 \sin\theta_2, \tag{12.77}$$

or,

$$n_1 \sin\theta_1 = n_2 \sin\theta_2. \tag{12.78}$$

Equation (12.78) is also known as *refraction's law* or *Snell's law*.
   The reflection's law and Snell's law determine the directions of the reflected and refracted (transmitted) waves; that is, using Eqs. (12.75) and (12.78), we can determine the reflected angle θ_1' and refracted angle θ_2, if the incident angle θ_1 is known, and if the index of refraction n is known for both dielectrics. Next, the fraction of the reflected or transmitted intensity is also of interest. The fraction of the incident intensity that is reflected is called *reflectance* (denoted by R), and the fraction of the incident intensity that is refracted is called *transmittance*, denoted by T.
   Note that the reflectance and transmittance are different if the polarization plane is oriented parallel or perpendicular to the incident plane. The incident electromagnetic plane wave can be considered as superposition of two linearly polarized waves with polarization's planes being orthogonal, where one of the polarization plane is parallel and the other perpendicular to the plane of the incidence. The polarization perpendicular to the plane of the incidence is denoted as *TE mode*, and the polarization parallel to that plane is called *TM mode*. In the TE mode (transverse electric field), the vector $\mathbf{E}$ is perpendicular to the plane of incidence, and in the TM mode

(transverse magnetic field), the vector **B** is perpendicular to the plane of incidence. Note that in the electromagnetic plane waves, the vectors **E** and **B** are both perpendicular to each other and perpendicular to the vector **k** (which indicates the direction of the wave propagation).

12.8   Fresnel Equations

Fresnel equations give the relationships between the amplitude reflection coefficients and the reflectance and transmittance with the incidence angle θ for the two polarization.

12.8.1   Boundary Conditions

Consider an electromagnetic plane wave intersects the interface between two linear medium, as depicted in Fig. 12.8. The boundary conditions on the electromagnetic field determine the characteristics of the reflected and transmitted waves. In the first linear medium where the incident wave is passing through, the electric and magnetic field vectors (**E**, **B**) are the sum of the incident and reflected waves. The boundary conditions to be satisfied are derived from Maxwell's equations.

The Gauss's law can be applied for the closed surface S. For electric field,

$$\oint_S \mathbf{D} \cdot d\mathbf{S} = q_f, \tag{12.79}$$

and for magnetic field (no-magnetic charges exist),

$$\oint_S \mathbf{B} \cdot d\mathbf{S} = 0. \tag{12.80}$$

Fig. 12.8 Boundary conditions on the electromagnetic plane wave

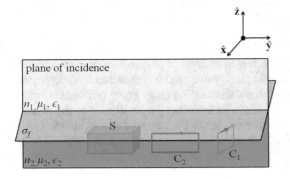

If the closed surface is taken infinitesimally small (such that volume charge density converges to surface charge density σ_f), and since there is no charges (that is, $\sigma_f = 0$), we obtain the following boundary conditions:

$$\epsilon_1 \left(\mathbf{E}_1 + \mathbf{E}_1' \right) \cdot \hat{\mathbf{z}} = \epsilon_2 \mathbf{E}_2 \cdot \hat{\mathbf{z}} \quad \text{for } \mathbf{D}, \tag{12.81}$$

$$\left(\mathbf{B}_1 + \mathbf{B}_1' \right) \cdot \hat{\mathbf{z}} = \mathbf{B}_2 \cdot \hat{\mathbf{z}} \quad \text{for } \mathbf{B}.$$

Using Ampére's law for the loop C_2 (see also Fig. 12.8), we can write

$$\oint_{C_2} \mathbf{H} \cdot d\mathbf{r} = \int_S \left(\mathbf{J}_f + \frac{\partial \mathbf{D}}{\partial t} \right) \cdot d\mathbf{S}, \tag{12.82}$$

and using the Faraday's law for the loop C_1, we obtain

$$\oint_{C_1} \mathbf{E} \cdot d\mathbf{r} = - \int_S \frac{\partial \mathbf{B}}{\partial t} \cdot d\mathbf{S}. \tag{12.83}$$

If we assume that the loops becomes infinitesimally thin such that contributions from $\partial \mathbf{D}/\partial t \approx 0$ and $\partial \mathbf{B}/\partial t \approx 0$, then only the free surface charge density contributes. Therefore,

$$\frac{1}{\mu_1} \left(\mathbf{B}_1 + \mathbf{B}_1' \right) \cdot \mathbf{x} = \frac{1}{\mu_2} \mathbf{B}_2 \mathbf{x} \quad \text{for } \mathbf{H}, \tag{12.84}$$

$$\frac{1}{\mu_1} \left(\mathbf{B}_1 + \mathbf{B}_1' \right) \cdot \mathbf{y} = \frac{1}{\mu_2} \mathbf{B}_2 \mathbf{y} \quad \text{for } \mathbf{H},$$

$$\left(\mathbf{E}_1 + \mathbf{E}_1' \right) \cdot \mathbf{x} = \mathbf{E}_2 \mathbf{x} \quad \text{for } \mathbf{E},$$

$$\left(\mathbf{E}_1 + \mathbf{E}_1' \right) \cdot \mathbf{y} = \mathbf{E}_2 \mathbf{y} \quad \text{for } \mathbf{E}.$$

Also, we can use the relationship between the magnitude of the electric and magnetic field: $B = nE/c$.

12.8.2 Perpendicular Polarization

Consider the perpendicular polarization of the electromagnetic plane wave with a geometry as depicted in Fig. 12.9. For this geometry, we have

$$\mathbf{E}_1 \cdot \hat{\mathbf{z}} = 0, \quad \mathbf{E}_1 \cdot \hat{\mathbf{y}} = 0, \quad \mathbf{B}_1 \cdot \hat{\mathbf{x}} = 0, \tag{12.85}$$

$$\mathbf{E}_1' \cdot \hat{\mathbf{z}} = 0, \quad \mathbf{E}_1' \cdot \hat{\mathbf{y}} = 0, \quad \mathbf{B}_1' \cdot \hat{\mathbf{x}} = 0,$$

$$\mathbf{E}_2 \cdot \hat{\mathbf{z}} = 0, \quad \mathbf{E}_2 \cdot \hat{\mathbf{y}} = 0, \quad \mathbf{B}_2 \cdot \hat{\mathbf{x}} = 0.$$

Therefore, there are three equations remaining only as follows:

Fig. 12.9 Perpendicular
polarization of the
electromagnetic plane wave

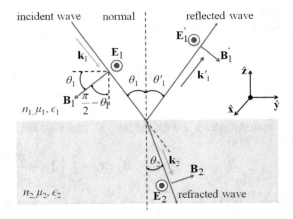

$$\left(E_1^\perp + E_1^{\perp,\prime}\right)\frac{n_1}{c}\cos\left(\frac{\pi}{2} - \theta_1\right) = E_2^\perp\frac{n_2}{c}\cos\left(\frac{\pi}{2} - \theta_2\right) \quad \text{for } \mathbf{B}, \tag{12.86}$$

$$\left(-E_1^\perp + E_1^{\perp,\prime}\right)\frac{n_1}{c\mu_m}\cos\theta_1 = -E_2^\perp\frac{n_2}{c\mu_m}\cos\theta_2 \quad \text{for } \mathbf{H},$$

$$\left(E_1^\perp + E_1^{\perp,\prime}\right) = E_2^\perp \quad \text{for } \mathbf{E}.$$

Solving Eq. (12.86) for $E_1^{\perp,\prime}$ and $E_2^\perp$, we obtain

$$E_1^{\perp,\prime} = \frac{\cos\theta_1 - \dfrac{n_2}{n_1}\cos\theta_2}{\cos\theta_1 + \dfrac{n_2}{n_1}\cos\theta_2}E_1^\perp \tag{12.87}$$

$$E_2^\perp = \frac{2\cos\theta_1}{\cos\theta_1 + \dfrac{n_2}{n_1}\cos\theta_2}E_1^\perp.$$

Using the Snell's law, Eq. (12.79), we obtain that

$$\cos\theta_2 = \sqrt{1 - \left(\frac{n_1}{n_2}\sin\theta_1\right)^2}. \tag{12.88}$$

Substituting Eq. (12.88) into (12.87), we get

$$E_1^{\perp,'} = \frac{\cos\theta_1 - \sqrt{\left(\dfrac{n_2}{n_1}\right)^2 - \sin^2\theta_1}}{\cos\theta_1 + \sqrt{\left(\dfrac{n_2}{n_1}\right)^2 - \sin^2\theta_1}} E_1^{\perp} \tag{12.89}$$

$$E_2^{\perp} = \frac{2\cos\theta_1}{\cos\theta_1 + \sqrt{\left(\dfrac{n_2}{n_1}\right)^2 - \sin^2\theta_1}} E_1^{\perp}.$$

Then, the amplitude of the reflection and transmission coefficient for perpendicular polarization are as follows:

$$R_{\perp} = \frac{E_1^{\perp,'}}{E_1^{\perp}} = \frac{\cos\theta_1 - \sqrt{\left(\dfrac{n_2}{n_1}\right)^2 - \sin^2\theta_1}}{\cos\theta_1 + \sqrt{\left(\dfrac{n_2}{n_1}\right)^2 - \sin^2\theta_1}}, \tag{12.90}$$

$$T_{\perp} = \frac{E_2^{\perp}}{E_1^{\perp}} = \frac{2\cos\theta_1}{\cos\theta_1 + \sqrt{\left(\dfrac{n_2}{n_1}\right)^2 - \sin^2\theta_1}}.$$

In Fig. 12.10 are plotted the amplitude coefficients, $R_{\perp}$ and $T_{\perp}$ versus the incident angle for crown glass ($n_2 = 1.52$) and water ($n_2 = 1.33$) in the case of perpendicular polarization of the electromagnetic plane wave. For the medium 1, the index of refraction was $n_1 = 1.00$.

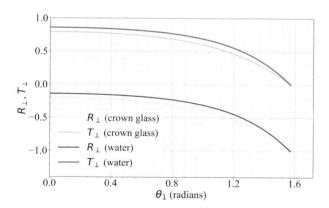

Fig. 12.10 The amplitude coefficients versus the incident angle for crown glass ($n_2 = 1.52$) and water ($n_2 = 1.33$) in the case of perpendicular polarization of the electromagnetic plane wave. For the medium 1, the index of refraction was $n_1 = 1.00$

12.8.3   Parallel Polarization

Figure 12.11 shows the geometry of a monochromatic electromagnetic plane wave with parallel polarization. In this case, the following relations hold:

$$\mathbf{B}_1 \cdot \hat{\mathbf{z}} = 0, \quad \mathbf{B}_1 \cdot \hat{\mathbf{y}} = 0, \quad \mathbf{E}_1 \cdot \hat{\mathbf{x}} = 0, \tag{12.91}$$
$$\mathbf{B}_1' \cdot \hat{\mathbf{z}} = 0, \quad \mathbf{B}_1' \cdot \hat{\mathbf{y}} = 0, \quad \mathbf{E}_1' \cdot \hat{\mathbf{x}} = 0,$$
$$\mathbf{B}_2 \cdot \hat{\mathbf{z}} = 0, \quad \mathbf{B}_2 \cdot \hat{\mathbf{y}} = 0, \quad \mathbf{E}_2 \cdot \hat{\mathbf{x}} = 0.$$

Then, we obtain the following three equations to be solved:

$$\left(\epsilon_1 E_1^{\parallel} + \epsilon_1 E_1^{\parallel,'}\right) \sin \theta_1 = \epsilon_2 E_2^{\parallel} \sin \theta_2 \quad \text{(for } \mathbf{D}\text{)}, \tag{12.92}$$
$$\left(E_1^{\parallel} + E_1^{\parallel,'}\right) n_1 = E_2^{\parallel} n_2 \quad \text{(for } \mathbf{H}\text{)},$$
$$\left(-E_1^{\parallel} + E_1^{\parallel,'}\right) \cos \theta_1 = -E_2^{\parallel} \cos \theta_2 \quad \text{(for } \mathbf{E}\text{)}.$$

Solving Eq. (12.92) for $E_1^{\parallel,'}$ and $E_2^{\parallel}$, we obtain

$$E_1^{\parallel,'} = \frac{\dfrac{n_2}{n_1} \cos \theta_1 - \cos \theta_2}{\dfrac{n_2}{n_1} \cos \theta_1 + \cos \theta_2} E_1^{\parallel}, \tag{12.93}$$

$$E_2^{\parallel} = \frac{2 \cos \theta_1}{\dfrac{n_2}{n_1} \cos \theta_1 + \cos \theta_2} E_1^{\parallel}.$$

Using the Snell's law, Eq. (12.89) is substituted in Eq. (12.93) to obtain the following:

$$E_1^{\parallel,'} = \frac{\left(\dfrac{n_2}{n_1}\right)^2 \cos \theta_1 - \sqrt{\left(\dfrac{n_2}{n_1}\right)^2 - \sin^2 \theta_1}}{\left(\dfrac{n_2}{n_1}\right)^2 \cos \theta_1 + \sqrt{\left(\dfrac{n_2}{n_1}\right)^2 - \sin^2 \theta_1}} E_1^{\parallel}, \tag{12.94}$$

$$E_2^{\parallel} = \frac{2 \left(\dfrac{n_2}{n_1}\right) \cos \theta_1}{\left(\dfrac{n_2}{n_1}\right)^2 \cos \theta_1 + \sqrt{\left(\dfrac{n_2}{n_1}\right)^2 - \sin^2 \theta_1}} E_1^{\parallel}.$$

From Eq. (12.94), we obtain the amplitude reflection and transmission coefficients for parallel polarization as follows:

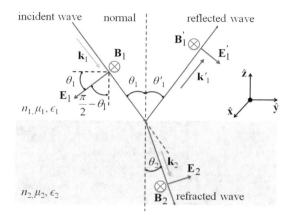

Fig. 12.11 Parallel polarization of the electromagnetic plane wave

$$R_\| = \frac{E_1^{\|,'}}{E_1^{\|}} = \frac{\left(\dfrac{n_2}{n_1}\right)^2 \cos\theta_1 - \sqrt{\left(\dfrac{n_2}{n_1}\right)^2 - \sin^2\theta_1}}{\left(\dfrac{n_2}{n_1}\right)^2 \cos\theta_1 + \sqrt{\left(\dfrac{n_2}{n_1}\right)^2 - \sin^2\theta_1}}, \tag{12.95}$$

$$T_\| = \frac{E_2^{\|}}{E_1^{\|}} = \frac{2\left(\dfrac{n_2}{n_1}\right)\cos\theta_1}{\left(\dfrac{n_2}{n_1}\right)^2 \cos\theta_1 + \sqrt{\left(\dfrac{n_2}{n_1}\right)^2 - \sin^2\theta_1}}.$$

In Fig. 12.12 are plotted the amplitude coefficients, $R_\|$ and $T_\|$ versus the incident angle for crown glass ($n_2 = 1.52$) and water ($n_2 = 1.33$) in the case of parallel polarization of the electromagnetic plane wave. For the incident wave medium, the index of refraction was $n_1 = 1.00$.

12.8.4 External and Internal Reflection

The external reflection is defined for $n_2 > n_1$. Therefore, in Eqs. (12.91) and (12.95), the term $\sqrt{(n_2/n_1)^2 - \sin^2\theta_1}$ is real for every $0° \le \theta_1 \le 90°$. This was also illustrated in Figs. 12.10 and 12.12. For the perpendicular polarization, the amplitude transmission coefficient, $T_\perp$, reduces smoothly from about $+0.86$ to 0.00 (see also Fig. 12.10) for $n_1 = 1.00$ and $n_2 = 1.52, 1.33$ (respectively, for crown glass and water). Furthermore, the amplitude reflection coefficient, $R_\perp$, reduces gradually from about -0.21 to -1.00. On the other hand, for the parallel polarization, the amplitude reflection coefficient, $R_\|$, goes smoothly from about $+0.21$ to -1.00, as shown in Fig. 12.12. By definition, the incident angle θ_1 for the parallel polarization for which $R_\|$ is zero

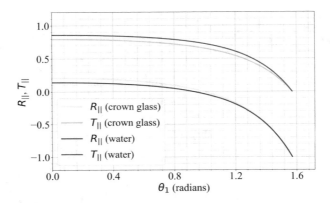

Fig. 12.12 The amplitude coefficients versus the incident angle for crown glass ($n_2 = 1.52$) and water ($n_2 = 1.33$) in the case of parallel polarization of the electromagnetic plane wave. For the medium 1, the index of refraction was $n_1 = 1.00$

is called *Brewster's angle*, denoted by θ_B. That is, at the angle of incidence waves θ_B only perpendicular polarized waves are reflected, and hence, creating linearly polarized light. Replacing $R_\parallel = 0$ in Eq. (12.95), we obtain the Brewster's angle θ_B as follows:

$$\cos \theta_B = \frac{1}{\sqrt{1 + \left(\dfrac{n_2}{n_1}\right)^2}}, \tag{12.96}$$

$$\tan \theta_B = \frac{n_2}{n_1}.$$

For $R_{\perp,\parallel} \geq 0$, the phase shift is zero, and for $R_{\perp,\parallel} < 0$, the phase shift on the reflection is π (or 180°) because $e^{\pm \pi} = -1$. In the case of perpendicular polarization $R_\perp \geq 0$, and hence the phase shift is zero. In the case of the parallel polarization, $R_\parallel \geq 0$ for $\theta_1 \geq \theta_B$, and hence the phase shift is zero; however, for $\theta_1 > \theta_B$, $R_\parallel > 0$, and hence there is a phase shift of π (or 180°).

In the case of the internal reflection, $n_1 > n_2$, and thus $n_2/n_1 < 1$. In Fig. 12.13 are shown the amplitude coefficients as a function of the incident angle θ_1 for the perpendicular polarization. The incident wave medium was crown glass ($n_1 = 1.52$) and water ($n_1 = 1.33$), and the refracted wave medium has an index of refraction $n_2 = 1.00$. Fig. 12.14 presents similar plot for parallel polarization. There is a critical angle θ_c such that for $\theta_1 > \theta_c$ the Snell's law does not apply because $\sin \theta_2$ becomes greater than one. For these incident angles (that is, $\theta_1 > \theta_c$), there will be no refracted waves (i.e., $T_{\perp,\parallel} = 0$); however, there is a total internal reflection such that $\mid R_{\perp,\parallel} \mid = 1$.

Using the Snell's law, Eq. (12.79), for $\theta_2 = 90°$, we can determine the critical angle as follows:

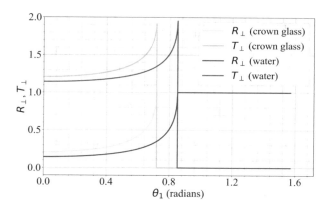

Fig. 12.13 The amplitude coefficients versus the incident angle for crown glass ($n_1 = 1.52$) and water ($n_1 = 1.33$) in the case of perpendicular polarization of the electromagnetic plane wave. For the refracted wave medium, the index of refraction was $n_2 = 1.00$. For $\theta_1 > \theta_c$, $| R_\perp |$ is plotted

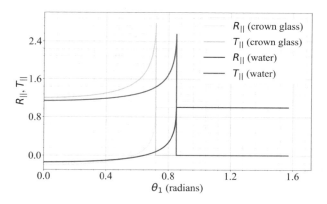

Fig. 12.14 The amplitude coefficients versus the incident angle for crown glass ($n_1 = 1.52$) and water ($n_1 = 1.33$) in the case of parallel polarization of the electromagnetic plane wave. For the refracted wave medium, the index of refraction was $n_2 = 1.00$. For $\theta_1 > \theta_c$, $| R_\parallel |$ is plotted

$$\sin \theta_c = \frac{n_2}{n_1}, \tag{12.97}$$

or,

$$\theta_c = \sin^{-1} \left(\frac{n_2}{n_1} \right). \tag{12.98}$$

Since for the internal reflection to occur, $n_2/n_1 < 1$, and thus

$$\left(\frac{n_2}{n_1} \right)^2 - \sin^2 \theta_1 = \begin{cases} > 0, \ \theta_1 < \theta_c \\ < 0, \ \theta_1 > \theta_c \end{cases} \tag{12.99}$$

Therefore, from Eqs. (12.90) and (12.95), the amplitude reflection coefficients are real numbers for $\theta_1 < \theta_c$; that is, for $\theta_1 < \theta_c$ the phase shift is $\phi = 0°$, if $R_{\perp,\parallel} > 0$, and $\phi = 180°$ if $R_{\perp,\parallel} < 0$, which is similar to the external reflection. On the other hand, for $\theta_1 > \theta_c$, the amplitude reflection coefficients are complex numbers. In that case, the amplitude reflection coefficient is written as

$$R_{\perp,\parallel} = \mid R_{\perp,\parallel} \mid e^{-i\phi}, \tag{12.100}$$

where ϕ is the phase shift. From Eqs. (12.90) and (12.95), since $\theta_1 > \theta_c$, the amplitude reflection coefficients can be written as follows:

$$R_{\perp,\parallel} = \frac{x - iy}{x + iy} = \frac{\sqrt{x^2 + y^2}e^{-i\psi}}{\sqrt{x^2 + y^2}e^{+i\psi}} = e^{-2i\psi} \equiv e^{-i\phi}, \tag{12.101}$$

where ϕ is the phase shift defined as

$$\phi = 2\psi = 2\tan^{-1}\left(\frac{y}{x}\right). \tag{12.102}$$

Here, for the perpendicular polarization of the electromagnetic plane wave, x and y are given as

$$x = \cos\theta_1, \tag{12.103}$$

$$y = \sqrt{\sin^2\theta_1 - \left(\frac{n_2}{n_1}\right)^2}.$$

For the parallel polarization of the electromagnetic plane wave, x and y are as follows:

$$x = \left(\frac{n_2}{n_1}\right)^2 \cos\theta_1, \tag{12.104}$$

$$y = \sqrt{\sin^2\theta_1 - \left(\frac{n_2}{n_1}\right)^2}.$$

The phase shift as a function of the incident angle for crown glass ($n_1 = 1.52$) and water ($n_1 = 1.33$) in the case of perpendicular polarization of the electromagnetic plane wave is plotted in Fig. 12.15, and for parallel polarization is shown in Fig. 12.16. In both cases, for the refracted wave medium, the index of refraction was $n_2 = 1.00$.

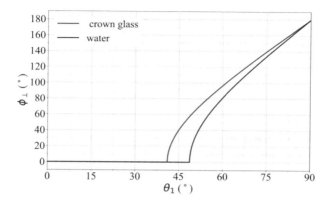

Fig. 12.15 The phase shift versus the incident angle for crown glass ($n_1 = 1.52$) and water ($n_1 = 1.33$) in the case of perpendicular polarization of the electromagnetic plane wave. For the refracted wave medium, the index of refraction was $n_2 = 1.00$

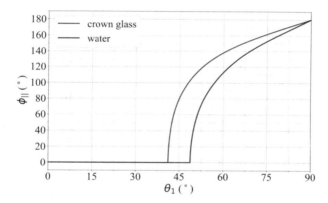

Fig. 12.16 The phase shift versus the incident angle for crown glass ($n_1 = 1.52$) and water ($n_1 = 1.33$) in the case of parallel polarization of the electromagnetic plane wave. For the refracted wave medium, the index of refraction was $n_2 = 1.00$

12.8.5 Normal Incidence of Electromagnetic Waves

Consider normal incidence of the electromagnetic plane waves to the boundary of the two linear media. Since the plane of incidence is not defined, the results for the perpendicular and parallel polarization are obtained as limiting case for the θ_1 going to zero ($\theta_1 \to 0$).

Thus, in the case of the perpendicular polarization, Eq. (12.90), for $\theta_1 \to 0$, becomes

$$R_\perp = \frac{E_1^{\perp,\prime}}{E_1^\perp} = \frac{1 - \dfrac{n_2}{n_1}}{1 + \dfrac{n_2}{n_1}}, \tag{12.105}$$

$$T_\perp = \frac{E_2^\perp}{E_1^\perp} = \frac{2}{1 + \dfrac{n_2}{n_1}}.$$

For the parallel polarization (see Eq. (12.95)), the amplitude coefficients in the case of normal incidence are as follows:

$$R_\parallel = \frac{E_1^{\parallel,\prime}}{E_1^\parallel} = \frac{\dfrac{n_2}{n_1} - 1}{1 + \dfrac{n_2}{n_1}}, \tag{12.106}$$

$$T_\parallel = \frac{E_2^\parallel}{E_1^\parallel} = \frac{2}{1 + \dfrac{n_2}{n_1}}.$$

Comparing Eqs. (12.105) and (12.106), we find that $R_\perp = -R_\parallel$ for $\theta_1 = 0°$; on the other hand, we expect that for normal incidence the perpendicular and parallel polarization to be identical. Note that the amplitude coefficients and the phase shifts are defined relative to the electric field directions of the incident, reflected, and refracted waves, as depicted in Figs. 12.9 and 12.11.

Figure 12.17 illustrates the direction of electric field for close to normal incidence in the case of internal reflection ($n_2 < n_1$), as in Fig. 12.17a, external reflection ($n_2 > n_1$), as in Fig. 12.17b for perpendicular polarization, and internal reflection (see Fig. 12.17c) and external reflection (see Fig. 12.17d) for parallel polarization. In the case of external reflection at the normal incidence, the electric field direction of the reflected wave is opposite to that of the incident wave (see also Fig. 12.17b, d), whereas in the case of the internal reflection (Fig. 12.17a, c), the electric field direction is the same as the incident wave. Therefore, for the reflection at normal incidence, there is a phase shift of 180° (or π) on the reflection for the external reflection; however, no phase shift occurs for the internal reflection for both polarization waves.

12.8.6 Reflectance and Transmittance

The reflectance and transmittance determine the rate of energy flow incident on the unit area of the interface: $\mathbf{S}_1 \cdot \hat{\mathbf{z}} = S_1 \cos \theta_1$. Then, the reflectance is given as

$$r = \frac{S_1' \cos \theta_1'}{S_1 \cos \theta_1} = \frac{S_1' \cos \theta_1}{S_1 \cos \theta_1} = \frac{S_1'}{S_1}, \tag{12.107}$$

where the reflection's law is used.

The transmittance is as follows:

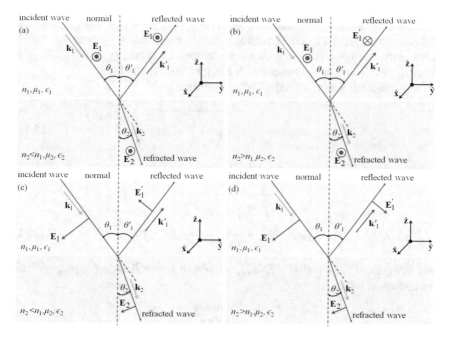

Fig. 12.17 The direction of electric field for close to normal incidence: **a** internal reflection ($n_2 < n_1$) and **b** external reflection ($n_2 > n_1$) for perpendicular polarization; **c** internal reflection and **d** external reflection for parallel polarization

$$t = \frac{S_2 \cos \theta_2}{S_1 \cos \theta_1}. \tag{12.108}$$

In Eqs. (12.107) and (12.108), S is the magnitude of the Poynting vector (for $\mu_m \approx 1$; that is, a dielectric medium):

$$S = EH = \frac{1}{\epsilon_m} E B = \frac{1}{\epsilon_m v_p} E^2 = \frac{n}{\epsilon_m c} E^2. \tag{12.109}$$

Here, the factor $\cos \theta$ considers the projected area of the interface as seen by the incoming or outgoing wave. Therefore, we obtain the reflectance and transmittance for the perpendicular and parallel polarization as follows:

$$r_{\perp,\parallel} = \mid R_{\perp,\parallel} \mid^2, \tag{12.110}$$

$$t_{\perp,\parallel} = \frac{\left(\dfrac{n_2}{n_1}\right) \cos \theta_2}{\cos \theta_1} \left(T_{\perp,\parallel}\right)^2 = 1 - r_{\perp,\parallel}.$$

In the case of the internal reflection, as $\theta_1 \to \theta_c$, $R_{\perp,\parallel} \to 1$, while the amplitude reflection coefficients $T_\perp$ and $T_\parallel$ remain finite. Because $\cos\theta_2 \to 0$ as $\theta_1 \to \theta_c$ (or equivalently, $\cos\theta_1 \to \cos\theta_c$), which is finite, then we obtain $t_{\perp,\parallel} \to 0$ as $\theta_1 \to \theta_c$. Also, $r_{\perp,\parallel} + t_{\perp,\parallel} = 1$ in agreement with energy conservation law.

Using the Snell's law (see Eq. (12.79)), we have

$$\cos\theta_2 = \sqrt{1 - \left(\frac{n_1}{n_2}\right)^2 \sin^2\theta_1}. \tag{12.111}$$

Substituting Eq. (12.111) into the second expression in Eq. (12.110), we obtain

$$t_{\perp,\parallel} = \frac{\sqrt{\left(\frac{n_2}{n_1}\right)^2 - \sin^2\theta_1}}{\cos\theta_1} \left(T_{\perp,\parallel}\right)^2. \tag{12.112}$$

Combining Eq. (12.90) with (12.112), we obtain transmittance for perpendicular polarization as follows:

$$t_\perp = \frac{4\cos\theta_1 \sqrt{\left(\frac{n_2}{n_1}\right)^2 - \sin^2\theta_1}}{\left(\cos\theta_1 + \sqrt{\left(\frac{n_2}{n_1}\right)^2 - \sin^2\theta_1}\right)^2}. \tag{12.113}$$

For the parallel polarization, the transmittance is obtained by substituting Eq. (12.95) into (12.112) as follows:

$$t_\parallel = \frac{4\left(\frac{n_2}{n_1}\right)^2 \cos\theta_1 \sqrt{\left(\frac{n_2}{n_1}\right)^2 - \sin^2\theta_1}}{\left(\left(\frac{n_2}{n_1}\right)^2 \cos\theta_1 + \sqrt{\left(\frac{n_2}{n_1}\right)^2 - \sin^2\theta_1}\right)^2}. \tag{12.114}$$

The reflectance is calculated as follows:

$$r_{\perp,\parallel} = 1 - t_{\perp,\parallel}. \tag{12.115}$$

The reflectance for the external reflection is shown in Fig. 12.18 for perpendicular and parallel polarization with $n_1 = 1.00$ and $n_2 = 1.52$ (crown glass). In Fig. 12.19, it is shown the reflectance for the internal reflection in the case of perpendicular and parallel polarization with $n_1 = 1.52$ (crown glass) and $n_1 = 1.00$. In the case of the

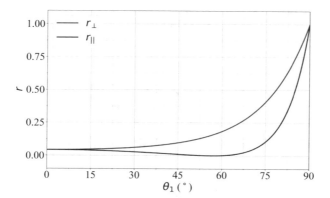

Fig. 12.18 The reflectance for the external reflection ($n_2 > n_1$) for perpendicular and parallel polarization with $n_1 = 1.00$ and $n_2 = 1.52$ (crown glass)

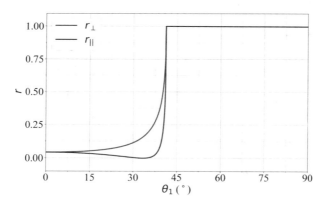

Fig. 12.19 The reflectance for internal reflection ($n_2 < n_1$) for perpendicular and parallel polarization with $n_1 = 1.52$ (crown glass) and $n_2 = 1.00$

external reflection (see also Fig. 12.18) when $\theta_1 \rightarrow 90°$, the reflectance approaches to one. This is a property used in the build of imaging X-ray space telescopes.

If $\theta_1 > \theta_c$, then total internal reflection occurs. However, because of the boundary conditions there is still an electromagnetic field present on the other side of the interface. The transmitted electromagnetic wave on the other side has a wave vector as follows:

$$\mathbf{k}_2 = \frac{\omega n_2}{c} \left(\sin\theta_2 \hat{\mathbf{y}} + \cos\theta_2 \hat{\mathbf{z}} \right). \tag{12.116}$$

For $\theta_1 > \theta_c$, using the Snell's law, we obtain

$$\cos \theta_2 = \sqrt{1 - \sin^2 \theta_2} \tag{12.117}$$

$$= \sqrt{1 - \left(\frac{n_1}{n_2}\right)^2 \sin^2 \theta_1}$$

$$= i\sqrt{\left(\frac{n_1}{n_2}\right)^2 \sin^2 \theta_1 - 1}.$$

Substituting Eq. (12.117) into (12.116) and using the Snell's law, we find that

$$\mathbf{k}_2 = \frac{\omega n_2}{c} \left(\frac{n_1}{n_2} \sin \theta_1 \hat{\mathbf{y}} + i\sqrt{\left(\frac{n_1}{n_2}\right)^2 \sin^2 \theta_1 - 1} \,\hat{\mathbf{z}} \right). \tag{12.118}$$

The refracted electromagnetic wave electric field vector is

$$\mathbf{E}_2(\mathbf{r}, t) = E_{02} \exp\left(i(\mathbf{k}_2 \cdot \mathbf{r} - \omega t)\right) \tag{12.119}$$

$$= E_{02} \exp\left(i \left(\frac{\omega n_2}{c} \left(\frac{n_1}{n_2} \sin \theta_1 \hat{\mathbf{y}} + i\sqrt{\left(\frac{n_1}{n_2}\right)^2 \sin^2 \theta_1 - 1} \,\hat{\mathbf{z}} \right) \cdot \mathbf{r} \right.\right.$$

$$\left.\left. - \omega t \right)\right)$$

$$= E_{02} \exp\left(-\frac{\omega n_2}{c} \sqrt{\left(\frac{n_1}{n_2}\right)^2 \sin^2 \theta_1 - 1} \,\hat{\mathbf{z}} \cdot \mathbf{r} \right)$$

$$\times \exp\left(i \left(\frac{\omega n_1}{c} \sin \theta_1 \mathbf{r} \cdot \hat{\mathbf{y}} - \omega t \right) \right),$$

which indicates that $\mathbf{E}_2(\mathbf{r}, t)$ is a wave propagating along the y-axis and its amplitude is

$$E_0 = E_{02} \exp\left(-\frac{\omega n_2}{c} \sqrt{\left(\frac{n_1}{n_2}\right)^2 \sin^2 \theta_1 - 1} \,\hat{\mathbf{z}} \cdot \mathbf{r} \right). \tag{12.120}$$

Therefore, the wave is exponentially decaying along the z-axis direction, which means in the direction perpendicular to the surface of dielectric; thus, this is a surface wave. This wave is called *evanescent wave*. Besides, Eq. (12.120) indicates that the decay distance of the electric field is of order c/ω, which of order of the wavelength; therefore, the wave can not propagate into the region outside the dielectric as a monochromatic plane wave. The evanescent wave may travel a tiny distance (such as in tunneling phenomena in quantum mechanics), if a slab of similar dielectric is brought within a few wavelength of the dielectric in which the wave is incident. If that occurs, a part of the incident wave tunnels through the gap and propagates in the slab as a monochromatic plane wave in the same direction as the original incident wave; however, its intensity reduces depending on the width of the gap.

This phenomena is called *frustrated total internal reflection*, and it can be applied to optics with beam-splitters and optical fiber junctions.

12.9 Exercises

Exercise 12.1 Assume that Fizeau's wheel has 360 teeth and is rotating at 27.5 rev/s when a pulse of light passing through opening A is blocked by tooth B on its return. If the distance to the mirror is 7 500 m, what is the speed of light?

Solution 12.1 The wheel has 360 teeth, and so it must have 360 openings. Therefore, because the light passes through opening A but is blocked by the tooth immediately adjacent to A, the wheel must rotate through an angular displacement of (1/720) rev in the time interval during which the light pulse makes its round trip. From the definition of angular speed, that time interval is

$$\Delta t = \frac{\Delta \theta}{\omega} \tag{12.121}$$
$$= \frac{(1/720) \text{ rev}}{27.5 \text{ rev/s}} = 5.05 \times 10^{-5} \text{ s}$$

Thus the speed of light calculated from this data is

$$c = \frac{2d}{\Delta t} \tag{12.122}$$
$$= \frac{2(7500 \text{ m})}{5.05 \times 10^{-5} \text{ s}} = 2.97 \times 10^8 \text{ m/s}$$

Exercise 12.2 A beam of light of wavelength 550 nm traveling in air is incident on a slab of transparent material. The incident beam makes an angle of 40.0° with the normal, and the refracted beam makes an angle of 26.0° with the normal. Find the index of refraction of the material.

Solution 12.2 Using the Snell's law of refraction

$$n_1 \sin \theta_1 = n_2 \sin \theta_2$$

with $n_1 = 1.00$ for air, we get

$$n_2 = n_1 \frac{\sin \theta_1}{\sin \theta_2} \tag{12.123}$$

$$= (1.00) \frac{\sin 40°}{\sin 26°} = 1.47$$

Exercise 12.3 A light ray of wavelength 589 nm traveling through air is incident on a smooth, flat slab of crown glass at an angle of 30.0° to the normal. Find the angle of refraction.

Solution 12.3 Using the Snell's law

$$n_1 \sin \theta_1 = n_2 \sin \theta_2$$

and thus,

$$\sin \theta_2 = \frac{n_1}{n_2} \sin \theta_1$$

for the air $n_1 = 1.00$ and for the crown glass $n_2 = 1.52$. Therefore,

$$\sin \theta_2 = \left(\frac{1.00}{1.52} \right) \sin 30° = 0.329$$

or,

$$\theta_2 = \sin^{-1}(0.329) = 19.2°$$

Because this is less than the incident angle of 30°, the refracted ray is bent toward the normal, as expected. Its change in direction is called the angle of deviation and is given by

$$\delta = | \theta_1 - \theta_2 | = 10.8°$$

Exercise 12.4 A laser in a compact disc player generates light that has a wavelength of 780 nm in air. (A) Find the speed of this light once it enters the plastic of a compact disc ($n = 1.55$). (B) What is the wavelength of this light in the plastic?

Solution 12.4 (A) We expect to find a value less than 3.00×10^8 m/s because $n > 1$. We can obtain the speed of light in the plastic by using Equation

$$v = \frac{c}{n} = \frac{3.00 \times 10^8 \text{ m/s}}{1.55} = 1.94 \times 10^8 \text{ m/s}$$

(B) We use Equation

$$\lambda_n = \frac{\lambda}{n}$$

to calculate the wavelength in plastic, noting that we are given the wavelength in air to be $\lambda = 780$ nm:

$$\lambda_n = \frac{\lambda}{n} = \frac{780 \text{ nm}}{1.55} = 503 \text{ nm}$$

Exercise 12.5 A light beam passes from medium 1 to medium 2, with the latter medium being a thick slab of material whose index of refraction is n_2. Show that the emerging beam is parallel to the incident beam.

Solution 12.5 Using the Snell's law of refraction to the upper surface, we get

$$n_1 \sin \theta_1 = n_2 \sin \theta_2 \rightarrow \sin \theta_2 = \frac{n_1}{n_2} \sin \theta_1$$

for the lower surface we get $\sin \theta_3 = \frac{n_2}{n_1} \sin \theta_2$. Combining these two equations, we get

$$\sin \theta_3 = \frac{n_2}{n_1} \left[\frac{n_1}{n_2} \sin \theta_1 \right] = \sin \theta_1 \rightarrow \theta_3 = \theta_1$$

Exercise 12.6 Find the critical angle for an air-water boundary. (The index of refraction of water is 1.33.)

Solution 12.6 The air above the water having index of refraction n_2 and the water having index of refraction n_1. Applying the equation

$$\sin \theta_c = \frac{n_2}{n_1} = \frac{1.00}{1.33} = 0.752$$

or

$$\theta_c = \sin^{-1}(0.752) = 48.8°$$

Exercise 12.7 A narrow beam of sodium yellow light, with wavelength 589 nm in vacuum, is incident from air onto a smooth water surface at an angle $\theta_1 = 35.0°$. Determine the angle of refraction θ_2 and the wavelength of the light in water. Find the critical angle for an air-water boundary. (The index of refraction of water is 1.33.)

Solution 12.7 Using the Snell's law:

$$n_1 \sin \theta_1 = n_2 \sin \theta_2$$

where $n_1 = 1.00$, $n_2 = 1.33$, and $\theta_1 = 35.0°$. Thus,

$$\sin \theta_2 = \frac{n_1 \sin \theta_1}{n_2} = \frac{(1.00)(\sin 35.0°)}{1.33} \approx 0.43126$$

Or,

$$\theta_2 = \sin^{-1} 0.43126 \approx 25.5°$$

The wavelength of light in water is:

$$\lambda_n = \frac{\lambda}{n_2} = \frac{589 \text{ nm}}{1.33} \approx 443 \text{ nm}$$

Exercise 12.8 The wavelength of red helium - neon laser light in air is $\lambda = 632.8$ nm. (a) What is its frequency? (b) What is its wavelength in glass that has an index of refraction of 1.50? (c) What is its speed in the glass?

Solution 12.8 We consider that the speed of light in air ($n_1 = 1.00$) is $c = 3.00 \times 10^8$ m/s.

(a) The frequency of light wave is:

$$f = \frac{1}{T} = \frac{c}{\lambda} = \frac{3.00 \times 10^8 \text{ m/s}}{632.8 \times 10^{-9} \text{ m}} \approx 4.74 \times 10^{14} \text{ Hz}$$

(b) The wavelength of light in glass is

$$\lambda_n = \frac{\lambda}{n} = \frac{632.8 \text{ nm}}{1.50} \approx 421.9 \text{ nm}$$

(c) The speed of light in glass is:

$$v = \frac{c}{n} = \frac{3.00 \times 10^8 \text{ m/s}}{1.50} = 2.00 \times 10^8 \text{ m/s}$$

Exercise 12.9 An underwater scuba diver sees the Sun at an apparent angle of $45.0°$ from the vertical. What is the actual direction of the Sun?

Solution 12.9 The indices of refraction for air and water are respectively: $n_{air} = 1.00$ and $n_{wat} = 1.33$. Using the Snell's law, we write

$$n_{air} \sin \theta_1 = n_{wat} \sin \theta_2$$

Here, $\theta_2 = 45.0°$. Therefore,

$$\theta_1 = \sin^{-1}\left(\frac{n_{wat} \sin \theta_2}{n_{air}}\right) = \sin^{-1}\left(\frac{(1.33) \sin 45.0°}{1.00}\right)$$

Or,

$$\theta_1 \approx \sin^{-1}(0.94) \approx 70.1°$$

which is the angle with vertical direction.

Exercise 12.10 A monochromatic plane wave with electric field vector

$$\mathbf{E}(\mathbf{r}, t) = \mathbf{E}_0 \exp\left(i\left(\mathbf{k} \cdot \mathbf{r} - \omega t\right)\right)$$

is traveling along the positive z axis through a linear medium with permittivity $\epsilon = 4$ and permeability $\mu = 1$, which is polarized along the x axis. The frequency is $f = 1$ GHz and E has a maximum value of $+10^{-3}$ V/m at $t = 5$ ns and $z = 1$ m. (a) Determine the angular frequency, phase velocity, wave number, wave vector, and wavelength. (b) Determine the expression for $\mathbf{E}(\mathbf{r}, t)$. (c) Determine the expression for $\mathbf{H}(\mathbf{r}, t)$. (d) Determine the Poynting vector and its time-averaged value. (e) Determine the position where E_x is maximum at $t = 0$ s.

Solution 12.10 (a) The angular frequency is

$$\omega = 2\pi f \approx 6.28 \times 10^6 \text{ rad/s}.$$

The phase velocity is

$$v_p = \frac{c}{\sqrt{\mu\epsilon}} = \frac{3.00 \times 10^8 \text{ m/s}}{\sqrt{4}} = 1.5 \times 10^8 \text{ m/s}.$$

The wave number magnitude is

$$k = \frac{\omega}{v_p} = \frac{6.28 \times 10^6 \text{ rad/s}}{1.5 \times 10^8 \text{ m/s}} = 4.19 \times 10^{-2} \text{ m}^{-1}.$$

The wave number vector is

$$\mathbf{k} = k\hat{\mathbf{z}} = 0.0419\hat{\mathbf{z}} \text{ m}^{-1}.$$

The wavelength is given as

$$\lambda = \frac{2\pi}{k} = \frac{2\pi}{0.0419 \text{ m}^{-1}} = 149.88 \text{ m}.$$

(b) The expression for the electric field vector is

$$\mathbf{E}(z, t) = E_0 \exp\left(i\left(0.0419z - 6.28 \times 10^6 t\right)\right) \hat{\mathbf{x}}.$$

At $t = 5 \text{ ns} = 5 \times 10^{-9}$ s and $z = 1$ m, we have

$$\mathbf{E}(1 \text{ m}, 5 \times 10^{-9} \text{ s}) = E_0 \exp\left(i\left(0.0419 - 0.0314\right)\right) \hat{\mathbf{x}} = E_0 \exp\left(i0.0105\right) \hat{\mathbf{x}}.$$

The real part is

$$\text{Re}\{\mathbf{E}(1 \text{ m}, 5 \times 10^{-9} \text{ s})\} = E_0 \cos(0.0105)\hat{\mathbf{x}},$$

and the maximum is given as

$$10^{-3} = E_0 \cos(0.0105)$$

or,

$$E_0 = \frac{10^{-3}}{0.99989} \approx 10^{-3} \text{ V/m}.$$

Thus,

$$\mathbf{E}(z, t) = \left(10^{-3} \text{ V/m}\right) \exp\left(i\left(0.0419z - 6.28 \times 10^6 t\right)\right) \hat{\mathbf{x}}.$$

(c) The magnetic field vector is

$$\mathbf{H}(\mathbf{r}, t) = H(\mathbf{r}, t)\hat{\mathbf{y}},$$

and its magnitude is

$$H(\mathbf{r}, t) = \frac{1}{Z} E = \frac{E(\mathbf{r}, t)}{\sqrt{\frac{\mu_m}{\epsilon_m}}} \tag{12.124}$$

$$= \frac{E(\mathbf{r}, t)}{\sqrt{\frac{\mu_0}{\epsilon_0 \epsilon}}} = \frac{E(\mathbf{r}, t)}{Z_0 \sqrt{\frac{1}{\epsilon}}}$$

$$\approx \frac{E(\mathbf{r}, t)}{188.5},$$

where

$$Z = \sqrt{\frac{\mu_m}{\epsilon_m}} = Z_0 \sqrt{\frac{1}{\epsilon}} = 188.5\,\Omega,$$

is the impedance of dielectric medium and $Z_0 \approx 377\,\Omega$ is the impedance of the vacuum. Therefore,

$$\mathbf{H}(\mathbf{r}, t) = (5.31 \times 10^{-6}\,\text{T}) \exp\left(i\left(0.0419z - 6.28 \times 10^6 t\right)\right) \hat{\mathbf{y}}.$$

(d) Using Eq. (12.39), the Poynting vector is

$$\mathbf{S} = \frac{E^2(\mathbf{r}, t)}{Z} \hat{\mathbf{z}} \tag{12.125}$$

$$= \left(5.31 \times 10^{-9}\,\frac{\text{V}^2}{(\text{m}^2 \cdot \Omega)}\right) \exp\left(i\left(0.0838z - 12.56 \times 10^6 t\right)\right) \hat{\mathbf{z}}.$$

The period is

$$T = \frac{2\pi}{\omega} = 1.00 \times 10^{-6}\,\text{s}$$

The time average of the Poynting vector is given by Eq. (12.41) as follows:

$$\langle \mathbf{S} \rangle = v_p \frac{\epsilon_m \mid E_0 \mid^2}{2} \tag{12.126}$$

$$= \sqrt{\frac{\epsilon_m}{\mu_m}} \frac{\mid E_0 \mid^2}{2}$$

$$= \frac{\mid E_0 \mid^2}{2Z} = \frac{10^{-6}}{377} \approx 2.65 \times 10^{-9}\,\frac{\text{V}^2}{(\text{m}^2 \cdot \Omega)}$$

(e) First, we write $E_x = \mathbf{E} \cdot \hat{\mathbf{x}}$ as

$$E_x(z, t) = \left(10^{-3}\,\text{V/m}\right) \exp\left(i\left(0.0419z - 6.28 \times 10^6 t\right)\right).$$

At $t = 0$, we obtain

$$E_x(z, 0) = \left(10^{-3}\,\text{V/m}\right) \exp\left(i0.0419z\right),$$

which has a maximum for

$$0.0419z = 2\pi.$$

This

$$z = \frac{2\pi}{0.0419} \approx 149.88\,\text{m}.$$

References

Altland A, Simons B (2010) Condensed matter field theory, 2nd edn. Cambridge University Press
Griffiths DJ (1999) Introduction to electrodynamics, 3rd edn. Prentice Hall
Holliday D, Resnick R, Walker J (2011) Fundamentals of physics. John Wiley and Sons
Jackson JD (1999) Classical electrodynamics, 3rd edn. John Wiley and Sons
Landau LD, Lifshitz EM (1971) The classical theory of fields. Pergamon Press
Protheroe RJ (2013) Essential electrodynamics, 1st edn. Bookboon
Sykja H (2006) Bazat e Elektrodinamikës. SHBUT

Chapter 13
Electromagnetic Waves in Dispersive Media

This chapter aims to derive electromagnetic wave equations in dispersive media. In particular, the aims of this chapter include the understanding the absorption, Lorentz's oscillator model of a dielectric, deriving the wave equation of a conductor, deriving the wave equation of a dilute plasma and understanding the magnetized plasma or dielectric.

In this chapter, we derive electromagnetic wave equations in dispersive media. This chapter describes the absorption, Lorentz's oscillator model of a dielectric, the wave equation of a conductor, the wave equation of a dilute plasma, and the magnetized plasma or dielectric. For further reading, one can also consider other available literature (Jackson 1999; Landau and Lifshitz 1971; Sykja 2006; Griffiths 1999; Altland and Simons 2010; Protheroe 2013).

13.1 Dispersion and Absorption

In Eq. (12.13) (Chap. 12), it was assumed that the wave number k is a constant; however, in general, the index of refraction $n(\omega)$ and the wave number $k(\omega)$ are functions of the angular frequency:

$$k(\omega) = \frac{\omega}{v_p} = n(\omega)\frac{\omega}{c}. \tag{13.1}$$

For a monochromatic plane wave in matter, we have

© The Author(s), under exclusive license to Springer Nature Switzerland AG 2022
H. Kamberaj, *Electromagnetism*, Undergraduate Texts in Physics,
https://doi.org/10.1007/978-3-030-96780-2_13

$$\mathbf{E}(\mathbf{r}, t) = \mathbf{E}_0 \exp\left(i\left(\mathbf{k} \cdot \mathbf{r} - \omega t\right)\right) \tag{13.2}$$

$$= \mathbf{E}_0 \exp\left(i\left(k_r \hat{\mathbf{k}} \cdot \mathbf{r} - \omega t\right)\right) \exp\left(-k_i \hat{\mathbf{k}} \cdot \mathbf{r}\right),$$

where $k_r = \mathrm{Re}\,(k)$ (real part) and $k_i = \mathrm{Im}\,(k)$ (imaginary part) such that $\mathbf{k} = (k_r + i k_i)\,\hat{\mathbf{k}}$. The intensity of the electromagnetic wave is given as the time average of the Poynting vector $\mathbf{S}(\mathbf{r}, t)$

$$I(\mathbf{r}) = \langle \mathbf{S}(\mathbf{r}, t) \rangle \propto \langle E^2(\mathbf{r}, t) \rangle \tag{13.3}$$

$$= E_0^2 \exp\left(-\alpha(\omega)\hat{\mathbf{k}} \cdot \mathbf{r}\right) \exp\left(2i k_r(\omega)\hat{\mathbf{k}} \cdot \mathbf{r}\right),$$

which indicates that intensity decays exponentially with distance, and $\alpha = 2\mathrm{Im}\,(k(\omega)) = 2k_i(\omega)$ is the so-called *absorption coefficient*.

13.1.1 Lorentz's Model of Oscillations in Dielectrics

The polarizability of atoms in molecules can be approximately modeled using the semi-classical approaches, which consider the atoms as nucleus with effective charge $+e$ surrounded by a spherical cloud of electrons with radius $a_0 \sim 10^{-10}$ m corresponding to a single valence electron of charge $-e$ (where e magnitude of the electron's charge $e = 1.6 \times 10^{-19}$ C). In this approximation, the nucleus is screened by all the electrons in the atom except one electron (the valence electron), and hence it has an effective charge of $+e$.

In the presence of an external electric field, the electron cloud displaces by an amount $\mathbf{r}_e$ relative to the nucleus, and hence it experiences a restoring force from the nucleus equal to $-\kappa \mathbf{r}_e$, where

$$\kappa \approx \frac{e^2}{4\pi\epsilon_0 a_0^3}. \tag{13.4}$$

Here, κ is a spring constant, which is a measure of the strength of restoring force of the nucleus. Applying the second law of Newton for the valence electron, we have

$$m_e \ddot{\mathbf{r}}_e = -e\mathbf{E} - \kappa \mathbf{r}_e, \tag{13.5}$$

where $\ddot{\mathbf{r}}_e$ is the second time derivative of the valence electron displacement $\mathbf{r}_e$ and $\mathbf{E}$ is the external electric field vector. In equilibrium, the right-hand side (which is the net force acting on the valence electron) is zero, and thus the equilibrium displacement for the valence electron $\mathbf{r}_e$ is

$$\mathbf{r}_e = -\frac{e\mathbf{E}}{\kappa}. \tag{13.6}$$

The dipole momentum is

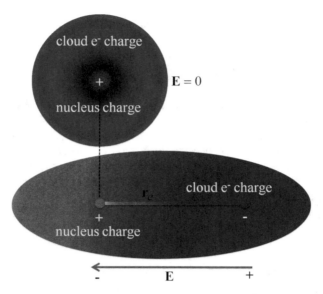

Fig. 13.1 Displacement $\mathbf{r}_e$ of the cloud electron charge ($-e$) surrounding the nucleus with a charge $+e$ under the external electric field $\mathbf{E}$ due to the electrical force $-e\mathbf{E}$

$$\mathbf{p} = e\mathbf{r}_e = -\frac{e^2\mathbf{E}}{\kappa}, \tag{13.7}$$

and the molecular polarizability is

$$\alpha_{\text{pol}} = \frac{e^2}{\kappa} = 4\pi\epsilon_0 a_0^3. \tag{13.8}$$

If the external electric field $\mathbf{E}$ oscillates in time, then the electrons will oscillate in response. Note there are various dragging forces (such as friction forces and radiation from stopping[1]) that apply to the valence electron, which are added as a dumping term proportional to the velocity that are added to the equation of motion as follows:

$$m_e\ddot{\mathbf{r}}_e = -e\mathbf{E}(t) - \kappa\mathbf{r}_e - m_e\gamma\dot{\mathbf{r}}_e, \tag{13.9}$$

where $\mathbf{v}_e = \dot{\mathbf{r}}_e = d\mathbf{r}_e/dt$ is the valence electron velocity and γ is a proportionality constant.

Figure 13.1 illustrates the displacement of the cloud electron charge under the external electric field $\mathbf{E}$.

---

[1] Sykja (2006).

The frequency of simple harmonic motion for the valence electrons of atoms is defined as

$$\omega_0 = \sqrt{\frac{\kappa}{m_e}} = \sqrt{\frac{e^2}{4\pi\epsilon_0 a_0^3 m_e}} \approx 2 \times 10^{16} \, \text{rad/s}. \tag{13.10}$$

This is also the resonance frequency and it corresponds to the frequencies in the ultraviolet range. When a monochromatic plane wave (or photons) with frequency ω_0 (or close to that value) strikes an atom, then the work done against the friction forces to allow the atom oscillating causes an absorption of photons. Note that the atom is not a classical system with just one resonance, but a quantum system with many resonant frequencies. Therefore, this semi-classical system is a good approximation to describe some important properties of dielectrics.

Thus, Eq. (13.9) takes the form

$$m_e \ddot{\mathbf{r}}_e = -e\mathbf{E}(t) - m_e \omega_0^2 \mathbf{r}_e - m_e \gamma \dot{\mathbf{r}}_e. \tag{13.11}$$

Consider a sinusoidal electric field of the electromagnetic wave; that is, $\mathbf{E}(t) \equiv \mathbf{E}_0 e^{-i\omega t}$. Then, the solution of Eq. (13.11) can be required in the following form:

$$\mathbf{r}_e(t) = \mathbf{r}_0 e^{\rho t}, \tag{13.12}$$

where ρ and $\mathbf{r}_0$ are constants to be determined. Replacing $\mathbf{r}_e$ from Eq. (13.12) into (13.11), we obtain

$$m_e \rho^2 e^{\rho t} \mathbf{r}_0 = -e\mathbf{E}_0 e^{-i\omega t} - m_e \omega_0^2 \mathbf{r}_0 e^{\rho t} - m_e \gamma \rho \mathbf{r}_0 e^{\rho t}, \tag{13.13}$$

or

$$m_e \left(\rho^2 + \omega_0^2 + \gamma\rho \right) e^{\rho t} \mathbf{r}_0 = -e\mathbf{E}_0 e^{-i\omega t}. \tag{13.14}$$

Equation (13.14) implies that $\rho = -i\omega$, and

$$m_e \left(\rho^2 + \omega_0^2 + \gamma\rho \right) \mathbf{r}_0 = -e\mathbf{E}_0. \tag{13.15}$$

Therefore, we obtain

$$\mathbf{r}_0 = -\frac{e}{m_e} \frac{\mathbf{E}_0}{\omega_0^2 - \omega^2 - i\gamma\omega}. \tag{13.16}$$

The displacement of the valence electrons is given as follows:

$$\mathbf{r}_e = -\frac{e}{m_e} \frac{\mathbf{E}_0}{\omega_0^2 - \omega^2 - i\gamma\omega} e^{-i\omega t} \tag{13.17}$$

$$= -\frac{e}{m_e} \frac{\omega_0^2 - \omega^2 + i\gamma\omega}{\left(\omega_0^2 - \omega^2\right)^2 + (\gamma\omega)^2} \mathbf{E}_0 e^{-i\omega t}$$

$$= -\frac{e}{m_e} \frac{\sqrt{(\omega_0^2 - \omega^2)^2 + (\gamma\omega)^2} e^{+i\phi}}{\left(\omega_0^2 - \omega^2\right)^2 + (\gamma\omega)^2} \mathbf{E}_0 e^{-i\omega t}$$

$$= -\frac{e/m_e}{\left(\left(\omega_0^2 - \omega^2\right)^2 + (\gamma\omega)^2\right)^{3/2}} \mathbf{E}_0 e^{-i(\omega t - \phi)},$$

where

$$\phi = \tan^{-1}\left(\frac{\gamma\omega}{\omega_0^2 - \omega^2}\right). \tag{13.18}$$

The minus sign in Eq. (13.17) indicates that the cloud electron displacement is opposite of the electric field. The dipole moment is as follows:

$$\mathbf{p}(t) = -e\mathbf{r}_e(t) = \frac{e^2}{m_e} \frac{\left(\omega_0^2 - \omega^2\right) + i\gamma\omega}{\left(\omega_0^2 - \omega^2\right)^2 + (\gamma\omega)^2} \mathbf{E}(t). \tag{13.19}$$

Consider a molecule of N atoms, n oscillators with frequency $\omega_{0,k}$, dumping coefficients γ_k, and f_k cloud electrons (for $k = 1, 2, \ldots, n$). The polarization of molecule is

$$\mathbf{P}(\mathbf{r}, t) = \frac{Ne^2}{m_e} \left(\sum_{k=1}^{n} f_k \frac{\left(\omega_{0,k}^2 - \omega^2\right) + i\gamma_k\omega}{\left(\omega_{0,k}^2 - \omega^2\right)^2 + (\gamma_k\omega)^2}\right) \mathbf{E}(t). \tag{13.20}$$

Using Eq. (12.25) (see Chap. 12), the refraction index is (for $\mu \approx 1$)

$$n(\omega) = \sqrt{\mu\epsilon} \approx \sqrt{\epsilon} = \sqrt{1 + \chi_e} = \sqrt{1 + \frac{\mathbf{P}}{\epsilon_0 \mathbf{E}}} \tag{13.21}$$

$$= \left(1 + \frac{Ne^2}{\epsilon_0 m_e} \left(\sum_{k=1}^{n} f_k \frac{\left(\omega_{0,k}^2 - \omega^2\right) + i\gamma_k\omega}{\left(\omega_{0,k}^2 - \omega^2\right)^2 + (\gamma_k\omega)^2}\right)\right)^{1/2}.$$

Equation (13.21) indicates that refractive index is a complex number, and thus there is an attenuation of the wave occurring in a dispersive media, which is more important close to the resonance frequencies ($\omega_k \approx \omega_0$), as described in the following.

First, considering that the second term in Eq. (13.21) is small, we can retain only the second-order terms, and thus:

$$n(\omega) \approx 1 + \frac{Ne^2}{2\epsilon_0 m_e}\left(\sum_{k=1}^{n} f_k \frac{(\omega_{0,k}^2 - \omega^2) + i\gamma_k\omega}{(\omega_{0,k}^2 - \omega^2)^2 + (\gamma_k\omega)^2}\right) \tag{13.22}$$

$$= \underbrace{\left(1 + \frac{Ne^2}{2\epsilon_0 m_e}\sum_{k=1}^{n} \frac{f_k (\omega_{0,k}^2 - \omega^2)}{(\omega_{0,k}^2 - \omega^2)^2 + (\gamma_k\omega)^2}\right)}_{n_r(\omega)\equiv\mathrm{Re}(n(\omega))}$$

$$+ i \underbrace{\frac{Ne^2}{2\epsilon_0 m_e}\sum_{k=1}^{n} \frac{f_k \gamma_k\omega}{(\omega_{0,k}^2 - \omega^2)^2 + (\gamma_k\omega)^2}}_{n_i(\omega)\equiv\mathrm{Im}(n(\omega))}.$$

Then, the wave number is complex number, and the real part $k_r(\omega)$ is

$$k_r(\omega) = \frac{\omega n_r(\omega)}{c} = \frac{\omega}{c}\left(1 + \frac{Ne^2}{2\epsilon_0 m_e}\sum_{k=1}^{n} \frac{f_k (\omega_{0,k}^2 - \omega^2)}{(\omega_{0,k}^2 - \omega^2)^2 + (\gamma_k\omega)^2}\right). \tag{13.23}$$

Here, $k_r(\omega)$ is the so-called *dispersion relation* for the medium. Using Eq. (13.23), the phase velocity $v_p(\omega)$ can be determined for a given frequency of the electromagnetic wave ω:

$$v_p(\omega) = \frac{\omega}{k_r(\omega)}. \tag{13.24}$$

Figure 13.2 shows the real part $n_r(\omega)$ and imaginary part $n_i(\omega)$ of the index of refraction as a function of frequency ω for illustration.

The absorption coefficient $\alpha(\omega)$ is determined from the imaginary part of $k(\omega)$ as follows:

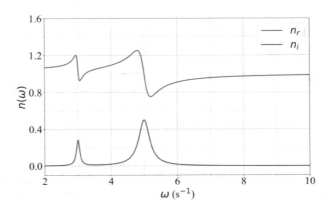

Fig. 13.2 The illustration of the real part $n_r(\omega)$ and imaginary part $n_i(\omega)$ of the index of refraction as a function of frequency ω for $n = 2$

$$\alpha(\omega) = 2k_i(\omega) = 2\frac{\omega n_i(\omega)}{c} = \frac{Ne^2}{\epsilon_0 m_e c} \sum_{k=1}^{n} \frac{f_k \gamma_k \omega^2}{\left(\omega_{0,k}^2 - \omega^2\right)^2 + (\gamma_k \omega)^2}. \tag{13.25}$$

Close to the resonance frequencies, $\omega_{0,k} \approx \omega$, and so $\omega_{0,k} + \omega \approx 2\omega$; therefore, the absorption coefficient can be approximated as

$$\alpha(\omega) = \frac{Ne^2}{\epsilon_0 m_e c} \sum_{k=1}^{n} \frac{f_k \gamma_k \omega^2}{\left(\omega_{0,k}^2 - \omega^2\right)^2 + (\gamma_k \omega)^2} \tag{13.26}$$

$$= \frac{Ne^2}{\epsilon_0 m_e c} \sum_{k=1}^{n} \frac{f_k \gamma_k \omega^2}{\left((\omega_{0,k} - \omega)(\omega_{0,k} + \omega)\right)^2 + (\gamma_k \omega)^2}$$

$$\approx \frac{Ne^2}{\epsilon_0 m_e c} \sum_{k=1}^{n} \frac{f_k \gamma_k}{4\left(\omega_{0,k} - \omega\right)^2 + \gamma_k^2}.$$

Replacing the Lorentzian function, given as

$$L(\omega, \omega_k, \gamma_k) = \frac{1}{\pi} \frac{(\gamma_k/2)}{\left(\omega_{0,k} - \omega\right)^2 + (\gamma_k/2)^2}, \tag{13.27}$$

Equation (13.26) can also be written as

$$\alpha(\omega) = \frac{\pi Ne^2}{\epsilon_0 m_e c} \sum_{k=1}^{n} f_k L(\omega, \omega_k, \gamma_k). \tag{13.28}$$

The Lorentz's function has its full width at half maximum (FWHM) of γ_k. It is normalized to one, and it gives the shape of absorption and emission lines, applied to driven resonant systems. The heavy damped oscillators have broad line widths and respond to a wider range of driving frequencies around the resonant frequency ω_k. The width of the spectral line depends on the quality factor η, which (see also Chap. 10) is a measure of the sharpness of the resonance and depends on γ_k:

$$\eta = \frac{\omega_k}{\Delta\omega} = \frac{\omega_k}{\gamma_k}. \tag{13.29}$$

13.2 Dispersion

The phase velocity is determined by Eq. (13.24), which can also be written as

$$v_p(\omega) = \frac{c}{\text{Re}(n(\omega))}. \tag{13.30}$$

Equation (13.29) indicates that the refraction index is a function of frequency $n(\omega)$, and thus there is a dispersion. Because of the dispersion, different colors present in a white light separate when passed through a glass prism. Besides, due to the dispersion, the wave packets travel with a velocity called the *group velocity*.

13.2.1   Wave Packets and Group Velocity

A wave packet is a superposition of a group of waves traveling together and localized in space at some initial time. Consider the general expression of a planar wave solution

$$\psi(\mathbf{r}, t) = \psi_0 \exp\left(i\left(\mathbf{k}(\omega) \cdot \mathbf{r} - \omega t\right)\right). \tag{13.31}$$

Let us consider a superposition of such plane waves as follows:

$$\psi(\mathbf{r}, t) = \int d\mathbf{k}\, \psi_0(\mathbf{k}) \exp\left(i\left(\mathbf{k} \cdot \mathbf{r} - \omega(\mathbf{k})t\right)\right), \tag{13.32}$$

where $\psi_0(\mathbf{k})$ is some arbitrary function of $\mathbf{k}$, which is chosen such that $\psi(\mathbf{r}, t = 0) = 0$ (i.e., it is localized in a finite region in space). This kind of configurations in space and time are also called *wave packets*. There are various choices for the spatially localized envelope functions; however, a convenient choice for the weighting function ψ_0 is the Gaussian shape function as follows:

$$\psi_0(\mathbf{k}) = \psi_0 \exp\left(-(\mathbf{k} - \mathbf{k}_0)^2 \frac{\xi^2}{4}\right), \tag{13.33}$$

where ψ_0 is an amplitude coefficient, ξ is a coefficient that has the units of lengths determining the spatial extent of the wave package at $t = 0$. For simplicity, we consider a wave packet traveling in the positive z-axis direction; that is, $\mathbf{k} = k\hat{\mathbf{z}}$, and $\mathbf{k} \cdot \mathbf{r} = kz$. Therefore, the spatial profile $\psi(z, 0)$ of the wave package at the initial time is

$$\psi(z, 0) = \psi_0 \int dk\, e^{-(k-k_0)^2 \frac{\xi^2}{4} + ikz} \tag{13.34}$$

$$= \psi_0 \int ds\, e^{-s^2 \frac{\xi^2}{4}} e^{i(s+k_0)z} \quad (s \equiv k - k_0)$$

$$= \psi_0 e^{ik_0 z} \int dk\, e^{-k^2 \frac{\xi^2}{4}} e^{ikz} \quad (k \equiv s)$$

$$= \psi_0 e^{ik_0 z} \int dk\, e^{-k^2 \frac{\xi^2}{4}} e^{-(z/\xi)^2}$$

$$= \frac{\sqrt{\pi}}{2} \psi_0 \xi e^{ik_0 z} e^{-(z/\xi)^2},$$

where the Gaussian integral is used

$$\int_0^{+\infty} dk\, e^{-\alpha k^2} = \sqrt{\frac{\pi}{\alpha}}. \tag{13.35}$$

Thus, $\psi(z, 0)$ is concentrated in a volume side ξ, and it describes the profile of a small wave packet at initial time $t = 0$. Next, we discuss the time evolution of the plane wave $\psi(z, t)$. For that, we assume that $w(k)$ varies slowly with k compared with the extension of the wave package in the wave number space k, which is considered the Fourier space of the real space z. Thus, we can expand $w(k)$ around k_0 as follows:

$$w(k) = w(k_0) + (k - k_0)\left(\frac{dw}{dk}\right)_{k_0} + \frac{1}{2}(k - k_0)^2\left(\frac{d^2w}{dk^2}\right)_{k_0} + \cdots. \tag{13.36}$$

Denoting $w_0 \equiv w(k_0)$, $v_g \equiv (dw/dk)_{k_0}$, and $\beta \equiv (d^2w/dk^2)_{k_0}$, we obtain

$$w(k) \approx w_0 + (k - k_0)v_g + \frac{\beta}{2}(k - k_0)^2. \tag{13.37}$$

Therefore,

$$\psi(z, t) = \psi_0 \int dk\, e^{-(k-k_0)^2\xi^2/4} e^{i(kz - w(k)t)} \tag{13.38}$$

$$\approx \psi_0 e^{ik_0(z - v_p t)} \int dk\, e^{-(\xi^2 + 2i\beta t)(k-k_0)^2/4} e^{i(k-k_0)(z - v_g t)}$$

$$= \frac{\sqrt{\pi}}{2}\psi_0\xi(t)e^{ik_0(z - v_p t)}e^{-((z - v_g t)/\xi(t))^2},$$

where $\xi(t) = \sqrt{\xi^2 + 2i\beta t}$, and $v_p = w_0/k_0$ (phase velocity at w_0). These finding indicate that the center of the wave package moves with group velocity v_g to the right along the z-axis:

$$v_g = \left(\frac{dw}{dk}\right)_{k_0}, \tag{13.39}$$

which is the group velocity at k_0 (or, at w_0). v_g determines the effective center velocity of a superposition of a continuum or group of plane waves. In vacuum where $c(w) = c$ is constant and hence independent of frequency w, we have $w = ck$, and thus $v_g = c$, which equals the speed of light in vacuum.

Note that the group velocity is physical velocity; that is, it describes the speed of propagation of the physical information. On the other hand, the phase velocity v_p is characteristic of an individual plane wave:

$$\exp\left(i\,(kz - w(k)t)\right) = \exp\left(ik\left(z - \frac{w(k)}{k}t\right)\right) = \exp\left(ik\,(z - v_p t)\right). \tag{13.40}$$

Thus, the phase velocity v_p is the velocity of the traveling of the wave. However, the wave extends over the entire system, and hence the phase velocity does not represent a physical information. Furthermore, from the definition $v_p = c/n(\omega)$, and since $n > 1$, it implies that $v_p < c$; however, the phase velocity can exceed the speed of light, and this is not in conflict with the Einstein's special theory of relativity postulate about the speed of light because the dynamics of a uniform wave train is not related to the transport of energy of any other form of physical information.

The group velocity, on the other hand, is such that

$$v_g = \frac{d\omega(k)}{dk} = \left(\frac{dk(\omega)}{d\omega}\right)^{-1} \tag{13.41}$$

$$= \left(\frac{d\omega n(\omega)}{d\omega}\right)^{-1} c = \left(n(\omega) + \omega \frac{dn(\omega)}{d\omega}\right)^{-1} c$$

$$= \frac{v_p}{1 + \dfrac{\omega}{n(\omega)} \dfrac{dn}{d\omega}}.$$

Close to the normal dispersion region, $dn/d\omega > 0$, which implies that $v_g < v_p$; for the anomalous cases, however, $dn/d\omega < 0$, and thus $v_g > v_p$ or even the speed of light in vacuum, c.[2]

13.2.2   Normal and Anomalous Dispersion

Note that the energy is transported with the wave packet at the group velocity v_g. Besides, since $v_g < c$, then the Einstein's special theory of relativity is satisfied. That is also called *normal dispersion*.

From Eq. (13.41), we have

$$v_g = \frac{c}{\text{Re}\,(n(\omega)) + \omega \dfrac{d\text{Re}\,(n(\omega))}{d\omega}}. \tag{13.42}$$

Equation (13.42) indicates that normal dispersion occurs for $d\text{Re}(n(\omega))/d\omega > 0$. However, for $d\text{Re}(n(\omega))/d\omega < 0$, the group speed v_g can exceed the speed of light in vacuum c, and this is the so-called *anomalous dispersion*. Note that the anomalous dispersion is related to the absorption, and hence the group velocity does not characterize any longer physical quantity; in that case, we can say that it does not represent the speed of transferring energy of information. Therefore, the Einstein's special theory of relativity is not violated.

The dispersive spreading of the electromagnetic wave packets is an important phenomenon, such as the dispersive deformation is one of the main factors that

---

[2] Jackson (1999).

limits the information load transported through the fiber optical cables; if the load is too high, the wave packets constituting the bit stream through the fibers begin to overload and thus loosing the identity. Construction of more sophisticated counter measures optimizing the data capacity of optical fibers represents an important area of research.

13.3 Refractive Index of a Conductor

If an electromagnetic wave propagates in a conductor, the current density is $\mathbf{J} = \sigma\mathbf{E}$ (where σ is the conductivity). Therefore, when deriving the wave equations from Maxwell's equations in a medium, the current density appears in these equations. Thus, taking the curl of the second Maxwell's equations in a conductor,

$$\nabla \times (\nabla \times \mathbf{E}) = -\frac{\partial}{\partial t}(\mu_0\mu\nabla \times \mathbf{H}).$$

(13.43)

Using the fourth Maxwell's equation for a conductor ($\nabla \times \mathbf{H} = \sigma\mathbf{E} + \partial\mathbf{D}/\partial t$), we have

$$-\nabla^2\mathbf{E} = -\left(\mu_m\sigma\frac{\partial\mathbf{E}}{\partial t} + \epsilon_m\mu_m\frac{\partial^2\mathbf{E}}{\partial t^2}\right),$$

(13.44)

where $\mathbf{D} = \epsilon_m\mathbf{E}$. Equation (13.44) can be re-arranged in the following form:

$$\nabla^2\mathbf{E} - \mu_m\sigma\frac{\partial\mathbf{E}}{\partial t} - \epsilon_m\mu_m\frac{\partial^2\mathbf{E}}{\partial t^2} = 0.$$

(13.45)

Similarly, using the curl of the fourth Maxwell's equation for $\mathbf{\Delta} \times \mathbf{B}$, we can obtain the wave equation for the magnetic field as follows:

$$\nabla^2\mathbf{B} - \mu_m\sigma\frac{\partial\mathbf{B}}{\partial t} - \epsilon_m\mu_m\frac{\partial^2\mathbf{B}}{\partial t^2} = 0.$$

(13.46)

Consider the monochromatic plane wave solutions, representing waves propagating to the right, as

$$\mathbf{E}(\mathbf{r}, t) = \mathbf{E}_0 \exp(i(\mathbf{k}\cdot\mathbf{r} - \omega t)),$$

(13.47)

$$\mathbf{B}(\mathbf{r}, t) = \mathbf{B}_0 \exp(i(\mathbf{k}\cdot\mathbf{r} - \omega t)).$$

Substituting expression for $\mathbf{E}$ from Eq. (13.47) into (13.45), we obtain

$$\left((i\mathbf{k})^2 - \mu_m\sigma(-i\omega) - \mu_m\epsilon_m(i\omega)^2\right)\mathbf{E} = 0.$$

(13.48)

From Eq. (13.48), we obtain the dispersion equation for the electromagnetic waves in a conductor as follows:

$$k(\omega) = \sqrt{i\mu_m\sigma\omega + \mu_m\epsilon_m\omega^2}.$$ (13.49)

If we consider a good conductor such that $\sigma \gg \epsilon_m\omega$, then from Eq. (13.49):

$$k(\omega) \approx \sqrt{i}\sqrt{\mu_m\sigma\omega} = \sqrt{e^{i\pi/2}}\sqrt{\mu_m\sigma\omega}$$ (13.50)

$$= e^{i\pi/4}\sqrt{\mu_m\sigma\omega}$$

$$= \frac{1+i}{\sqrt{2}}\sqrt{\mu_m\sigma\omega}.$$

Equation (13.50) indicates that wave number k is complex number with real and imaginary part as follows:

$$k_r(\omega) \equiv \text{Re}(k(\omega)) = \frac{1}{\sqrt{2}}\sqrt{\mu_m\sigma\omega},$$ (13.51)

$$k_i(\omega) \equiv \text{Im}(k(\omega)) = \frac{1}{\sqrt{2}}\sqrt{\mu_m\sigma\omega}.$$

The time average of the magnitude of the electric field vector is

$$\langle E(\mathbf{r}, t)\rangle \propto \exp\left(-k_i\hat{\mathbf{k}} \cdot \mathbf{r}\right) \equiv \exp\left(-\hat{\mathbf{k}} \cdot \mathbf{r}/\delta(\omega)\right),$$ (13.52)

where $\delta(\omega)$ is the so-called *skin depth* defined as

$$\delta(\omega) = \sqrt{\frac{2}{\mu_m\sigma\omega}}.$$ (13.53)

Therefore, when the electromagnetic waves propagate through a conductor, the intensity decreases as follows:

$$I = I_{\text{max}} \exp\left(-2\hat{\mathbf{k}} \cdot \mathbf{r}/\delta(\omega)\right),$$ (13.54)

where the absorption coefficient is given as

$$\alpha(\omega) = \frac{2}{\delta(\omega)}.$$ (13.55)

Figure 13.3 presents a log plot of the skin depth versus linear frequency for soft iron ($\sigma = 11.2 \times 10^6 \ \Omega^{-1}m^{-1}$), silver ($\sigma = 62.1 \times 10^6 \ \Omega^{-1}m^{-1}$), lead ($\sigma = 4.7 \times 10^6 \ \Omega^{-1}m^{-1}$), and seawater at 20 °C and a salt concentration of 3 % ($\sigma = 4.1 \ \Omega^{-1}m^{-1}$), and $\mu_m = 2 \times 10^{-5}$. At high electromagnetic wave frequencies, the

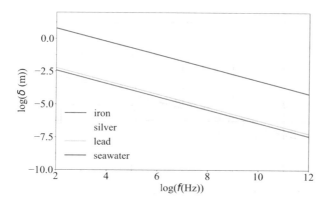

Fig. 13.3 Log plot of the skin depth versus linear frequency for soft iron ($\sigma = 11.2 \times 10^6 \, \Omega^{-1}\mathrm{m}^{-1}$), silver ($\sigma = 62.1 \times 10^6 \, \Omega^{-1}\mathrm{m}^{-1}$), lead ($\sigma = 4.7 \times 10^6 \, \Omega^{-1}\mathrm{m}^{-1}$), and seawater at $20\,^\circ$C and a salt concentration of 3 % ($\sigma = 4.1 \, \Omega^{-1}\mathrm{m}^{-1}$). $\mu_m = 2 \times 10^{-5}$

waves can only penetrate small distances inside the conductor. For example, in seawater the skin depth is about $\delta = 30$ cm at $f = 1$ MHz and much longer of $\delta = 10$ m at $f = 1$ kHz for a temperature of $20\,^\circ$C and salt concentration of 3 %. That is why, there is a communication problem of radio waves at frequency range 3–30 kHz. In the case of a conductor at high-frequency oscillating currents, magnetic fields are created that induce electric fields opposing the change in magnetic field; these induced electric fields produce the so-called *eddy currents*, and hence a resistive power loss. Furthermore, the induced electric fields are stronger at the center of the conductor, producing a high resistance at the center of the conductor, and thus the conduction current is confined to a thin outer layer with thickness the skin depth. That is equivalent to using a hollow wire, which is lighter than a conducting wire.

The reflectance of a good conductor is high; therefore, the parabolic or spherical glass surfaces of a reflecting telescope are coated with a thin silver or aluminum, typically $0.1\ \mu$m. Also, the index of refraction of a conductor is complex number. The amplitude reflection coefficient for the external reflection (for example, air/metal) at the normal incidence at $\theta_1 = 0^\circ$ (for both perpendicular and parallel polarization), and obtain a reflectance as follows:

$$r_{\perp,\parallel}(\theta_1 = 0^\circ) = \frac{(1 - n_r(\omega))^2 + (n_i(\omega))^2}{(1 + n_r(\omega))^2 + (n_i(\omega))^2}. \tag{13.56}$$

Using expressions given by Eq. (13.51), we derive the index of the refraction (both the real and imaginary parts) as follows:

$$n_r(\omega) = \frac{k_r(\omega)c}{\omega} = c\sqrt{\frac{\mu_m\sigma}{2\omega}}, \tag{13.57}$$

$$n_i(\omega) = \frac{k_i(\omega)c}{\omega} = c\sqrt{\frac{\mu_m\sigma}{2\omega}}.$$

Knowing the conductivity σ (or equivalently, the resistivity $\rho = 1/\sigma$) and μ_m, one can calculate the reflectance of a conductor depending on the frequency ω for normal incidence waves from air to metal.

13.4 Wave Propagation in a Dilute Plasma

Consider the wave propagation in an ionized plasma consisting of n_e electrons and n_i ions per unit of volume. Besides, we will ignore the distant ions as they will provide just a little damping to the electrons motion, and because of their large mass, their contribution to the current density can be omitted.

Applying the second law of Newton for the motion of an individual electron, we have

$$m_e\frac{d\mathbf{v}}{dt} = -e\mathbf{E}(t), \tag{13.58}$$

where $\mathbf{E}(t)$ is the time-varying external electric field of an electromagnetic wave. The current density arriving from all bulk electrons in the plasma is

$$\mathbf{J}(t) = n_e(-e)\mathbf{v}(t). \tag{13.59}$$

The time derivative of the current density is

$$\frac{d\mathbf{J}(t)}{dt} = \frac{n_e e^2}{m_e}\mathbf{E}(t). \tag{13.60}$$

Consider an oscillating electric field such that $\mathbf{E}(t) = \mathbf{E}_0 e^{-i\omega t}$, then from Eq. (13.60), we obtain

$$\frac{d\mathbf{J}(t)}{dt} = \frac{n_e e^2}{m_e}\mathbf{E}_0 e^{-i\omega t}. \tag{13.61}$$

Or,

$$J(t) = \frac{n_e e^2}{m_e} E_0 \int dt\, e^{-i\omega t} \tag{13.62}$$

$$= \frac{n_e e^2}{-i\omega m_e} E_0 \int d(-i\omega t)\, e^{-i\omega t}$$

$$= i\frac{n_e e^2}{\omega m_e} E.$$

Equation (13.62) implies that conductivity is

$$\sigma = i\frac{n_e e^2}{\omega m_e}, \tag{13.63}$$

which is purely imaginary quantity. Using Eq. (13.62) and $i = e^{i\pi/2}$, we write

$$J(t) = \frac{n_e e^2}{\omega m_e} E_0 e^{-i(\omega t - \pi/2)}. \tag{13.64}$$

Therefore, the current density and the electric field have a phase shift of $\pi/2$, or $90°$. If we calculate the average power, then it equals to zero. This is analogue to the LC circuit discussed in Chap. 10 where the energy was continuously exchanging between the electric field energy of the capacitor C and magnetic field energy of the inductor L.

13.4.1 Electromagnetic Waves in a Dilute Plasma

Let us consider a dilute (i.e., low density) plasma and neural (i.e., the charge density is $\rho = 0$). We may also assume that $\mu_m = \mu_0$ and $\epsilon_m = \epsilon_0$ (i.e., the vacuum). Then, the wave equation for the electric field vector E can be obtained by taking the curl of the second Maxwell's equation (Faraday's law) as usually

$$\nabla \times (\nabla \times E) = -\frac{\partial}{\partial t}(\nabla \times B), \tag{13.65}$$

$$-\nabla^2 E = -\frac{\partial}{\partial t}\left(\mu_0 J + \mu_0 \epsilon_0 \left(\frac{\partial E}{\partial t}\right)\right), \tag{13.66}$$

$$\nabla^2 E = \mu_0 \frac{\partial J}{\partial t} + \mu_0 \epsilon_0 \left(\frac{\partial^2 E}{\partial t^2}\right), \tag{13.67}$$

$$\nabla^2 E = \mu_0 \frac{n_e e^2}{m_e} E + \mu_0 \epsilon_0 \left(\frac{\partial^2 E}{\partial t^2}\right), \tag{13.68}$$

where Eq. (13.61) is used. Therefore, the wave equation for E becomes

$$\nabla^2 \mathbf{E} - \mu_0 \left(\frac{n_e e^2}{m_e} \right) \mathbf{E} - \frac{1}{c^2} \left(\frac{\partial^2 \mathbf{E}}{\partial t^2} \right) = 0. \tag{13.69}$$

Substituting the monochromatic electromagnetic plane wave solution, $\mathbf{E}(\mathbf{r}, t) = \mathbf{E}_0 \exp(i(\mathbf{k} \cdot \mathbf{r} - \omega t))$ into Eq. (13.69), we get

$$(i\mathbf{k})^2 \mathbf{E} - \mu_0 \left(\frac{n_e e^2}{m_e} \right) \mathbf{E} - \frac{1}{c^2} (-i\omega)^2 \mathbf{E} = 0 \tag{13.70}$$

or,

$$\left(-k^2 - \frac{n_e e^2}{c^2 \epsilon_0 m_e} + \frac{\omega^2}{c^2} \right) \mathbf{E} = 0. \tag{13.71}$$

Therefore, the following dispersion relation is obtained:

$$k^2 c^2 = \omega^2 - \omega_p^2, \tag{13.72}$$

where ω_p is the so-called *plasma frequency*:

$$\omega_p = \sqrt{\frac{n_e e^2}{\epsilon_0 m_e}}. \tag{13.73}$$

Equation (13.72) implies that for $\omega > \omega_p$, $\mathbf{k}$ is purely real number, and hence the electromagnetic wave propagates in the plasma, whereas for $\omega < \omega_p$, wave number vector $\mathbf{k}$ is purely imaginary, the electromagnetic waves can not propagate in plasma.

13.4.2   Phase and Group Velocity in a Dilute Plasma

Using Eq. (13.72), the index of refraction n, group velocity v_g and phase velocity v_p are given as follows:

$$n(\omega) = \frac{kc}{\omega} = \sqrt{1 - \frac{\omega_p^2}{\omega^2}}, \tag{13.74}$$

$$v_g(\omega) = \left(\frac{dk}{d\omega} \right)^{-1} = c\sqrt{1 - \frac{\omega_p^2}{\omega^2}}, \tag{13.75}$$

$$v_p(\omega) = \frac{\omega}{k} = \frac{c}{\sqrt{1 - \frac{\omega_p^2}{\omega^2}}}. \tag{13.76}$$

From Eqs. (13.75) and (13.76),

$$v_g v_p = c^2. \tag{13.77}$$

Thus,

$$n(\omega) \equiv \frac{c}{v_p} = \frac{v_g}{c}. \tag{13.78}$$

Therefore, for $\omega < \omega_p$, there is no electromagnetic wave propagation in plasma, which is also called *cutoff frequency* of a dilute plasma. For $\omega > \omega_p$ the phase velocity magnitude exceeds the speed of light in vacuum c; however, $v_g < c$.

13.4.3   Plasma and Dielectric at High Frequency

Consider $\omega \gg \omega_p$, then the index of refraction $n(\omega)$ of a dilute plasma from Eq. (13.74) becomes

$$n(\omega) = \sqrt{1 - \frac{\omega_p^2}{\omega^2}} \approx 1 - \frac{\omega_p^2}{2\omega^2}. \tag{13.79}$$

On the other hand, using Eq. (13.22), the real part of the index of refraction for a dielectric using the Lorentz's oscillator model is

$$n_r(\omega) \approx 1 + \frac{Ne^2}{2\epsilon_0 m_e} \sum_{k=1}^{n} f_k \frac{\omega_{0,k}^2 - \omega^2}{(\omega_{0,k}^2 - \omega^2)^2 + (\gamma_k \omega)^2} \tag{13.80}$$

$$\approx 1 - \frac{Ne^2}{2\epsilon_0 m_e} \sum_{k=1}^{n} \frac{f_k}{\omega^2}.$$

Therefore,

$$n_r(\omega) \approx 1 - \frac{\omega_p^2}{2\omega^2}, \tag{13.81}$$

where

$$\omega_p = \sqrt{\frac{N'e^2}{\epsilon_0 m_e}}, \tag{13.82}$$

and N' is the total number density of electron oscillators:

$$N' = N \sum_{k=1}^{n} f_k. \tag{13.83}$$

Comparing Eq. (13.79) with (13.83), we find that the index of refraction for the dielectric and plasma at high frequency are identical.

13.5 Exercises

Exercise 13.1 Show that the time-averaged power density $\langle \mathbf{E} \cdot \mathbf{J} \rangle$ in a dilute plasma is zero.

Solution 13.1 Consider $\mathbf{E}(t) = \mathbf{E}_0 e^{-i\omega t}$, then using Eq. (13.62), we have

$$\mathbf{E} \cdot \mathbf{J} = i \frac{n_e e^2}{\omega m_e} \mathbf{E}^2 = \frac{n_e e^2}{\omega m_e} \mathbf{E}_0^2 e^{-i(2\omega t - \pi/2)}, \tag{13.84}$$

where $i = e^{\pi/2}$ is used. Thus,

$$\mathbf{E} \cdot \mathbf{J} = \frac{n_e e^2}{\omega m_e} \mathbf{E}_0^2 \left(\cos(2\omega t - \pi/2) - i \sin(2\omega t - \pi/2) \right) \tag{13.85}$$

$$= \frac{n_e e^2}{\omega m_e} \mathbf{E}_0^2 \left(\sin(2\omega t) + i \cos(2\omega t) \right).$$

The time average of the following functions are

$$\langle \sin(2\omega t) \rangle = \frac{1}{T} \int_0^T \sin(2\omega t) dt = \frac{1}{2T\omega} \int_0^T \sin(2\omega t) d(2\omega t) = 0, \tag{13.86}$$

$$\langle \cos(2\omega t) \rangle = \frac{1}{T} \int_0^T \cos(2\omega t) dt = \frac{1}{2T\omega} \int_0^T \cos(2\omega t) d(2\omega t) = 0. \tag{13.87}$$

Therefore,

$$\langle \mathbf{E} \cdot \mathbf{J} \rangle = 0. \tag{13.88}$$

Exercise 13.2 The resistivity of silver is $\rho = 1.6 \times 10^{-8}\ \Omega\text{m}$, and its permeability is $\mu_m = 0.9998\mu_0$, find the reflectance for light with wavelength 500 nm at normal incidence from air.

Solution 13.2 Using Eq. (13.56) for the reflectance:

$$r_{\perp,\parallel}(\theta_1 = 0°) = \frac{(1 - n_r(\omega))^2 + (n_i(\omega))^2}{(1 + n_r(\omega))^2 + (n_i(\omega))^2}. \tag{13.89}$$

For both n_r and n_i, we can use Eq. (13.57) as follows:

$$n_r = n_i = \frac{k_r c}{\omega} = \frac{c}{\omega}\sqrt{\frac{\mu_m \sigma \omega}{2}} = \sqrt{\frac{c^2 \mu_m}{2\omega\rho}}. \tag{13.90}$$

$$\mu_m = 4\pi \times 10^{-7}\,(\text{T}\cdot\text{m/A})0.9998 \approx 12.56 \times 10^{-7}\,(\text{T}\cdot\text{m/A})$$

Furthermore,

$$\omega = \frac{2\pi}{T} = \frac{2\pi c}{\lambda},$$

and thus

$$n_r = n_i = \sqrt{\frac{c\lambda\mu_m}{4\pi\rho}} = \sqrt{\frac{(3 \times 10^8)(500 \times 10^{-9})(12.56 \times 10^{-7})}{4\pi(1.6 \times 10^{-8})}} \approx 30.6.$$

Therefore,

$$r_{\perp,\|} = 0.937.$$

Exercise 13.3 The density of electrons in a dilute plasma is $n_e = 1.0 \times 10^{20}$ electrons/m^3. Determine the plasma frequency ω_p.

Solution 13.3 Using Eq. (13.78), $m_e = 9.11 \times 10^{-31}$ kg, $e = 1.6 \times 10^{-19}$ C, then the plasma frequency is

$$\omega_p = \left(\frac{n_e e^2}{\epsilon_0 m_e}\right)^{1/2} \tag{13.91}$$

$$= \left(\frac{(10^{20})(1.6 \times 10^{-19})^2}{(8.8542 \times 10^{-12})(9.11 \times 10^{-31})}\right)^{1/2}$$

$$\approx 0.563 \times 10^{12}\text{ Hz}.$$

Exercise 13.4 Determine the absorption coefficient of seawater knowing that at low frequency, seawater has an electrical conductivity $\sigma = 4.0\,\Omega^{-1}\text{m}^{-1}$. The frequency is $f = 10^8$ Hz.

Solution 13.4 Using Eqs. (13.53) and (13.55) for $\mu \approx 1$, the absorption coefficient of seawater is

$$\alpha = \sqrt{4\pi\mu_0\sigma f} \tag{13.92}$$
$$= \sqrt{4\pi(4\pi \times 10^{-7})(4.0)(10^8)}$$
$$\approx 79.5\,\mathrm{m}^{-1}.$$

Exercise 13.5 Determine the skin length of seawater knowing that at low frequency, seawater has an electrical conductivity $\sigma = 4.0\,\Omega^{-1}\mathrm{m}^{-1}$. The frequency is $f = 100$ Hz.

Solution 13.5 Using Eqs. (13.53) and (13.55) for $\mu \approx 1$, the absorption coefficient of seawater is

$$\alpha = \sqrt{4\pi\mu_0\sigma f} \tag{13.93}$$
$$= \sqrt{4\pi(4\pi \times 10^{-7})(4.0)(100)}$$
$$\approx 0.079\,\mathrm{m}^{-1}.$$

Therefore, the skin length is

$$\delta = \frac{1}{\alpha} \approx 12.6\,\mathrm{m}.$$

References

Altland A, Simons B (2010) Condensed matter field theory, 2nd edn. Cambridge University Press
Griffiths DJ (1999) Introduction to electrodynamics, 3rd edn. Prentice Hall
Jackson JD (1999) Classical electrodynamics, 3rd edn. John Wiley and Sons
Landau LD, Lifshitz EM (1971) The classical theory of fields. Pergamon Press
Protheroe RJ (2013) Essential electrodynamics, 1st edn. Bookboon
Sykja H (2006) Bazat e Elektrodinamikës. SHBUT

Appendix
Vectorial Analysis

Some mathematical background on vector analysis, useful to electromagnetism.

In this appendix, we discuss some applications of vectorial calculus and vector differential operators to electromagnetism. In addition, we present some useful expressions of the gradient operator acting on a scalar and vector field in Cartesian, spherical, and cylindrical coordinates. Moreover, Laplacian expressions in Cartesian, spherical, and cylindrical coordinates are provided.

A.1 Vector Calculus

Consider $\mathbf{A}$ a vector, which is a function of a scalar quantity s. We can express $\mathbf{A}$ in terms of its components as

$$\mathbf{A}(s) = A_x(s)\mathbf{i} + A_y(s)\mathbf{j} + A_z(s)\mathbf{k}. \tag{A.1}$$

Definition A.1 (*Derivative*) The derivative of $\mathbf{A}$ for the variable s is defined as

$$\frac{d\mathbf{A}(s)}{ds} = \lim_{\Delta s \to 0} \frac{\mathbf{A}(s + \Delta s) - \mathbf{A}(s)}{\Delta s}. \tag{A.2}$$

In terms of the components of $\mathbf{A}$, we can write

$$\frac{d\mathbf{A}(s)}{ds} = \frac{dA_x}{ds}\mathbf{i} + \frac{dA_y}{ds}\mathbf{j} + \frac{dA_z}{ds}\mathbf{k}. \tag{A.3}$$

© The Editor(s) (if applicable) and The Author(s), under exclusive license to Springer Nature Switzerland AG 2022
H. Kamberaj, *Electromagnetism*, Undergraduate Texts in Physics,
https://doi.org/10.1007/978-3-030-96780-2

Fig. A.1 Graphical interpretation of the derivative

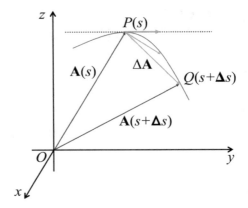

A graphical interpretation of derivative of vector $\mathbf{A}$ is shown in Fig. A.1. As it can be seen that as $\Delta s \to 0$, the vector difference $\nabla \mathbf{A} = \mathbf{A}(s + \Delta s) - \mathbf{A}(s)$ approaches along the line tangent to the curve $\mathbf{A}(s)$ at point s. Therefore, $d\mathbf{A}/ds$ is a vector, and it is tangent to the curve.

The magnitude of the vector $d\mathbf{A}/ds$ is given as

$$\left| \frac{d\mathbf{A}(s)}{ds} \right| \equiv \frac{dA}{ds} = \sqrt{\left(\frac{dA_x}{ds} \right)^2 + \left(\frac{dA_y}{ds} \right)^2 + \left(\frac{dA_z}{ds} \right)^2}. \tag{A.4}$$

The following rules hold for the derivative:

$$\frac{d}{ds}(\mathbf{A}(s) \pm \mathbf{B}(s)) = \frac{d\mathbf{A}}{ds} \pm \frac{d\mathbf{B}}{ds}, \tag{A.5}$$

$$\frac{d}{ds}(f(s)\mathbf{A}(s)) = \frac{df}{ds}\mathbf{A}(s) + f(s)\frac{d\mathbf{A}}{ds},$$

$$\frac{d}{ds}(\mathbf{A}(s) \cdot \mathbf{B}(s)) = \frac{d\mathbf{A}}{ds} \cdot \mathbf{B}(s) + \mathbf{A} \cdot \frac{d\mathbf{B}}{ds},$$

$$\frac{d}{ds}(\mathbf{A}(s) \times \mathbf{B}(s)) = \frac{d\mathbf{A}}{ds} \times \mathbf{B}(s) + \mathbf{A} \times \frac{d\mathbf{B}}{ds}.$$

Note that in the last expression of Eq. (A.5) the order of the vectors $\mathbf{A}$ and $\mathbf{B}$ matters. In Eq. (A.5), $f(s)$ is a scalar function of s.

Usually, in three-dimensional space, we have to distinguish two types of functions or fields. Namely, the scalar point functions of the form:

$$f(\mathbf{r}) \equiv f(x, y, z) \tag{A.6}$$

and the vector point functions

$$\mathbf{A}(\mathbf{r}) \equiv \mathbf{A}(x, y, z) = \left(A_x(x, y, z), \ A_y(x, y, z), \ A_z(x, y, z) \right). \tag{A.7}$$

Fig. A.2 A line $\mathcal{L}$ in
three-dimensional space, and
a vector **A** defined at every
point along the curve $\mathcal{L}$

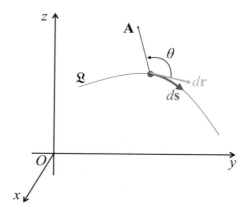

For example, a scalar function is the electric potential, $\phi(\mathbf{r}) = \phi(x, y, z)$, and a vector function is electric field vector $\mathbf{E}(\mathbf{r}) = \mathbf{E}(x, y, z)$. These functions can also be explicit function of some other scalar, such as s, which is often time t.

Consider a line $\mathcal{L}$ in three-dimensional space, and a vector **A** defined at every point along the curve $\mathcal{L}$, as shown in Fig. A.2.

Definition A.2 (*Line integral*) The line integral of **A** along the curve $\mathcal{L}$ is defined as

$$\int_{\mathcal{L}} \mathbf{A}(\mathbf{r}) \cdot d\mathbf{s}. \tag{A.8}$$

If we consider the displacement vector $d\mathbf{r}$ such that $|\, d\mathbf{r} \,| = ds$, then the line integral can also be written as

$$\int_{\mathcal{L}} \mathbf{A}(\mathbf{r}) \cdot d\mathbf{s} = \int_{\mathcal{L}} \mathbf{A}(\mathbf{r}) \cdot d\mathbf{r} = \int_{\mathcal{L}} A(\mathbf{r}) \cos\theta\, ds. \tag{A.9}$$

Furthermore, if $\mathbf{r} = x\mathbf{i} + y\mathbf{j} + z\mathbf{k}$, then $d\mathbf{r} = dx\mathbf{i} + dy\mathbf{j} + dz\mathbf{k}$, and hence, from Eq. (A.9), we can write

$$\int_{\mathcal{L}} \mathbf{A}(\mathbf{r}) \cdot d\mathbf{s} = \int_{\mathcal{L}} \left(A_x dx + A_y dy + A_z dz \right) \tag{A.10}$$

$$= \int_{\mathcal{L}} A_x dx + \int_{\mathcal{L}} A_y dy + \int_{\mathcal{L}} A_z dz.$$

If we denote ds a distance measured along the line, which parametrizes the line, then Eq. (A.9) can also be written as

$$\int_{\mathcal{L}} \mathbf{A}(\mathbf{r}) \cdot d\mathbf{r} = \int_{\mathcal{L}} \left(\mathbf{A} \cdot \frac{d\mathbf{r}}{ds} \right) ds \tag{A.11}$$

$$= \int_{\mathcal{L}} \left(A_x \frac{dx}{ds} + A_y \frac{dy}{ds} + A_z \frac{dz}{ds} \right) ds .$$

A.2 Vector Differential Operators

Definition A.3 (*Gradient operator*) The vector gradient operator is denoted by **grad** or ∇, and it is not a vector but it is a vector operator, which in Cartesian coordinates is defined as

$$\nabla = \mathbf{i} \frac{\partial}{\partial x} + \mathbf{j} \frac{\partial}{\partial y} + \mathbf{k} \frac{\partial}{\partial z} \tag{A.12}$$

$$= \left(\frac{\partial}{\partial x}, \frac{\partial}{\partial y}, \frac{\partial}{\partial z} \right) .$$

When this operator acts on a scalar function ϕ, the result is a vector; for example, let $\phi(x, y, z)$ be a scalar function of (x, y, z), then $\nabla \phi$ is a vector given as

$$\nabla \phi = \mathbf{i} \frac{\partial \phi}{\partial x} + \mathbf{j} \frac{\partial \phi}{\partial y} + \mathbf{k} \frac{\partial \phi}{\partial z} \tag{A.13}$$

$$= \left(\frac{\partial \phi}{\partial x}, \frac{\partial \phi}{\partial y}, \frac{\partial \phi}{\partial z} \right) .$$

A practical aspect of the gradient is as the following. Suppose we want to calculate a small change on the scalar function ϕ due to a change on the variables $d\mathbf{r}$. For that, using Eq. (A.13), we write

$$d\phi = \nabla \phi \cdot d\mathbf{r} = | \nabla \phi | | d\mathbf{r} | \cos \theta , \tag{A.14}$$

where θ is the angle between $\nabla \phi$ and $d\mathbf{r}$.

When this operator acts on a vector function $\mathbf{A}$, it performs a scalar product and hence the result is a scalar; for example, let $\mathbf{A}(x, y, z)$ be a vector function of (x, y, z), then $\nabla \cdot \mathbf{A}$ is a scalar given as dot product:

$$\nabla \cdot \mathbf{A} = \frac{\partial}{\partial x} (\mathbf{i} \cdot \mathbf{A}) + \frac{\partial}{\partial y} (\mathbf{j} \cdot \mathbf{A}) + \frac{\partial}{\partial z} (\mathbf{k} \cdot \mathbf{A}) \tag{A.15}$$

$$= \frac{\partial A_x}{\partial x} + \frac{\partial A_y}{\partial y} + \frac{\partial A_z}{\partial z} ,$$

which is also called divergence of $\mathbf{A}$, and it is denoted by div$\mathbf{A}$.

Moreover, when the gradient operator performs a cross-product with another vector function, it gives as a result the curl of $\mathbf{A}$ or the rot (rotation) of $\mathbf{A}$, denoted by $\nabla \times \mathbf{A}$ or curl$\mathbf{A}$, or rot$\mathbf{A}$. This is a vector, and it is defined as

$$
\nabla \times \mathbf{A} = \begin{bmatrix} \mathbf{i} & \mathbf{j} & \mathbf{k} \\ \dfrac{\partial}{\partial x} & \dfrac{\partial}{\partial y} & \dfrac{\partial}{\partial z} \\ A_x & A_y & A_z \end{bmatrix} \tag{A.16}
$$

$$
= \det\left(\begin{bmatrix} \dfrac{\partial}{\partial y} & \dfrac{\partial}{\partial z} \\ A_y & A_z \end{bmatrix}\right)\mathbf{i} - \det\left(\begin{bmatrix} \dfrac{\partial}{\partial x} & \dfrac{\partial}{\partial z} \\ A_x & A_z \end{bmatrix}\right)\mathbf{j} + \det\left(\begin{bmatrix} \dfrac{\partial}{\partial x} & \dfrac{\partial}{\partial y} \\ A_x & A_y \end{bmatrix}\right)\mathbf{k}
$$

$$
= \left(\frac{\partial A_z}{\partial y} - \frac{\partial A_y}{\partial z}\right)\mathbf{i} + \left(\frac{\partial A_x}{\partial z} - \frac{\partial A_z}{\partial x}\right)\mathbf{j} + \left(\frac{\partial A_y}{\partial x} - \frac{\partial A_x}{\partial y}\right)\mathbf{k}.
$$

A.3 Stokes' Formula

Consider the vector field lines $\mathbf{J}$ passing through the cross-sectional surface area $\mathbf{A}$ enclosed by the closed line $\mathcal{L}$, as shown in Fig. A.3. Here, the surface area vector $\mathbf{A} = A\mathbf{n}$, where $\mathbf{n}$ is a unit vector perpendicular to surface area. Then, according to Stokes' formula, the line integral of $\mathbf{F}$ along the curve $\mathcal{L}$ is

$$
\oint_{\mathcal{L}} \mathbf{F} \cdot d\mathbf{s} = \int_{A} (\nabla \times \mathbf{F}) \cdot d\mathbf{A}. \tag{A.17}
$$

In Eq. (A.17), the left-hand side gives the line integral of the vector field $\mathbf{F}$, and the right-hand side gives the flux of the vector $\mathbf{J} \equiv \nabla \times \mathbf{F}$ (the rot of $\mathbf{F}$) through the surface area enclosed by the line $\mathcal{L}$.

Fig. A.3 Vector field lines $\mathbf{J}$ passing through the cross-sectional surface area $\mathbf{A}$ enclosed by the closed line $\mathcal{L}$

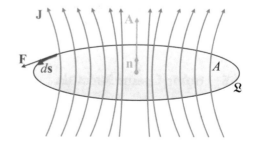

Fig. A.4 Vector field lines **F** passing through the closed surface area **A** enclosing the volume V. dA is a small surface element and **n** is an outward unit vector normal to the surface element

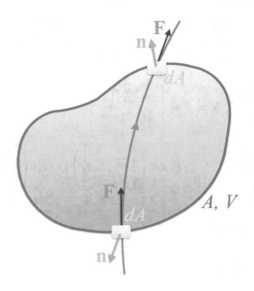

A.4 Gauss's Formula

Consider the vector field lines **F** passing through the closed surface area **A** enclosing the volume V, as shown in Fig. A.4. dA is a small surface element and **n** is an outward unit vector normal to the surface element. The vector **F** is tangent to the line at every point of the line. Then, according to Gauss's formula, the closed surface integral of **F** along the surface A is

$$\oint_A \mathbf{F} \cdot d\mathbf{A} = \int_V (\nabla \cdot \mathbf{F})\, dV , \qquad (A.18)$$

where $d\mathbf{A} = dA\mathbf{n}$. In Eq. (A.18), the left-hand side gives the net number of lines leaving the surface A and the right-hand side gives the total amount of source creating the field **F** (e.g., charge) inside the volume V (Fig. A.4).

A.5 Some Useful Formula

Let ϕ and **A** be a scalar field and vector field, respectively. Then,

$$\nabla \cdot (\phi \mathbf{A}) = \frac{\partial(\phi A_x)}{\partial x} + \frac{\partial(\phi A_y)}{\partial y} + \frac{\partial(\phi A_z)}{\partial z} \qquad (A.19)$$

$$= \frac{\partial \phi}{\partial x} A_x + \phi \frac{\partial A_x}{\partial x} + \frac{\partial \phi}{\partial y} A_y + \phi \frac{\partial A_y}{\partial y} + \frac{\partial \phi}{\partial z} A_z + \phi \frac{\partial A_z}{\partial z}$$

$$= (\nabla \phi) \cdot \mathbf{A} + \phi (\nabla \cdot \mathbf{A}) .$$

Furthermore, the following formula can be proved:

$$\nabla \times (\phi \mathbf{A}) = \left(\frac{\partial (\phi A_z)}{\partial y} - \frac{\partial (\phi A_y)}{\partial z} \right) \mathbf{i} \tag{A.20}$$

$$+ \left(\frac{\partial (\phi A_x)}{\partial z} - \frac{\partial (\phi A_z)}{\partial x} \right) \mathbf{j}$$

$$+ \left(\frac{\partial (\phi A_y)}{\partial x} - \frac{\partial (\phi A_x)}{\partial y} \right) \mathbf{k}$$

$$= \left(\frac{\partial \phi}{\partial y} A_z + \phi \frac{\partial A_z}{\partial y} - \frac{\partial \phi}{\partial z} A_y - \phi \frac{\partial A_y}{\partial z} \right) \mathbf{i}$$

$$+ \left(\frac{\partial \phi}{\partial z} A_x + \phi \frac{\partial A_x}{\partial z} - \frac{\partial \phi}{\partial x} A_z - \phi \frac{\partial A_z}{\partial x} \right) \mathbf{j}$$

$$+ \left(\frac{\partial \phi}{\partial x} A_y + \phi \frac{\partial A_y}{\partial x} - \frac{\partial \phi}{\partial y} A_x - \phi \frac{\partial A_x}{\partial y} \right) \mathbf{k}$$

$$= (\nabla \phi) \times \mathbf{A} + \phi (\nabla \times \mathbf{A}) .$$

Moreover, since $\mathbf{i}$, $\mathbf{j}$, and $\mathbf{k}$ are unit and constant vectors, then

$$\nabla \cdot \mathbf{i} = \nabla \cdot \mathbf{j} = \nabla \cdot \mathbf{k} = 0, \tag{A.21}$$
$$\nabla \times \mathbf{i} = \nabla \times \mathbf{j} = \nabla \times \mathbf{k} = 0 .$$

It is easy to show that

$$\nabla \cdot \mathbf{r} = \frac{\partial x}{\partial x} + \frac{\partial y}{\partial y} + \frac{\partial z}{\partial z} = 3 \tag{A.22}$$

and

$$\nabla \times \mathbf{r} = \begin{bmatrix} \mathbf{i} & \mathbf{j} & \mathbf{k} \\ \frac{\partial}{\partial x} & \frac{\partial}{\partial y} & \frac{\partial}{\partial z} \\ x & y & z \end{bmatrix} \tag{A.23}$$

$$= \det \left(\begin{bmatrix} \frac{\partial}{\partial y} & \frac{\partial}{\partial z} \\ y & z \end{bmatrix} \right) \mathbf{i} - \det \left(\begin{bmatrix} \frac{\partial}{\partial x} & \frac{\partial}{\partial z} \\ x & z \end{bmatrix} \right) \mathbf{j} + \det \left(\begin{bmatrix} \frac{\partial}{\partial x} & \frac{\partial}{\partial y} \\ x & y \end{bmatrix} \right) \mathbf{k}$$

$$= \left(\frac{\partial z}{\partial y} - \frac{\partial y}{\partial z} \right) \mathbf{i} + \left(\frac{\partial x}{\partial z} - \frac{\partial z}{\partial x} \right) \mathbf{j} + \left(\frac{\partial y}{\partial x} - \frac{\partial x}{\partial y} \right) \mathbf{k} = 0 .$$

If $\hat{\mathbf{r}}$ is a unit vector along the direction $\mathbf{r}$ such that $\hat{\mathbf{r}} = \mathbf{r}/r$, then using Eq. (A.19) with $\phi = 1/r$ and $\mathbf{A} = \mathbf{r}$, we get

$$\nabla \cdot \hat{\mathbf{r}} = \nabla \cdot \left(\frac{\mathbf{r}}{r} \right) \tag{A.24}$$

$$= \nabla \left(\frac{1}{r} \right) \cdot \mathbf{r} + \frac{1}{r} (\nabla \cdot \mathbf{r})$$

$$= \left[\mathbf{i} \frac{\partial}{\partial x} \left(\frac{1}{r} \right) + \mathbf{j} \frac{\partial}{\partial y} \left(\frac{1}{r} \right) + \mathbf{k} \frac{\partial}{\partial z} \left(\frac{1}{r} \right) \right] \cdot \mathbf{r} + \frac{3}{r}$$

$$= -\frac{\mathbf{r} \cdot \mathbf{r}}{r^3} + \frac{3}{r}$$

$$= \frac{2}{r},$$

and

$$\nabla \times \hat{\mathbf{r}} = 0. \tag{A.25}$$

A.6   Laplacian

Laplacian formulas can be obtained by applying the vector gradient operator twice.

Definition A.4 (*Laplacian*) Let ϕ be a scalar field function and $\mathbf{A}$ a vector field. The Laplacian are defined as follows:

$$\Delta\phi = \nabla \cdot (\nabla\phi) \tag{A.26}$$

$$\Delta\mathbf{A} = \nabla(\nabla \cdot \mathbf{A}) - \nabla \times \nabla \times \mathbf{A}.$$

In Eq. (A.26), $\Delta\phi$ is called scalar Laplacian and $\Delta\mathbf{A}$ is called vector Laplacian. It can be seen that the scalar Laplacian is given in Cartesian coordinates as

$$\Delta\phi = \frac{\partial^2\phi}{\partial x^2} + \frac{\partial^2\phi}{\partial y^2} + \frac{\partial^2\phi}{\partial z^2}. \tag{A.27}$$

Furthermore,

$$\nabla \times \nabla \times \mathbf{A} = \mathbf{i} \left[\frac{\partial}{\partial y} \left(\frac{\partial A_y}{\partial x} - \frac{\partial A_x}{\partial y} \right) + \frac{\partial}{\partial z} \left(\frac{\partial A_z}{\partial x} - \frac{\partial A_x}{\partial z} \right) \right] \tag{A.28}$$

$$+ \mathbf{j} \left[\frac{\partial}{\partial x} \left(\frac{\partial A_x}{\partial y} - \frac{\partial A_y}{\partial x} \right) + \frac{\partial}{\partial z} \left(\frac{\partial A_z}{\partial y} - \frac{\partial A_y}{\partial z} \right) \right]$$

$$+ \mathbf{k} \left[\frac{\partial}{\partial x} \left(\frac{\partial A_x}{\partial z} - \frac{\partial A_z}{\partial x} \right) + \frac{\partial}{\partial y} \left(\frac{\partial A_y}{\partial z} - \frac{\partial A_z}{\partial y} \right) \right],$$

and

$$\nabla(\nabla \cdot \mathbf{A}) = \mathbf{i}\frac{\partial}{\partial x}\left[\frac{\partial A_x}{\partial x} + \frac{\partial A_y}{\partial y} + \frac{\partial A_z}{\partial z}\right] \tag{A.29}$$
$$+ \mathbf{j}\frac{\partial}{\partial y}\left[\frac{\partial A_x}{\partial x} + \frac{\partial A_y}{\partial y} + \frac{\partial A_z}{\partial z}\right]$$
$$+ \mathbf{k}\frac{\partial}{\partial z}\left[\frac{\partial A_x}{\partial x} + \frac{\partial A_y}{\partial y} + \frac{\partial A_z}{\partial z}\right].$$

Therefore, substituting Eqs. (A.28) and (A.29) into (A.26), we write that

$$\Delta\mathbf{A} = \mathbf{i}\left[\frac{\partial^2}{\partial x^2} + \frac{\partial^2}{\partial y^2} + \frac{\partial^2}{\partial z^2}\right]A_x \tag{A.30}$$
$$+ \mathbf{j}\left[\frac{\partial^2}{\partial x^2} + \frac{\partial^2}{\partial y^2} + \frac{\partial^2}{\partial z^2}\right]A_y$$
$$+ \mathbf{k}\left[\frac{\partial^2}{\partial x^2} + \frac{\partial^2}{\partial y^2} + \frac{\partial^2}{\partial z^2}\right]A_z.$$

A.7 Curvilinear Coordinates

In the following, we introduce two other coordinate systems, namely *spherical coordinate system* (r, θ, ϕ) and *cylindrical coordinate system* (ρ, ϕ, z), as presented in Fig. A.5. Furthermore, we introduce a metric for each geometry, defined as follows:

$$\mathbf{g} \equiv (1, 1, 1), \quad \text{Cartesian coordinate system} \tag{A.31}$$
$$\mathbf{g} \equiv (1, r, r\sin\theta), \quad \text{Spherical coordinate system}$$
$$\mathbf{g} \equiv (1, \rho, 1), \quad \text{Cylindrical coordinate system}.$$

An interval length ds relates to the change on variables in these coordinate systems in the following general form:

$$(ds)^2 = \sum_{i=1}^{3} g_i^2(dx_i)^2. \tag{A.32}$$

Therefore,

$$(ds)^2 = (dx)^2 + (dy)^2 + (dz)^2, \quad \text{Cartesian coordinate system,} \tag{A.33}$$
$$(ds)^2 = (dr)^2 + r^2(d\theta)^2 + r^2\sin^2\theta(d\phi)^2, \quad \text{Spherical coordinate system,}$$
$$(ds)^2 = (d\rho)^2 + \rho^2(d\phi)^2 + (dz)^2, \quad \text{Cylindrical coordinate system}.$$

Moreover, the generals form of the vector gradient operator acting on a scalar field ψ or vector field $\mathbf{A}$ are as follows:

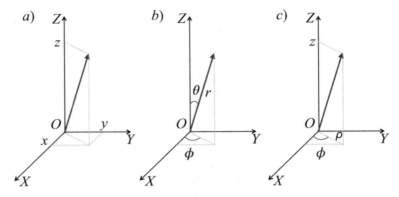

Fig. A.5 **a** A Cartesian coordinate system (x, y, z); **b** A spherical coordinate system (r, θ, ϕ); **c** A cylindrical coordinate system (ρ, ϕ, z)

$$\nabla \psi = \sum_{i=1}^{3} \hat{\mathbf{x}}_i \frac{1}{g_i} \frac{\partial \psi}{\partial x_i}, \tag{A.34}$$

$$\nabla \cdot \mathbf{A} = \frac{1}{g_1 g_2 g_3} \sum_{i=1}^{3} \frac{\partial}{\partial x_i} \left(\frac{g_1 g_2 g_3}{g_i} A_i \right),$$

$$\nabla \times \mathbf{A} = \frac{1}{2} \sum_{i \neq j \neq k = 1}^{3} \hat{\mathbf{x}}_i \epsilon_{ijk} \frac{1}{g_j g_k} \left[\frac{\partial}{\partial x_j} (g_k A_k) - \frac{\partial}{\partial x_k} (g_j A_j) \right]$$

$$= \sum_{i \neq j > k = 1}^{3} \hat{\mathbf{x}}_i \epsilon_{ijk} \frac{1}{g_j g_k} \left[\frac{\partial}{\partial x_j} (g_k A_k) - \frac{\partial}{\partial x_k} (g_j A_j) \right],$$

where $\hat{\mathbf{x}}_i$ is a unit vector along the coordinate x_i, and ϵ_{ijk} is Levi-Civita number:

$$\epsilon_{ijk} = \begin{cases} +1, & \text{if } (i \to j \to k) \text{ rotates clockwise,} \\ 0, & \text{if } i = j, \text{ or } i = k, \text{ or } j = k, \\ -1, & \text{if } (i \to j \to k) \text{ rotates counterclockwise.} \end{cases} \tag{A.35}$$

For example, for any scalar field and vector field functions, Eqs. (A.13) and (A.15) give the expressions of the gradient and divergence in Cartesian coordinates. In spherical coordinates, then for a scalar field ψ and vector field $\mathbf{A}$, we find

$$\nabla \psi = \hat{\mathbf{r}} \frac{\partial \psi}{\partial r} + \hat{\theta} \frac{1}{r} \frac{\partial \psi}{\partial \theta} + \hat{\phi} \frac{1}{r \sin \theta} \frac{\partial \psi}{\partial \phi}, \tag{A.36}$$

and

$$\mathbf{\nabla} \cdot \mathbf{A} = \frac{1}{g_1 g_2 g_3} \left[\frac{\partial}{\partial x_1} \left(\frac{g_1 g_2 g_3}{g_1} A_1 \right) + \frac{\partial}{\partial x_2} \left(\frac{g_1 g_2 g_3}{g_2} A_2 \right) \right. \tag{A.37}$$
$$\left. + \frac{\partial}{\partial x_3} \left(\frac{g_1 g_2 g_3}{g_3} A_3 \right) \right]$$
$$= \frac{1}{g_1 g_2 g_3} \left[\frac{\partial}{\partial x_1} (g_2 g_3 A_1) + \frac{\partial}{\partial x_2} (g_1 g_3 A_2) + \frac{\partial}{\partial x_3} (g_1 g_2 A_3) \right]$$
$$= \frac{1}{r^2 \sin \theta} \left[\sin \theta \frac{\partial}{\partial r} (r^2 A_r) + r \frac{\partial}{\partial \theta} (\sin \theta A_\theta) + r \frac{\partial}{\partial \phi} (A_\phi) \right]$$
$$= \frac{1}{r^2} \frac{\partial}{\partial r} (r^2 A_r) + \frac{1}{r \sin \theta} \frac{\partial}{\partial \theta} (\sin \theta A_\theta) + \frac{1}{r \sin \theta} \frac{\partial A_\phi}{\partial \phi} \, .$$

Furthermore, $\mathbf{\nabla} \times \mathbf{A}$ in spherical coordinates is given as

$$\mathbf{\nabla} \times \mathbf{A} = \hat{\mathbf{x}}_1 \epsilon_{123} \frac{1}{g_2 g_3} \left[\frac{\partial}{\partial x_2} (g_3 A_3) - \frac{\partial}{\partial x_3} (g_2 A_2) \right] \tag{A.38}$$
$$+ \hat{\mathbf{x}}_2 \epsilon_{213} \frac{1}{g_1 g_3} \left[\frac{\partial}{\partial x_1} (g_3 A_3) - \frac{\partial}{\partial x_3} (g_1 A_1) \right]$$
$$+ \hat{\mathbf{x}}_3 \epsilon_{312} \frac{1}{g_1 g_2} \left[\frac{\partial}{\partial x_1} (g_2 A_2) - \frac{\partial}{\partial x_2} (g_1 A_1) \right]$$
$$= \hat{\mathbf{r}} \frac{1}{r \sin \theta} \left[\frac{\partial}{\partial \theta} (\sin \theta A_\phi) - \frac{\partial A_\theta}{\partial \phi} \right]$$
$$+ \hat{\boldsymbol{\theta}} \frac{1}{r \sin \theta} \left[\frac{\partial A_r}{\partial \phi} - \sin \theta \frac{\partial}{\partial r} (r A_\phi) \right]$$
$$+ \hat{\boldsymbol{\phi}} \frac{1}{r} \left[\frac{\partial}{\partial r} (r A_\theta) - \frac{\partial A_r}{\partial \theta} \right] \, .$$

Laplacian of a scalar field ψ in spherical coordinates is calculated as

$$\mathbf{\Delta} \psi = \mathbf{\nabla} \cdot (\mathbf{\nabla} \psi) \tag{A.39}$$
$$= \mathbf{\nabla} \cdot \left[\hat{\mathbf{r}} \frac{\partial \psi}{\partial r} + \hat{\boldsymbol{\theta}} \frac{1}{r} \frac{\partial \psi}{\partial \theta} + \hat{\boldsymbol{\phi}} \frac{1}{r \sin \theta} \frac{\partial \psi}{\partial \phi} \right] \tag{A.40}$$
$$= \underbrace{\mathbf{\nabla} \cdot \left(\hat{\mathbf{r}} \frac{\partial \psi}{\partial r} \right)}_{T_1} + \underbrace{\mathbf{\nabla} \cdot \left(\hat{\boldsymbol{\theta}} \frac{1}{r} \frac{\partial \psi}{\partial \theta} \right)}_{T_2} + \underbrace{\mathbf{\nabla} \cdot \left(\hat{\boldsymbol{\phi}} \frac{1}{r \sin \theta} \frac{\partial \psi}{\partial \phi} \right)}_{T_3}$$
$$= T_1 + T_2 + T_3 \, ,$$

where the terms T_1, T_2, and T_3 are calculated using Eq. (A.34) as

$$T_1 = \frac{1}{g_1 g_2 g_3}\left[\frac{\partial}{\partial x_1}(g_2 g_3 A_1) + \frac{\partial}{\partial x_2}(g_1 g_3 A_2) + \frac{\partial}{\partial x_3}(g_1 g_2 A_3)\right] \quad\text{(A.41)}$$

$$= \frac{1}{r^2\sin\theta}\left[\frac{\partial}{\partial r}\underbrace{\left(r^2\sin\theta\, A_r\right)}_{A_r = \partial\psi/\partial r} + \frac{\partial}{\partial\theta}\underbrace{\left(r\sin\theta\, A_\theta\right)}_{A_\theta = 0} + \frac{\partial}{\partial\phi}\underbrace{\left(r A_\phi\right)}_{A_\phi = 0}\right]$$

$$= \frac{1}{r^2\sin\theta}\frac{\partial}{\partial r}\left(r^2\sin\theta\frac{\partial\psi}{\partial r}\right)$$

$$= \frac{2}{r}\frac{\partial\psi}{\partial r} + \frac{\partial\psi^2}{\partial r^2},$$

$$T_2 = \frac{1}{g_1 g_2 g_3}\left[\frac{\partial}{\partial x_1}(g_2 g_3 A_1) + \frac{\partial}{\partial x_2}(g_1 g_3 A_2) + \frac{\partial}{\partial x_3}(g_1 g_2 A_3)\right] \quad\text{(A.42)}$$

$$= \frac{1}{r^2\sin\theta}\left[\frac{\partial}{\partial r}\underbrace{\left(r^2\sin\theta\, A_r\right)}_{A_r = 0} + \frac{\partial}{\partial\theta}\underbrace{\left(r\sin\theta\, A_\theta\right)}_{A_\theta = \frac{1}{r}\frac{\partial\psi}{\partial\theta}} + \frac{\partial}{\partial\phi}\underbrace{\left(r A_\phi\right)}_{A_\phi = 0}\right]$$

$$= \frac{1}{r^2\sin\theta}\frac{\partial}{\partial\theta}\left(\sin\theta\frac{\partial\psi}{\partial\theta}\right)$$

$$= \frac{1}{r^2\tan\theta}\frac{\partial\psi}{\partial\theta} + \frac{1}{r^2}\frac{\partial^2\psi}{\partial\theta^2},$$

and

$$T_3 = \frac{1}{g_1 g_2 g_3}\left[\frac{\partial}{\partial x_1}(g_2 g_3 A_1) + \frac{\partial}{\partial x_2}(g_1 g_3 A_2) + \frac{\partial}{\partial x_3}(g_1 g_2 A_3)\right] \quad\text{(A.43)}$$

$$= \frac{1}{r^2\sin\theta}\left[\frac{\partial}{\partial r}\underbrace{\left(r^2\sin\theta\, A_r\right)}_{A_r = 0} + \frac{\partial}{\partial\theta}\underbrace{\left(r\sin\theta\, A_\theta\right)}_{A_\theta = 0} + \frac{\partial}{\partial\phi}\underbrace{\left(r A_\phi\right)}_{A_\phi = \frac{1}{r\sin\theta}\frac{\partial\psi}{\partial\phi}}\right]$$

$$= \frac{1}{r^2\sin\theta}\frac{\partial}{\partial\phi}\left(\frac{1}{\sin\theta}\frac{\partial\psi}{\partial\phi}\right)$$

$$= \frac{1}{r^2\sin^2\theta}\frac{\partial^2\psi}{\partial\phi^2}.$$

Therefore,

$$\Delta \psi = \frac{\partial \psi^2}{\partial r^2} + \frac{1}{r^2} \frac{\partial^2 \psi}{\partial \theta^2} + \frac{1}{r^2 \sin^2 \theta} \frac{\partial^2 \psi}{\partial \phi^2} + \frac{2}{r} \frac{\partial \psi}{\partial r} + \frac{1}{r^2 \tan \theta} \frac{\partial \psi}{\partial \theta} . \tag{A.44}$$

For Laplacian of a vector field (the second expression in Eq. (A.26)), we first evaluate

$$\mathbf{\nabla}(\mathbf{\nabla} \cdot \mathbf{A}) = \underbrace{\hat{\mathbf{r}} \frac{\partial}{\partial r}(\mathbf{\nabla} \cdot \mathbf{A})}_{I_1} \tag{A.45}$$

$$+ \underbrace{\hat{\boldsymbol{\theta}} \frac{1}{r} \frac{\partial}{\partial \theta}(\mathbf{\nabla} \cdot \mathbf{A})}_{I_2}$$

$$+ \underbrace{\hat{\boldsymbol{\phi}} \frac{1}{r \sin \theta} \frac{\partial}{\partial \phi}(\mathbf{\nabla} \cdot \mathbf{A})}_{I_3}$$

$$= I_1 + I_2 + I_3 ,$$

where $\mathbf{\nabla} \cdot \mathbf{A}$ is given by Eq. (A.37), and thus we have

$$I_1 = \frac{\hat{\mathbf{r}}}{\sin \theta} \frac{\partial}{\partial r} \left\{ \frac{1}{r^2} \left[\sin \theta \frac{\partial}{\partial r}\left(r^2 A_r\right) + r \frac{\partial}{\partial \theta}(\sin \theta A_\theta) + r \frac{\partial A_\phi}{\partial \phi} \right] \right\} . \tag{A.46}$$

Then, I_2 is

$$I_2 = \frac{\hat{\boldsymbol{\theta}}}{r^3} \frac{\partial}{\partial \theta} \left\{ \frac{1}{\sin \theta} \left[\sin \theta \frac{\partial}{\partial r}\left(r^2 A_r\right) + r \frac{\partial}{\partial \theta}(\sin \theta A_\theta) + r \frac{\partial A_\phi}{\partial \phi} \right] \right\} , \tag{A.47}$$

and I_3 is

$$I_3 = \frac{\hat{\boldsymbol{\phi}}}{r^3 \sin^2 \theta} \frac{\partial}{\partial \phi} \left\{ \left[\sin \theta \frac{\partial}{\partial r}\left(r^2 A_r\right) + r \frac{\partial}{\partial \theta}(\sin \theta A_\theta) + r \frac{\partial A_\phi}{\partial \phi} \right] \right\} . \tag{A.48}$$

Furthermore, we calculate

$$\mathbf{\nabla} \times \mathbf{\nabla} \times \mathbf{A} = \hat{\mathbf{r}} \frac{1}{r^2 \sin \theta} \left[\frac{\partial}{\partial \theta}\left(r \sin \theta (\mathbf{\nabla} \times \mathbf{A})_\phi\right) - \frac{\partial}{\partial \phi}\left(r(\mathbf{\nabla} \times \mathbf{A})_\theta\right) \right] \tag{A.49}$$

$$+ \hat{\boldsymbol{\theta}} \frac{1}{r \sin \theta} \left[\frac{\partial}{\partial \phi}\left((\mathbf{\nabla} \times \mathbf{A})_r\right) - \frac{\partial}{\partial r}\left(r \sin \theta (\mathbf{\nabla} \times \mathbf{A})_\phi\right) \right]$$

$$+ \hat{\boldsymbol{\phi}} \frac{1}{r} \left[\frac{\partial}{\partial r}\left(r(\mathbf{\nabla} \times \mathbf{A})_\theta\right) - \frac{\partial}{\partial \theta}\left((\mathbf{\nabla} \times \mathbf{A})_r\right) \right] ,$$

where

$$(\nabla \times \mathbf{A})_r = \frac{1}{r \sin \theta} \left[\frac{\partial}{\partial \theta} (\sin \theta A_\phi) - \frac{\partial A_\theta}{\partial \phi} \right], \tag{A.50}$$

$$(\nabla \times \mathbf{A})_\theta = \frac{1}{r \sin \theta} \left[\frac{\partial A_r}{\partial \phi} - \sin \theta \frac{\partial}{\partial r} (r A_\phi) \right],$$

$$(\nabla \times \mathbf{A})_\phi = \frac{1}{r} \left[\frac{\partial}{\partial r} (r A_\theta) - \frac{\partial A_r}{\partial \theta} \right].$$

Substituting Eq. (A.49) and (A.50) into second expression of Eq. (A.26), we calculate Laplacian of a vector field **A**.

In cylindrical coordinates, for the scalar field ψ, we can calculate the gradient as

$$\nabla \psi = \hat{\boldsymbol{\rho}} \frac{\partial \psi}{\partial \rho} + \hat{\boldsymbol{\phi}} \frac{1}{\rho} \frac{\partial \psi}{\partial \phi} + \hat{\mathbf{z}} \frac{\partial \psi}{\partial z}, \tag{A.51}$$

where the first expression of Eq. (A.34) is used. Then, for a vector field function **A**, we calculate the divergence in cylindrical coordinates as follows:

$$\begin{aligned} \nabla \cdot \mathbf{A} &= \frac{1}{g_1 g_2 g_3} \left[\frac{\partial}{\partial x_1} \left(\frac{g_1 g_2 g_3}{g_1} A_1 \right) + \frac{\partial}{\partial x_2} \left(\frac{g_1 g_2 g_3}{g_2} A_2 \right) \right. \\ &\quad \left. + \frac{\partial}{\partial x_3} \left(\frac{g_1 g_2 g_3}{g_3} A_3 \right) \right] \\ &= \frac{1}{g_1 g_2 g_3} \left[\frac{\partial}{\partial x_1} (g_2 g_3 A_1) + \frac{\partial}{\partial x_2} (g_1 g_3 A_2) + \frac{\partial}{\partial x_3} (g_1 g_2 A_3) \right] \\ &= \frac{1}{\rho} \left[\frac{\partial}{\partial \rho} (\rho A_\rho) + \frac{\partial A_\phi}{\partial \phi} + \frac{\partial}{\partial z} (\rho A_z) \right] \\ &= \frac{1}{\rho} \left[\frac{\partial}{\partial \rho} (\rho A_\rho) + \frac{\partial A_\phi}{\partial \phi} + \rho \frac{\partial A_z}{\partial z} \right]. \end{aligned} \tag{A.52}$$

The rotor $\nabla \times \mathbf{A}$ in cylindrical coordinates is given as

$$\begin{aligned} \nabla \times \mathbf{A} &= \hat{\mathbf{x}}_1 \epsilon_{123} \frac{1}{g_2 g_3} \left[\frac{\partial}{\partial x_2} (g_3 A_3) - \frac{\partial}{\partial x_3} (g_2 A_2) \right] \\ &\quad + \hat{\mathbf{x}}_2 \epsilon_{213} \frac{1}{g_1 g_3} \left[\frac{\partial}{\partial x_1} (g_3 A_3) - \frac{\partial}{\partial x_3} (g_1 A_1) \right] \\ &\quad + \hat{\mathbf{x}}_3 \epsilon_{312} \frac{1}{g_1 g_2} \left[\frac{\partial}{\partial x_1} (g_2 A_2) - \frac{\partial}{\partial x_2} (g_1 A_1) \right] \\ &= \hat{\boldsymbol{\rho}} \frac{1}{\rho} \left[\frac{\partial A_z}{\partial \phi} - \frac{\partial}{\partial z} (\rho A_\phi) \right] \\ &\quad + \hat{\boldsymbol{\phi}} \left[\frac{\partial A_\rho}{\partial z} - \frac{\partial A_z}{\partial \rho} \right] \\ &\quad + \hat{\mathbf{z}} \frac{1}{\rho} \left[\frac{\partial}{\partial \rho} (\rho A_\phi) - \frac{\partial A_\rho}{\partial \phi} \right]. \end{aligned} \tag{A.53}$$

Similarly, Laplacian of the scalar field ψ in cylindrical coordinates is calculated as

$$\Delta\psi = \nabla \cdot (\nabla\psi) \tag{A.54}$$

$$= \nabla \cdot \left[\hat{\rho}\frac{\partial\psi}{\partial\rho} + \hat{\phi}\frac{1}{\rho}\frac{\partial\psi}{\partial\phi} + \hat{z}\frac{\partial\psi}{\partial z} \right] \tag{A.55}$$

$$= \underbrace{\nabla \cdot \left(\hat{\rho}\frac{\partial\psi}{\partial\rho} \right)}_{T_1} + \underbrace{\nabla \cdot \left(\hat{\phi}\frac{1}{\rho}\frac{\partial\psi}{\partial\phi} \right)}_{T_2} + \underbrace{\nabla \cdot \left(\hat{z}\frac{\partial\psi}{\partial z} \right)}_{T_3}$$

$$= T_1 + T_2 + T_3 .$$

Each of the terms T_1, T_2, and T_3 is calculated using Eq. (A.34) as follows:

$$T_1 = \frac{1}{g_1 g_2 g_3} \left[\frac{\partial}{\partial x_1}(g_2 g_3 A_1) + \frac{\partial}{\partial x_2}(g_1 g_3 A_2) + \frac{\partial}{\partial x_3}(g_1 g_2 A_3) \right] \tag{A.56}$$

$$= \frac{1}{\rho} \left[\frac{\partial}{\partial\rho}\underbrace{(\rho A_\rho)}_{A_\rho = \frac{\partial\psi}{\partial\rho}} + \frac{\partial}{\partial\phi}\underbrace{(A_\phi)}_{A_\phi = 0} + \frac{\partial}{\partial z}\underbrace{(\rho A_z)}_{A_z = 0} \right]$$

$$= \frac{1}{\rho}\frac{\partial}{\partial\rho}\left(\rho\frac{\partial\psi}{\partial\rho} \right)$$

$$= \frac{1}{\rho}\frac{\partial\psi}{\partial\rho} + \frac{\partial\psi^2}{\partial\rho^2} ,$$

$$T_2 = \frac{1}{g_1 g_2 g_3} \left[\frac{\partial}{\partial x_1}(g_2 g_3 A_1) + \frac{\partial}{\partial x_2}(g_1 g_3 A_2) + \frac{\partial}{\partial x_3}(g_1 g_2 A_3) \right] \tag{A.57}$$

$$= \frac{1}{\rho} \left[\frac{\partial}{\partial\rho}\underbrace{(\rho A_\rho)}_{A_\rho = 0} + \frac{\partial}{\partial\phi}\underbrace{(A_\phi)}_{A_\phi = \frac{1}{\rho}\frac{\partial\psi}{\partial\phi}} + \frac{\partial}{\partial z}\underbrace{(\rho A_z)}_{A_z = 0} \right]$$

$$= \frac{1}{\rho}\frac{\partial}{\partial\phi}\left(\frac{1}{\rho}\frac{\partial\psi}{\partial\phi} \right)$$

$$= \frac{1}{\rho^2}\frac{\partial^2\psi}{\partial\phi^2} .$$

The last term is

$$T_3 = \frac{1}{g_1 g_2 g_3} \left[\frac{\partial}{\partial x_1} (g_2 g_3 A_1) + \frac{\partial}{\partial x_2} (g_1 g_3 A_2) + \frac{\partial}{\partial x_3} (g_1 g_2 A_3) \right] \qquad (A.58)$$

$$= \frac{1}{\rho} \left[\frac{\partial}{\partial \rho} \underbrace{(\rho A_\rho)}_{A_\rho = 0} + \frac{\partial}{\partial \phi} \underbrace{(A_\phi)}_{A_\phi = 0} + \frac{\partial}{\partial z} \underbrace{(\rho A_z)}_{A_z = \frac{\partial \psi}{\partial z}} \right]$$

$$= \frac{1}{\rho} \frac{\partial}{\partial z} \left(\rho \frac{\partial \psi}{\partial z} \right)$$

$$= \frac{\partial^2 \psi}{\partial z^2} \, .$$

Thus, combining T_1, T_2, and T_3, we obtain

$$\boldsymbol{\Delta} \psi = \frac{\partial \psi^2}{\partial \rho^2} + \frac{1}{\rho^2} \frac{\partial^2 \psi}{\partial \phi^2} + \frac{\partial^2 \psi}{\partial z^2} + \frac{1}{\rho} \frac{\partial \psi}{\partial \rho} \, . \qquad (A.59)$$

For Laplacian of a vector field in cylindrical coordinates (the second expression in Eq. (A.26)), we first calculate

$$\boldsymbol{\nabla} (\boldsymbol{\nabla} \cdot \mathbf{A}) = \hat{\boldsymbol{\rho}} \underbrace{\frac{\partial}{\partial \rho} (\boldsymbol{\nabla} \cdot \mathbf{A})}_{I_1} \qquad (A.60)$$

$$+ \hat{\boldsymbol{\phi}} \underbrace{\frac{1}{\rho} \frac{\partial}{\partial \phi} (\boldsymbol{\nabla} \cdot \mathbf{A})}_{I_2}$$

$$+ \hat{\mathbf{z}} \underbrace{\frac{\partial}{\partial z} (\boldsymbol{\nabla} \cdot \mathbf{A})}_{I_3}$$

$$= I_1 + I_2 + I_3 \, ,$$

with $\boldsymbol{\nabla} \cdot \mathbf{A}$ given by Eq. (A.52). Hence, we have

$$I_1 = \hat{\boldsymbol{\rho}} \frac{\partial}{\partial \rho} \left(\frac{1}{\rho} \frac{\partial}{\partial \rho} (\rho A_\rho) + \frac{1}{\rho} \frac{\partial A_\phi}{\partial \phi} + \frac{\partial A_z}{\partial z} \right) \, . \qquad (A.61)$$

The term, I_2 is

$$I_2 = \frac{\hat{\boldsymbol{\phi}}}{\rho} \frac{\partial}{\partial \phi} \left(\frac{1}{\rho} \frac{\partial}{\partial \rho} (\rho A_\rho) + \frac{1}{\rho} \frac{\partial A_\phi}{\partial \phi} + \frac{\partial A_z}{\partial z} \right) \, , \qquad (A.62)$$

and the term I_3 is

$$I_3 = \hat{\mathbf{z}} \frac{\partial}{\partial z} \left(\frac{1}{\rho} \frac{\partial}{\partial \rho} (\rho A_\rho) + \frac{1}{\rho} \frac{\partial A_\phi}{\partial \phi} + \frac{\partial A_z}{\partial z} \right). \tag{A.63}$$

Furthermore, we calculate

$$\nabla \times \nabla \times \mathbf{A} = \hat{\boldsymbol{\rho}} \frac{1}{\rho} \left[\frac{\partial (\nabla \times \mathbf{A})_z}{\partial \phi} - \frac{\partial}{\partial z} \left(\rho (\nabla \times \mathbf{A})_\phi \right) \right] \tag{A.64}$$
$$+ \hat{\boldsymbol{\phi}} \left[\frac{\partial (\nabla \times \mathbf{A})_\rho}{\partial z} - \frac{\partial (\nabla \times \mathbf{A})_z}{\partial \rho} \right]$$
$$+ \hat{\mathbf{z}} \frac{1}{\rho} \left[\frac{\partial}{\partial \rho} \left(\rho (\nabla \times \mathbf{A})_\phi \right) - \frac{\partial (\nabla \times \mathbf{A})_\rho}{\partial \phi} \right],$$

where

$$(\nabla \times \mathbf{A})_\rho = \frac{1}{\rho} \left[\frac{\partial A_z}{\partial \phi} - \frac{\partial}{\partial z} \left(\rho A_\phi \right) \right] \tag{A.65}$$
$$(\nabla \times \mathbf{A})_\phi = \frac{\partial A_\rho}{\partial z} - \frac{\partial A_z}{\partial \rho}$$
$$(\nabla \times \mathbf{A})_z = \frac{1}{\rho} \left[\frac{\partial}{\partial \rho} \left(\rho A_\phi \right) - \frac{\partial A_\rho}{\partial \phi} \right].$$

Some useful applications of the spherical and cylindrical coordinates include the cases when there is some symmetry of the problem such that either the scalar field function ψ or vector field function $\mathbf{A}$ depends on fewer coordinates. For example, in problems with cylindrical symmetry scalar electric potential function depends only on ρ, and it is independent on ϕ and z, then the formulas simplify significantly compare with those using the Cartesian coordinate system.

Printed in the United States
by Baker & Taylor Publisher Services